Quantization, Coherent States, and Complex Structures

Quantization, Coherent States, and Complex Structures

Edited by

J.-P. Antoine
Catholic University of Louvain
Louvain-la-Neuve, Belgium

S. Twareque Ali
Concordia University
Montréal, Québec, Canada

W. Lisiecki
Warsaw University
Białystok, Poland

I. M. Mladenov
Bulgarian Academy of Sciences
Sofia, Bulgaria

and

A. Odzijewicz
Warsaw University
Białystok, Poland

Plenum Press • New York and London

Library of Congress Cataloging-in-Publication Data

On file

Proceedings of the 13th Workshop on Geometric Methods in Physics,
held July 9–15, 1994, in Bialowieza, Poland

ISBN 0-306-45214-6

A Division of Plenum Publishing Corporation
233 Spring Street, New York, N. Y. 10013

10 9 8 7 6 5 4 3 2 1

Printed in the United States of America

PREFACE

The XIIIth Białowieża Summer Workshop was held from July 9 to 15, 1994. While still within the general framework of Differential Geometric Methods in Physics, the XIIIth Workshop was expanded in scope to include quantum groups, q-deformations and non-commutative geometry. It is expected that lectures on these topics will now become an integral part of future workshops. In the more traditional areas, lectures were devoted to topics in quantization, field theory, group representations, coherent states, complex and Poisson structures, the Berry phase, graded contractions and some infinite-dimensional systems.

Those of us who have taken part in the evolution of the workshops over the years, feel a good measure of satisfaction with the excellent quality of the papers presented, in particular the mathematical rigour and novelty. Each year a significant number of new results are presented and future directions of research are discussed. Their freshness and immediacy inevitably leads to intense discussions and an exchange of ideas in an informal and physically charming environment.

The present workshop also had a higher attendance than its predecessors, with approximately 65 registered participants. As usual, there was a large number of graduate students and young researchers among them.

The editors would like to thank all members of the local organizing committee for their painstaking efforts in ensuring the smooth running of the meeting and for organizing a delightful array of social events. Secondly, they would like to record their indebtedness to all the people who have contributed to this volume. Finally they would like to thank Warsaw University, Białystok Branch, for generous financial support.

J-P. Antoine
S. Twareque Ali
W. Lisiecki
I. M. Mladenov
A. Odzijewicz

CONTENTS

PART I: QUANTIZATION, FIELD THEORY AND REPRESENTATION THEORY

PART II: COHERENT STATES, COMPLEX AND POISSON STRUCTURES

PART III: Q-DEFORMATIONS AND QUANTUM GROUPS, NONCOMMUTATIVE GEOMETRY

PART IV: MISCELLANEOUS PROBLEMS OF QUANTUM DYNAMICS

PART I

QUANTIZATION, FIELD THEORY AND REPRESENTATION THEORY

ON QUANTUM MECHANICS IN A CURVED SPACETIME WITH ABSOLUTE TIME

Daniel Canarutto,[1] Arkadiusz Jadczyk[2] and Marco Modugno[1]

[1] Dipartimento di Matematica Applicata 'G. Sansone'
Via S. Marta 3, 50139 Firenze, Italia

[2] Institute of Theoretical Physics
pl. Maksa Borna 9, 50-204 Wrocław, Poland

Abstract

We present a new covariant approach to the quantum mechanics of a charged 1/2-spin particle in given electromagnetic and gravitational fields. The background space is assumed to be a curved Galileian spacetime, that is a curved space-time with absolute time. This setting is intended both as a suitable approximation for the case of low speeds and feeble gravitational fields, and as a guide for eventual extension to fully Einsteinian spacetime. Moreover, in the flat spacetime case one completely recovers standard non-relativistic quantum mechanics.

1. INTRODUCTION

Recently Jadczyk and Modugno[1,2] have proposed a new covariant formulation of the quantum mechanics of scalar charged particles interacting with given classical gravitational and electromagnetic fields, in the framework of a general relativistic Galileian spacetime. In this paper we extend this formulation to the quantum mechanics of a particle with spin.

Our work is related to a wide literature on the classical and quantum Galilei theory starting from E. Cartan (several quotations on this subject may be found in the above papers by Jadczyk and Modugno; in particular we recall the works by C. Duval, H. P. Künzle, P. Havas, K. Kuchař, J. Ehlers, J. M. Lévy-Leblond, M. Mangiarotti, M. Modugno, W.Pauli, E. Prugovečki, E. Schmutzer and J. Plebanski, A. Trautman, W. M. Tulczyjew). Moreover, our theory has evident relations but also important differences with the geometric quantisation. We stress that the touchstone of our approach is standard quantum mechanics.

The scheme of our approach is briefly the following.

First, we sketch the essential features of our background classical spacetime. We assume a 4-dimensional spacetime fibred over time and equipped with a space-like Euclidean metric, a time-preserving linear connection (the gravitational field) and a 2-form (the electromagnetic field). We couple the gravitational and electromagnetic fields and obtain "total" geometric objects, including a cosymplectic 2-form. We postulate the closure of this form thus yielding a link between the above geometrical structures and the first Maxwell equation; moreover, we postulate a kind of "reduced" Einstein and second Maxwell equations expressing the interaction of the above fields with their matter sources. Eventually we present a distinguished Lie algebra of functions, which are called "quantisable" in view of their role in the theory of quantum operators.

Then, we develop the quantum theory starting from the quantum bundle over spacetime obtained as tensor product of a Hermitian complex 1-dimensional scalar quantum bundle and a Hermitian complex 2-dimensional spin bundle equipped with a 2-form. On the scalar quantum bundle we assume a connection which is Hermitian "universal" and whose curvature is proportional to the cosymplectic form; in a sense, this connection is parametrised by all classical observers. We postulate a Pauli map which links the bundle of spacelike vectors with the spin bundle and yields a Hermitian connection on it. The tensor product of these two connections is a connection on the quantum bundle. This is our only primitive quantum structure; all other objects will be derived from it getting free from observers through a "principle of projectability", which is our implementation of covariance. Namely, we obtain a distinguished Lagrangian, which yields the generalised Pauli equation and conserved quantities.

Acknowledgements. This research has been supported by Italian MURST (national and local funds), by GNFM of Consiglio Nazionale delle Ricerche and by the EEC contract N. ERB CHRXCT 930096. Thanks are due to Andrzej Trautman for stimulating discussion.

2. PRELIMINARIES

2.1. A Few Recalls on Fibred Manifolds

We start with a few general notation and remarks concerning our geometrical language.

Remark 2.1 Let $\boldsymbol{M}$ be a manifold. We denote the $\mathbb{R}$-Lie algebra of functions $f : \boldsymbol{M} \to \mathbb{R}$ by $\mathcal{F}\boldsymbol{M}$, the *tangent bundle* of $\boldsymbol{M}$ by $T\boldsymbol{M} \to \boldsymbol{M}$ and the $\mathbb{R}$-Lie algebra of vector fields $X : \boldsymbol{M} \to T\boldsymbol{M}$ by $\mathcal{T}\boldsymbol{M}$. A local chart (x^λ) of $\boldsymbol{M}$ induces the local chart $(x^\lambda, \dot{x}^\lambda)$ of $T\boldsymbol{M}$, the local basis of vector fields $(\partial_\lambda) := (\partial x_\lambda)$ and the dual local basis of forms $(d^\lambda) := (dx^\lambda)$. The *tangent prolongation* of a map $f : \boldsymbol{M} \to \boldsymbol{N}$ is the map $Tf : T\boldsymbol{M} \to T\boldsymbol{N}$ with coordinate expresssion $Tf = \partial_\lambda f^i d^\lambda \otimes (\partial_i \circ f)$. •

Remark 2.2 A manifold $\boldsymbol{F}$ is said to be *fibred* over the *base space* $\boldsymbol{B}$ if it is equipped with a surjective map $p : \boldsymbol{F} \to \boldsymbol{B}$ whose rank equals the dimension of $\boldsymbol{B}$. A fibred manifold can be covered by local trivializations defined on open subsets $\boldsymbol{F}' \in \boldsymbol{F}$. Thus the concept of fibred manifold is more general than that of *bundle* (which can be covered by local trivializations defined on open subsets of the type $\boldsymbol{F}' = p^{-1}(\boldsymbol{U})$, where $\boldsymbol{U} \in \boldsymbol{B}$ is an open subset).

A chart (x^λ, y^i) of $\boldsymbol{F}$ is said to be *fibred* if x^λ's depend only on the base space. A fibred chart of $\boldsymbol{F}$ induces the local basis of vector fields $(\partial_\lambda, \partial_i)$ and the dual local basis of forms (d^λ, d^i) of $\boldsymbol{F}$. Hence, we obtain also the chart $(x^\lambda, y^i; \dot{x}^\lambda, \dot{y}^i)$ of $T\boldsymbol{F}$, the local basis of vector fields $(\partial_\lambda, \partial_i; \dot{\partial}_\lambda, \dot{\partial}_j)$ and the dual local basis of forms $(d^\lambda, d^i, \dot{d}^\lambda; \dot{d}^i)$.

We have a natural projection $T\boldsymbol{F} \to T\boldsymbol{B}$. A vector field $X : \boldsymbol{F} \to T\boldsymbol{F}$ is said to be *projectable* if it admits a projection $\underline{X} : \boldsymbol{B} \to T\boldsymbol{B}$ on the base space, i.e. if its coordinate expression is of the type $X = X^\lambda \partial_\lambda + X^i \partial_i$, with $X^\lambda \in \mathcal{F}\boldsymbol{B}$.

The *vertical subbundle* $V\boldsymbol{F} \subset T\boldsymbol{F}$ of $\boldsymbol{F}$ is constituted by all vectors tangent to the fibres and is characterised by the equation $(\dot{x}^\lambda = 0)$. Thus, a vector field X is vertical iff it is projectable over 0, i.e. iff $X^\lambda = 0$. The subset $\mathcal{V}\boldsymbol{F} \in \mathcal{T}\boldsymbol{F}$ is an ideal. We have a natural projection $V^*\boldsymbol{F} \to T^*\boldsymbol{F}$. We shall denote the vertical restrictions of forms by a check (' ˇ '). •

Remark 2.3 The *jet space at* $x \in \boldsymbol{B}$ of $\boldsymbol{F} \to \boldsymbol{B}$ is defined to be the set $J_{1x}\boldsymbol{F}$ of equivalence classes of sections $s : \boldsymbol{B} \to \boldsymbol{F}$ which have the same derivatives $\partial_\lambda s^i(x)$. Moreover, the *jet space* $J_1\boldsymbol{F}$ of $\boldsymbol{F}$ is the union of all $J_{1x}\boldsymbol{F}$ for $x \in \boldsymbol{B}$. We obtain the natural fibred charts (x^λ, y^i_λ) of $J_1\boldsymbol{F}$ and the *jet prolongation* $j_1 s : \boldsymbol{B} \to J_1\boldsymbol{F}$, charcterised by the coordinate expression $(y^i_\lambda \circ j_1 s = \partial_\lambda s^i)$. We can identify $j_1 s$ with $Ts : T\boldsymbol{B} \to T\boldsymbol{F}$, which projects over $\mathbf{1_B}$. Accordingly, we can regard $J_1\boldsymbol{F}$ as the subbundle of $T^*\boldsymbol{B} \otimes_{\mathbf{F}} T\boldsymbol{F}$, whose elments are projectable over $\mathbf{1_B}$. We denote this inclusion by the map

$$д : J_1\boldsymbol{F} \to T^*\boldsymbol{B} \underset{\mathbf{F}}{\otimes} T\boldsymbol{F} ,$$

with coordinate expression

$$д = d^\lambda \otimes д_\lambda = u^\lambda \otimes (\partial_\lambda + y^j_\lambda \partial_j) .$$

We obtain also the complementary map $\vartheta : J_{\mathbf{F}} \to T^*\boldsymbol{F} \otimes_{\mathbf{F}} V\boldsymbol{F}$ with coordinate expression

$$\vartheta = \vartheta^j \otimes \partial_j = (d^j - y^j_\lambda d^\lambda) \otimes \partial_j .$$

The vertical bundle of $J_1\boldsymbol{F}$ over the base space $\boldsymbol{F}$ turns out to be

$$V_{\mathbf{F}}\, J_1\boldsymbol{F} = J_1\boldsymbol{F} \underset{\mathbf{F}}{\times} (T^*\boldsymbol{B} \underset{\mathbf{F}}{\otimes} V\boldsymbol{F}) .$$

•

Connections will play an essential role in our approach. There are several equivalent ways to define the concept of a (possibly non linear) connection.

Remark 2.4 In general, we present a connection on a fibred manifold $\boldsymbol{F} \to \boldsymbol{B}$ as a section $c : \boldsymbol{F} \to J_1\boldsymbol{F}$ which, via the natural inclusion $д : J_1\boldsymbol{F} \to T^*\boldsymbol{B} \otimes_{\mathbf{F}} T\boldsymbol{F}$, can be seen as a *horizontal prolongation* $c : \boldsymbol{F} \to T^*\boldsymbol{B} \otimes_{\mathbf{F}} T\boldsymbol{F}$, whose coordinate expression is of the type $c = d^\lambda \otimes (\partial_\lambda + c_\lambda{}^j \partial_j)$, with $c_\lambda{}^j \in \mathcal{F}\boldsymbol{F}$. The associated *vertical projection* is $\nu_c : \boldsymbol{F} \to T^*\boldsymbol{F} \otimes_{\mathbf{F}} V\boldsymbol{F}$, with coordinate expression $\nu_c = (d^j - c_\lambda{}^j d^\lambda) \otimes \partial_j$.

Then, the *covariant differential* of a section $c : \boldsymbol{B} \to \boldsymbol{F}$ is defined to be the section $\nabla c := j_1 s - c \circ s = Ts \lrcorner\, \nu_c : \boldsymbol{B} \to T^*\boldsymbol{B} \otimes_{\mathbf{F}} T\boldsymbol{F}$, with coordinate expression $\nabla_\lambda s^i = \partial_\lambda s^i - c_\lambda{}^j \circ s$. •

We shall often use the following few facts.

Remark 2.5 If $\pi_{\mathbf{B}} : \boldsymbol{F} \to \boldsymbol{B}$ is any vector bundle, then one has the natural identification $V\boldsymbol{F} = \boldsymbol{F} \times_{\mathbf{B}} \boldsymbol{F}$. This fact yields some interesting consequences. First, any section $s : \boldsymbol{B} \to \boldsymbol{F}$ can be regarded as the *basic* vertical vector field $\boldsymbol{F} \to V\boldsymbol{F} : \varphi \mapsto (\varphi, s(\pi_{\mathbf{B}}(\varphi)))$. Hence, if $v : \boldsymbol{F} \to T\boldsymbol{F}$ is a linear vector field, projectable over $\underline{v} : \boldsymbol{B} \to T\boldsymbol{B}$, then the Lie bracket $[v, s]$ is a basic vertical vector field, i.e. it determines the section $v.s : \boldsymbol{B} \to \boldsymbol{F}$ with coordinate expression $(v.s)^j = v^\lambda \partial_\lambda s^j - v^j_k s^k$. Moreover, any linear map $f : \boldsymbol{F} \to \boldsymbol{F}$ fibred over $\boldsymbol{B}$ can be regarded as the vertical vector field $\boldsymbol{F} \to V\boldsymbol{F} : \varphi \mapsto (\varphi, f(\varphi))$. In particular, the *Liouville* vector field is defined to be the vertical vector field $и : \boldsymbol{F} \to V\boldsymbol{F} : \varphi \mapsto (\varphi, \varphi)$ associated with $\mathbf{1_F}$. •

2.2 Units of Measurement

Our theory is to be manifestly invariant with respect to any choice of measurement units; this is just an aspect of the general covariance. Then we need a few technical concepts.

We observe that homogeneous units can be added and multiplied by real numbers; however, in some cases, no zero unit exists and only multiplication by positive real numbers is allowed. This fact leads us to the following concept.

We can define algebraically a *semi-vector space* as a semi-field $\mathbb{U}$ associated with the semi-ring $\mathbb{R}^+$ (the axioms are analogous to those of vector spaces, with the only difference that $\mathbb{U}$ and $\mathbb{R}^+$ are additive semi-groups and not groups). Moreover, a semi-vector space is said to be *positive* if the multiplication by numbers can be extended neither to $\mathbb{R}^+ \cup \{0\}$ nor to $\mathbb{R}$. Each vector space is also a semi-vector space; moreover, a vector space and a basis yield a positive semi-vector space. Thus, a semi-vector space is a vector space, or a positive semi-vector space, or a positive semi-vector space extended by the zero element.

In order to write formulas which resemble the standard ones used by physicists, we adopt a 'number-wise' notation for 1-dimensional semi-vector spaces. So, if $\mathbb{U}$ and $\mathbb{V}$ are semi-vector spaces and $u \in \mathbb{U}, v \in \mathbb{V}$, then we write $uv \equiv u \otimes v$; if $\mathbb{U}$ is a 1-dimensional semi-vector space which does not contain 0, then we write $\mathbb{U}^{-1} = \mathbb{U}^*$ and denote by $1/u \in \mathbb{U}^{-1}$ the dual element of $u \in \mathbb{U}$.

We define a *unit space* to be a 1-dimensional semi-vector space.

We assume the following fundamental unit spaces: the oriented vector space of *time units* $\mathbb{T}$, the positive spaces of *lengths* $\mathbb{L}$ and of *masses* $\mathbb{M}$.

A time unit of measurement is denoted by $u_0 \in \mathbb{T}^+$ or $u^0 \in \mathbb{T}^{+*}$. We also set $u^{00} := u^0 \otimes u^0$ and $u_{00} := u_0 \otimes u_0$. For any $v \in \mathbb{T}, w \in \mathbb{T}^*$, according to our conventions, we shall often write $u^0 v$, $u_0 w \in \mathbb{R}$.

We attach to each particle a *mass*, a *charge* and a *magnetic constant*

$$m \in \mathbb{M} \,, \qquad q \in \mathbb{T}^* \otimes \mathbb{L}^{3/2} \otimes \mathbb{M}^{1/2} \,, \qquad \mu \in \mathbb{T}^* \otimes \mathbb{L}^{3/2} \otimes \mathbb{M}^{-1/2} \,.$$

We can also write μ as $\mu = G \frac{q}{2m}$, where $G \in \mathbb{R}$ is the *gyromagnetic ratio*; in particular, if $q = e$ is the positron's charge, then $\mu \hbar / G = e\hbar/2m$ is the *Bohr magneton*.

Moreover, we postulate two universal coupling constants, namely the *Newton's gravitational constant* and the *Planck's constant*

$$\kappa \in \mathbb{T}^{*2} \otimes \mathbb{L}^3 \otimes \mathbb{M}^* \,, \qquad \hbar \in (\mathbb{T}^+)^* \otimes \mathbb{L}^2 \otimes \mathbb{M} \,.$$

2.3. Classical Spacetime

We start with a brief sketch of the basic structures on the classical spacetime that are needed by the quantum theory as background; further details can be found in Jadcyk and Modugno[2].

Postulate C1 *Classical spacetime* is assumed to be a 4-dimensional oriented fibred manifold $t : \boldsymbol{E} \to \boldsymbol{T}$, where the base space $\boldsymbol{T}$ (*time*) is a 1-dimensional oriented affine space associated with the vector space $\mathbb{T}$. ♣

We shall not assume any distinguished splitting of spacetime into space and time, that is no distinguished observer. Actually our theory is observer-independent, namely it fulfills the general relativity principle in a 'Galileian' sense (with absolute time).

We shall use fibred spacetime charts, denoted by $(x^\lambda) := (x^0, y^i)$, where the coordinate x^0 is defined through a time unit u^0 and a time origin $\tau \in \boldsymbol{T}$ by $x^0(e) := u^0(t(e) - \tau)$. In general greek and latin indices will refer to spacetime and fibre coordinates, respectively.

We then have the scaled *time form* $dt : \boldsymbol{E} \to \mathbb{T} \otimes T^*\boldsymbol{E}$, with coordinate expression $dt = u_0 \otimes dx^0$.

Each fibre $\boldsymbol{E}_\tau$ of $\boldsymbol{E}$ represents the 'space at a given time' $\tau \in \boldsymbol{T}$; by analogy with Einstein relativity, we say that the vertical space $V\boldsymbol{E}$ is constituted by all 'spacelike' vectors on $\boldsymbol{E}$ (while we are not allowed to use the term 'timelike' in the present context).

The *phase space* of our theory is the jet bundle $J_1\boldsymbol{E} \to \boldsymbol{E}$, whose induced fibred coordinates are denoted by (x^0, y^j, y^j_0). From the general theory of jet spaces (see remark 2.3) we recall that $J_1\boldsymbol{E}$ can be naturally regarded as a subbundle $J_1\boldsymbol{E} \subset \mathbb{T}^*\otimes T\boldsymbol{E}$ over $\boldsymbol{E}$ constituted by tensors v whose time component is $v^0_0 = 1$. In other words, chosen a time unit u_0, the phase space $J_1\boldsymbol{E}$ can be identified with the affine subbundle of $T\boldsymbol{E}$ constituted by vectors v whose time component is $v^0 = 1$. Hence, the tangent space is unsufficient to represent the phase space of a theory explicitly independent of units of measurement.

A classical *particle motion* is defined to be a section

$$s : \boldsymbol{T} \to \boldsymbol{E} \ ,$$

and its (observer-independent) *velocity* is the jet prolongation

$$j_1 s : \boldsymbol{T} \to J_1\boldsymbol{E} \subset \mathbb{T}^*\otimes T\boldsymbol{E} \ .$$

We have the coordinate expression:

$$j_1 s = u^0\otimes\Big((\partial_0 \circ s) + \partial_0 s^j (\partial_j \circ s)\Big) \ .$$

Thus the jet space $J_1\boldsymbol{E}$ can be seen as the space of all particle 4-velocities. We stress that a (observer independent) 4-velocity v has no norm $\|v\|$ and its dimension is given just by $\mathbb{T}^*$ and not by $\mathbb{T}^*\otimes\mathbb{L}$.

An *observer* is defined to be a section $o : \boldsymbol{E} \to J_1\boldsymbol{E}$, i.e. just a field of particle velocities. By the way we observe that an observer can be regarded as a (possibly non linear) connection on $\boldsymbol{E} \to \boldsymbol{T}$.

Postulate C2 The fibres of $\boldsymbol{E}$ are assumed to be scaled Riemannian manifolds, i.e. spacetime is assumed to be equipped with a *scaled vertical Riemannian metric*

$$g : \boldsymbol{E} \to \mathbb{L}^2\otimes(V^*\boldsymbol{E} \underset{\mathbf{E}}{\otimes} V^*\boldsymbol{E}).$$

♣

The coordinate expression of the metric is $g = g_{hj}\check{d}y^h\otimes\check{d}y^j$. We stress that, differently form the Einstein case, we do not have a full spacetime metric: this is a weak feature of the Galileian theory. The metric yields vertical 'index-lowering' and 'index-raising' isomorphisms, $g^\flat : V\boldsymbol{E} \to \mathbb{L}^2\otimes V^*\boldsymbol{E}$ and $g^\# : \mathbb{L}^2\otimes V^*\boldsymbol{E} \to V\boldsymbol{E}$, but no similar isomorphisms between $T\boldsymbol{E}$ and $T^*\boldsymbol{E}$.

The metric and the time-form, along with the chosen orientation, yield the scaled *spacetime* and *spacelike volume forms*

$$\upsilon : \boldsymbol{E} \to (\mathbb{T}\otimes\mathbb{L}^3)\otimes \wedge^4 T^*\boldsymbol{E} \ , \quad \eta : \boldsymbol{E} \to \mathbb{L}^3\otimes \wedge^3 V^*\boldsymbol{E} \ ,$$

with coordinate expressions

$$\upsilon = \sqrt{|g|}\, u_0\otimes\omega \ , \qquad \eta = \sqrt{|g|}\, \check{\omega} \ ,$$

where we have set

$$\omega := d^0 \wedge d^1 \wedge d^2 \wedge d^3 \ , \qquad \check{\omega} := \check{d}^1 \wedge \check{d}^2 \wedge \check{d}^3 \ .$$

Differently from the Einstein case, the metric g does not characterise a unique spacetime connection; in order to fully appreciate the question we need to examine spacetime connections in some detail.

We first remark that there is a natural bijection between dt-preserving torsion-free linear connections on the tangent bundle $\boldsymbol{TE} \to \boldsymbol{E}$ and torsion-free affine connections on the jet bundle $J_1\boldsymbol{E} \to \boldsymbol{E}$ (see remark 2.4)

$$K : \boldsymbol{TE} \to T^*\boldsymbol{E} \underset{T\mathbf{E}}{\otimes} T\boldsymbol{TE} \,, \qquad \Gamma : J_1\boldsymbol{E} \to T^*\boldsymbol{E} \underset{J_1\mathbf{E}}{\otimes} TJ_1\boldsymbol{E} \,,$$

respectively. The coordinate expressions of such connections are:

$$K = d^\lambda \otimes \big(\partial_\lambda + (K_\lambda{}^j{}_h \dot{y}^h + K_\lambda{}^j{}_0 \dot{x}^0)\dot{\partial}_j\big) \,, \quad \Gamma = d^\lambda \otimes \big(\partial_\lambda + (\Gamma_\lambda{}^j{}_h y^h_0 + \Gamma_\lambda{}^j{}_0)\partial^0_j\big) \,,$$

with

$$K_\lambda{}^j{}_\mu = K_\mu{}^j{}_\lambda = \Gamma_\lambda{}^j{}_\mu = \Gamma_\mu{}^j{}_\lambda \,.$$

Then a *spacetime connection* is defined to be any of such equivalent connections. We shall be involved with both viewpoints; in fact, K and Γ will be suitable for the classical field theory and the classical and quantum mechanics, respectively.

A spacetime connection yields, by vertical restriction, a linear connection

$$K' : V\boldsymbol{E} \to T^*\boldsymbol{E} \underset{V\mathbf{E}}{\otimes} TV\boldsymbol{E}$$

on the bundle $V\boldsymbol{E} \to \boldsymbol{E}$, with coordinate expression $K' = \check{d}^\lambda \otimes (\partial_\lambda + K_\lambda{}^j{}_h \dot{y}^h \dot{\partial}_j)$. This connection will play a central role in the classical and quantum theory of spin.

A further vertical restriction gives the *vertical connection*

$$\check{K} : V\boldsymbol{E} \to V^*\boldsymbol{E} \underset{V\mathbf{E}}{\otimes} V_{\mathbf{E}} V\boldsymbol{E}$$

(which, more properly, is a family of connections: for each $\tau \in \boldsymbol{T}$, $\check{K}_\tau$ is a connection on the manifold $\boldsymbol{E}_\tau := t^{-1}(\tau)$). Its coordinate expression is $\check{K} = \check{d}^h \otimes (\partial_h + K_h{}^j{}_k \dot{y}^k \dot{\partial}_j)$.

A spacetime connection is said to be *metrical* if it preserves the vertical metric, i.e. if $\nabla[K']g = 0$. If K is metrical, then $\check{K}$ is exactly the Riemannian connection on the spacetime fibres; however, if $\check{K}$ is the Riemannian connection then K is not necessarily metrical, since $\nabla[K']g$ involves the covariant derivatives of g also along non-spacelike directions.

By recalling that (see remark 2.3)

$$V_{\mathbf{E}}\, J_1\boldsymbol{E} = J_1\boldsymbol{E} \underset{\mathbf{E}}{\times} (\mathbb{T}^* \otimes V\boldsymbol{E}) \,,$$

the vertical-valued 1-form associated with a spacetime connection Γ can be seen as a map

$$\nu_\Gamma : J_1\boldsymbol{E} \to \mathbb{T}^* \otimes (T^*J_1\boldsymbol{E} \underset{J_1\mathbf{E}}{\otimes} V\boldsymbol{E})$$

with coordinate expression $\nu_\Gamma = \big(d^j_0 - (\Gamma_\lambda{}^j{}_h y^h_0 + \Gamma_\lambda{}^j{}_0) d^\lambda\big) \otimes \partial_j$.

A spacetime connection yields the following two important objects: the (non linear) connection

$$\gamma := \text{д} \lrcorner \Gamma : J_1\boldsymbol{E} \to \mathbb{T}^* \otimes TJ_1\boldsymbol{E}$$

on the fibred manifold $J_1\boldsymbol{E} \to \boldsymbol{T}$ and the scaled 2-form (Janyška has proved that this form is essentially the unique natural object of this kind in the present framework)

$$\Omega := \nu_\Gamma \bar{\wedge} \vartheta : J_1\boldsymbol{E} \to (\mathbb{T}^* \otimes \mathbb{L}^2) \otimes \wedge^2 T^*J_1\boldsymbol{E}$$

on the manifold $J_1\boldsymbol{E}$ (here $\bar{\wedge}$ indicates exterior product followed by a metric contraction). These are called the *second order connection* and the *cosymplectic form* associated with Γ. Their coordinate expressions are

$$\gamma = u^0\otimes\big(\partial_0 + y^j_0\partial_j + \gamma^j\partial^0_j\big)\ , \qquad \Omega = g_{jk}u^0\otimes(d^j_0 - \gamma^j d^0 - \Gamma_h{}^j\vartheta^h)\wedge\vartheta^k\ ,$$

where

$$\gamma^j := \Gamma_{h\,k}^{\,j}y^h_0 y^k_0 + 2\Gamma_{h\,0}^{\,j}y^h_0 + \Gamma_{0\,0}^{\,j}\ , \qquad \Gamma_h{}^j = (\Gamma_{\lambda\,h}^{\,j}y^h_0 + \Gamma_{\lambda\,0}^{\,j})d^\lambda\ .$$

These objects fulfill the equality $\gamma\lrcorner\Omega = 0$, and it can be seen that they characterise Γ itself.

For any motion s the map

$$\nabla[\gamma]j_1 s := j_2 s - \gamma\circ j_1 s : \boldsymbol{T}\to(\mathbb{T}^*\otimes\mathbb{T}^*)\otimes V\boldsymbol{E}$$

is called the (observer-independent) *acceleration* of s.

Moreover

$$dt\wedge\Omega\wedge\Omega\wedge\Omega : J_1\boldsymbol{E}\to(\mathbb{T}^{*2}\otimes\mathbb{L}^6)\otimes\wedge^7 T^*J_1\boldsymbol{E}$$

is a scaled volume form on $J_1\boldsymbol{E}$. Also, if $o:\boldsymbol{E}\to J_1\boldsymbol{E}$ is any observer, we have the *observed scaled 2-form*

$$\Phi := 2o^*\Omega : \boldsymbol{E}\to(\mathbb{T}^*\otimes\mathbb{L}^2)\otimes\wedge^2 T^*\boldsymbol{E}$$

which, in a coordinate system adapted to o (i.e. such that $y^j_0\circ o = 0$), has the expression $\Phi = -2u^0\otimes(\Gamma_{0j0}d^0\wedge d^j + \Gamma_{hj0}d^h\wedge d^j)$.

It can be seen that, given an observer, a spacetime connection is characterised by $\nabla[K']g$ and Φ. Then these objects can be seen (in a sense) as the symmetric and antisymmetric parts of Γ with respect to a splitting determined by o. This is the keypoint for understanding how to characterise distinguished spacetime connections. In fact, a complex theorem[2] states that the condition that Ω is closed, i.e.

$$d\Omega = 0\ , \tag{1}$$

is equivalent to the couple of conditions that K is metrical and, for every observer, Φ is closed; a connection that satisfies this equation is then determined by g and a local potential of Φ, that is a 1-form

$$a : \boldsymbol{E}\to(\mathbb{T}^*\otimes\mathbb{L}^2)\otimes T^*\boldsymbol{E}$$

such that $\Phi = 2da$. Then a distinguished spacetime connection obeying eq.(1) is determined, similarly to the Einstein case, by ten scalar potentials: here, these are the six components of g and the four components of a.

Postulate C3 We assume that the *gravitational* and *electromagnetic fields* are represented, respectively, by a spacetime connection and a scaled 2-form

$$\Gamma^\natural : J_1\boldsymbol{E}\to T^*\boldsymbol{E}\underset{J_1\boldsymbol{E}}{\otimes}TJ_1\boldsymbol{E}\ , \qquad F:\boldsymbol{E}\to(\mathbb{L}\otimes\mathbb{M})^{1/2}\otimes\wedge^2 T^*\boldsymbol{E}\ .$$

♣

These two objects can be coupled in a natural way through any constant $c\in\mathbb{T}^*\otimes\mathbb{L}^{3/2}\otimes\mathbb{M}^{-1/2}$. Actually, we consider three possible natural choices for the coupling constant c. In the gravitational equation we are involved with $c=\sqrt{\kappa}$, where κ is Newton's gravitational constant (see §2.2). In the classical and quantum mechanics of a charged scalar particle, we take $c = q/m$; in the classical and quantum mechanics of a particle with spin, we take $c = \mu$ with reference to spin (see §2.2).

Namely, consider the *total cosymplectic form*

$$\Omega_c := \Omega^\natural + \tfrac{1}{2} cF ,$$

where $\Omega^\natural$ is the cosymplectic form of $\Gamma^\natural$. Then one sees that Ω_c characterises, in a natural way, a spacetime connection; namely there is a unique spacetime connection Γ_c such that $\Omega_c = \nu_{\Gamma_c} \bar{\wedge} \vartheta$ (that is, Ω_c is exactly the cosymplectic form associated with Γ_c). Actually, we can write

$$\Gamma_c = \Gamma^\natural + \Gamma^e_c ,$$

where

$$\Gamma^e_c : J_1 \boldsymbol{E} \to \mathbb{T}^* \otimes T^* \boldsymbol{E} \underset{\boldsymbol{E}}{\otimes} V\boldsymbol{E} .$$

We have the coordinate expression

$$(\Gamma_c)^j_{hk} = \Gamma^{\natural j}_{hk} , \quad (\Gamma_c)^j_{0k} = \Gamma^{\natural j}_{0k} + \tfrac{1}{2} u_0 c F^j{}_k , \quad (\Gamma_c)^j_{00} = \Gamma^{\natural j}_{00} + u_0 c F^j{}_0 .$$

Furthermore, the second order connection $\gamma_c := \text{л} \lrcorner \Gamma_c$ associated with Γ_c fulfills the condition $\gamma_c \lrcorner \Omega_c = 0$ and splits as

$$\gamma = \gamma^\natural + \gamma^e_c ,$$

where

$$\gamma^e_c : J_1 \boldsymbol{E} \to \mathbb{T}^* \otimes \mathbb{T}^* \otimes V\boldsymbol{E}$$

has the coordinate expression

$$\gamma^e_c = c(F^j{}_0 + F^j{}_h y^0_h) u^0 \otimes \partial^0_j .$$

Next we postulate the field equations.

Postulate C4 We postulate that the total connection Γ_c obeys, for all c's, the 'first field equation'

$$d\Omega_c = 0 .$$

♣

The closure of Ω_c implies that it is locally exact, but we cannot exhibit any distinguished potential. Clearly, this postulate is equivalent to the couple of conditions $d\Omega^\natural = 0$ and $dF = 0$ (first Maxwell equation). Also the observed cosymplectic form splits as $\Phi = \Phi^\natural + cF$.

Postulate C5 We postulate the *second gravitational* and *electromagnetic field equations*:

$$r^\natural = \mathrm{T} \qquad \operatorname{div}^\natural F = \rho \, dt ,$$

where $r^\natural$ is the Ricci tensor of $K^\natural$; T is the timelike energy tensor, which involves κ and contains source matter and electromagnetic terms; $\operatorname{div}^\natural$ is the spacelike divergence operator; ρ is the charge density of source matter. ♣

These equations yield the following synthetic formula

$$r_{\sqrt{\kappa}} = \mathrm{T}_{\sqrt{\kappa}} ,$$

where $r_{\sqrt{\kappa}}$ is the Ricci tensor of $K_{\sqrt{\kappa}}$, and $\mathrm{T}_{\sqrt{\kappa}} := \mathrm{T} + \sqrt{\kappa}\, \rho \, dt \otimes dt$.

Eventually, we postulate the fundamental law of motion of a charged particle.

Postulate C6 The equation of motion of a charged particle is

$$\nabla[\gamma_c] j_1 s = \nabla^\natural [\gamma_c] j_1 s - \gamma^e_c \circ j_1 s = 0 ,$$

with $c = q/m$. ♣

Then γ^e_c turns out to be just the *Lorentz force*.

2.4. Classical Spin

For a particle with spin, the fundamental bundle will be the vector bundle $\mathbb{L}^*\otimes V\boldsymbol{E} \to \boldsymbol{E}$ and the fundamental connection the total linear connection

$$C := K'_{2\mu} : V\boldsymbol{E} \to T^*\boldsymbol{E} \underset{V\boldsymbol{E}}{\otimes} TV\boldsymbol{E} \ .$$

We first note that g can be seen as a (non-scaled) metric on the vector bundle $\mathbb{L}^*\otimes V\boldsymbol{E} \to \boldsymbol{E}$. The induced 'index-lowering' and 'index-raising' maps will be indicated, respectively, by

$$g^\flat : \mathbb{L}^*\otimes V\boldsymbol{E} \to \mathbb{L}\otimes V^*\boldsymbol{E} \ , \qquad g^\# : \mathbb{L}\otimes V^*\boldsymbol{E} \to \mathbb{L}^*\otimes V\boldsymbol{E} \ .$$

We shall denote by (e_r) a positively-oriented orthonormal frame of $\mathbb{L}^*\otimes V\boldsymbol{E}$. The dual frame (ϵ^r) of $\mathbb{L}\otimes V^*\boldsymbol{E}$ determines a linear fibred chart (x^λ, ϵ^r) on $\mathbb{L}^*\otimes V\boldsymbol{E}$.

Clearly, C can be regarded also as a connection

$$C : \mathbb{L}^*\otimes V\boldsymbol{E} \to T^*\boldsymbol{E} \underset{\mathbb{L}^*\otimes V\boldsymbol{E}}{\otimes} T(\mathbb{L}^*\otimes V\boldsymbol{E}) \ ,$$

with coordinate expression

$$C = dx^\lambda \otimes (\partial x_\lambda + C_{\lambda}{}^{p}{}_{r} \epsilon^r e_p) \ ,$$

where

$$C_{h}{}^{r}{}_{s} = \tilde{\Gamma}^{\natural}{}_{h}{}^{r}{}_{s} \qquad C_{0}{}^{r}{}_{s} = \tilde{\Gamma}^{\natural}{}_{0}{}^{r}{}_{s} + u_0 \mu \tilde{F}^{r}{}_{s} = \tilde{\Gamma}^{\natural}{}_{0}{}^{r}{}_{s} + 2u_0 \mu \varepsilon^{r}{}_{sp} \tilde{B}^p \ ,$$

and

$$B := \tfrac{1}{2} * \check{F} : \boldsymbol{E} \to \mathbb{L}^{-5/2}\otimes\mathbb{M}^{1/2}\otimes V\boldsymbol{E}$$

is the (observer-independent) *magnetic field*. The tilde over the component of $\Gamma^\natural$, $\check{F}$ and B indicates that these are components in the frame (e_r).

Note that here λ is an index of spacetime coordinates, while the latin indices appearing in this formula are related to the linear coordinates ϵ^r, on the fibres of $\mathbb{L}^*\otimes V\boldsymbol{E}$, that are not induced by spacetime coordinates.

We shall indicate by $U\boldsymbol{E} \to \boldsymbol{E}$ the subbundle of $\mathbb{L}^*\otimes V\boldsymbol{E}$ whose fibres are unit 3-spheres. The history of a classical spinning particle will be described by a section $\upsilon : \boldsymbol{T} \to U\boldsymbol{E}$. Its projection $s : \boldsymbol{T} \to \boldsymbol{E}$ is a particle motion in the usual way, while the vertical vector field over it represents the particle's spin; more precisely, the classical intrinsic angular momentum of the particle is $\frac{1}{2}\hbar\upsilon$.

We postulate the following equation of motion for υ:

$$\nabla^\natural_{j_1 s}\upsilon = \mu\upsilon \times B = 0 \ .$$

For a classical charged particle in the flat case it is known that the interaction between spin and magnetic field yields an energy

$$-\mu\hbar\, g(\upsilon, B) = -\mu\hbar *(\upsilon^\flat \wedge \check{F}) = -\tfrac{1}{2}\mu\hbar\, \varepsilon_p{}^{rs} \upsilon^p \tilde{F}_{rs} \ .$$

This function is well-defined also in the general curved case. In order to see that it has the same meaning we should postulate the effect of spin on the electromagnetic field, through a suitable current to be coupled to the field via the Maxwell equations, and study the energy balance in the present context. We omit such an analysis, and just assume that the classical *spin Hamiltonian* $H^{\mathrm{s}} : J_1\boldsymbol{E} \times_{\boldsymbol{E}} V\boldsymbol{E} \to \mathbb{T}^{*2}\otimes\mathbb{L}^2\otimes\mathbb{M}$ is given by

$$H^{\mathrm{s}}[\upsilon] := H[s] - \mu\hbar\, g(\upsilon, B) \ ,$$

that is, in coordinates,

$$H^{\mathrm{s}} = \tfrac{1}{2} m g_{jk} y_0^j y_0^k - m a_0 - \tfrac{1}{2}\mu\hbar\varepsilon_p{}^{rs}\epsilon^p \tilde{F}_{rs} \ .$$

2.5. Quantisable Functions

Our phase space $J_1\boldsymbol{E}$ is odd-dimensional, thus there is no symplectic structure on it. Instead, we have the cosymplectic form Ω, which yields the linear morphism over $J_1\boldsymbol{E}$:

$$\Omega^\flat : TJ_1\boldsymbol{E} \to T^*J_1\boldsymbol{E} : v \mapsto \frac{m}{\hbar}\Omega(v) \ .$$

This is not an isomorphism. In fact, from $\gamma \lrcorner \Omega = 0$ it follows that $\Omega^\flat$ vanishes on any $v \in TJ_1\boldsymbol{E}$ which is in the image of $\gamma : \boldsymbol{E} \to \mathbb{T}^*\otimes TJ_1\boldsymbol{E}$. However, consider the vector subbundles over $J_1\boldsymbol{E}$:

$$T^*_\gamma J_1\boldsymbol{E} := \{\phi \in T^*J_1\boldsymbol{E} \,|\, \gamma \lrcorner \phi = 0\} \ , \qquad T_\tau J_1\boldsymbol{E} := \{v \in TJ_1\boldsymbol{E}, \,|\, v^0 = \tau(\pi(v))\} \ ,$$

where $\tau : J_1\boldsymbol{E} \to \mathbb{T}$ is *any* given smooth map (called a *time scale*) and $\pi : TJ_1\boldsymbol{E} \to J_1\boldsymbol{E}$ denotes the natural projection. Then one sees easily that $\Omega^\flat$ is an isomorphism $T_\tau J_1\boldsymbol{E} \to T^*_\gamma J_1\boldsymbol{E}$, whose inverse will be denoted by $\Omega^\#_\tau$.

Now, with any function $f : J_1\boldsymbol{E} \to \mathbb{R}$ we can associate the 1-form

$$d_\gamma f := df - \gamma \lrcorner df : J_1\boldsymbol{E} \to T^*_\gamma J_1\boldsymbol{E} \ ,$$

hence, for any time scale $\tau : J_1\boldsymbol{E} \to \mathbb{T}$, the vector field

$$f^\#_\tau := \Omega^\#_\tau(d_\gamma f) : J_1\boldsymbol{E} \to T_\tau J_1\boldsymbol{E} \ ;$$

the latter is called the *Hamiltonian lift* of f.

In particular, by taking $\tau = 0$ we can define the generalised Poisson Lie bracket

$$\{f_1\,,f_2\} := \frac{m}{\hbar}\Omega\big((f_1)^\#_0\,,(f_2)^\#_0\big) \ ,$$

which has the property

$$\{f_1\,,f_2\}^\#_0 = [(f_1)^\#_0\,,(f_2)^\#_0] \ .$$

In the quantum theory we will be involved with projectable Hamiltonian lifts. Actually, one can prove that the vector field $f^\#_\tau$ is projectable over a vector field $\boldsymbol{E} \to T\boldsymbol{E}$ iff f is, with respect to the fibres of $J_1\boldsymbol{E} \to \boldsymbol{E}$, is a polynomial of degree 2, whose second derivative equals $\tau\frac{m}{\hbar}g$. Namely, the coordinate expression of f must be of the type

$$f = u^{00}f''\frac{m}{2\hbar}g_{jk}y^j_0y^k_0 + f_jy^j_0 + f_\circ$$

with $f_j, f_\circ : \boldsymbol{E} \to \mathbb{R}$, $f'' : \boldsymbol{E} \to \mathbb{T}$, and τ must be equal to f''. Functions of this kind will be called *quantisable phase functions*. We shall indicate by $\mathcal{A}^{\mathrm{P}}$ the space of phase quantisable functions. The classical time, position, momentum, Hamiltonian and Lagrangian functions turn out to be of this kind.

Thus, if for any quantisable phase function f we choose $\tau = f''$, then we obtain the vector field

$$f^\# := \Omega^\#_{f''} : J_1\boldsymbol{E} \to T_{f''}J_1\boldsymbol{E} \ .$$

Its projection

$$X[f] : \boldsymbol{E} \to T\boldsymbol{E} \ ,$$

with coordinate expression

$$X[f] = u^0f''\partial_0 - u_0\frac{\hbar}{m}g^{jk}f_k\partial_j \ ,$$

is called the *tangent lift* of f.

Let now f_1 and f_2 be quantisable phase functions, and set

$$[f_1\,,f_2] := \{f_1\,,f_2\} + (f''_1\gamma).f_2 - (f''_2\gamma).f_1 \ .$$

Proposition 2.1 *The tangent lift*

$$\mathcal{A}^{\mathrm{P}} \to \mathcal{T}\boldsymbol{E} : f \mapsto X[f]$$

is an $\mathcal{F}\boldsymbol{E}$-linear epimorphism, with kernel $\mathcal{F}\boldsymbol{E} \subset \mathcal{A}^{\mathrm{P}}$, and an $\mathbb{R}$-Lie algebra morphism. Namely, we have

$$X[[f_1, f_2]] = \big[X[f_1], X[f_2]\big] .$$

The space of *quantisable spin functions* is defined to be the space $\mathcal{A}^{\mathrm{S}} := \mathcal{A}^{\mathrm{SQ}} \oplus \mathcal{A}^{\mathrm{SL}}$ of all functions $\phi : \mathbb{L}^{*} \otimes V\boldsymbol{E} \to \mathbb{R}$ of the type $\phi = \phi^{\mathrm{Q}} + \phi^{\mathrm{L}}$, where $\phi^{\mathrm{L}} \in \mathcal{A}^{\mathrm{SL}}$ is linear, $\phi^{\mathrm{Q}} \in \mathcal{A}^{\mathrm{SQ}}$ is quadratic and proportional to g. Namely, the expression of $\phi \in \mathcal{A}^{\mathrm{S}}$ in an orthonormal frame is of the type

$$\phi = \phi'' \tilde{g}_{rs} \epsilon^{r} \epsilon^{s} + \phi_{r} \epsilon^{r} ,$$

with $\phi'', \phi_r : \boldsymbol{E} \to \mathbb{R}$.

By means of the vertical isomorphism $g^{\#}$ any $\phi \in \mathcal{A}^{\mathrm{S}}$ yields the section

$$X[\phi] =: \phi^{\mathrm{Q}\#} + \phi^{\mathrm{L}\#} : \boldsymbol{E} \to \otimes^{2}(\mathbb{L}^{*} \otimes V\boldsymbol{E}) \underset{\boldsymbol{E}}{\oplus} \mathbb{L}^{*} \otimes V\boldsymbol{E} ,$$

Its orthonormal frame expression is

$$X[\phi] = \tilde{g}^{rs}(\phi'' e_{r} \otimes e_{s} + \phi_{s} e_{r}) := \tilde{g}^{rs} \phi'' e_{r} \otimes e_{s} + \phi^{r} e_{r} .$$

By analogy with phase functions we call $X[\phi]$ the *tangent lift* of ϕ.

We recall that $\mathbb{L}^{*} \otimes V\boldsymbol{E}$ is naturally equipped with the $\mathcal{F}\boldsymbol{E}$-Lie algebra structure given by cross-product. Since the map $\mathcal{A}^{\mathrm{SL}} \to \mathbb{L}^{*} \otimes V\boldsymbol{E} : \phi^{\mathrm{L}} \mapsto X[\phi]$ is an $\mathcal{F}\boldsymbol{E}$-linear isomorphism, it induces an $\mathcal{F}\boldsymbol{E}$-Lie algebra structure on $\mathcal{A}^{\mathrm{SL}}$. Moreover we define an $\mathcal{F}\boldsymbol{E}$-Lie algebra structure on $\mathcal{A}^{\mathrm{S}}$ by assuming $\mathcal{A}^{\mathrm{SQ}}$ to be an Abelian ideal. Then we have

$$[\phi, \theta] := (\phi^{\mathrm{L}\#} \times \theta^{\mathrm{L}\#})^{\flat} = \varepsilon_{p}{}^{rs} \phi_{r} \theta_{s} \epsilon^{p} .$$

Namely, only the linear parts of ϕ and θ contribute to $[\phi, \theta]$.

Eventually we note that $\mathcal{A}^{\mathrm{P}} \cap \mathcal{A}^{\mathrm{S}} = \{0\}$, and set

$$\mathcal{A} := \mathcal{A}^{\mathrm{P}} \oplus \mathcal{A}^{\mathrm{S}} .$$

We are going to define a bracket on $\mathcal{A}$. Since we have brackets on $\mathcal{A}^{\mathrm{P}}$ and $\mathcal{A}^{\mathrm{S}}$, it suffices to define the bracket between any $f \in \mathcal{A}^{\mathrm{P}}$ and any $\phi \in \mathcal{A}^{\mathrm{S}}$. Then we set

$$[f, \phi] := \nabla[C]_{X[f]} \phi^{\mathrm{L}} \in \mathcal{A}^{\mathrm{SL}} ,$$

and $[\phi, f] := -[f, \phi]$. Then $\mathcal{A}^{\mathrm{S}}$ and $\mathcal{A}^{\mathrm{SL}}$ are ideals of $\mathcal{A}$.

We have the coordinate expression

$$[f, \phi]_{s} = (u^{0} f'' \partial_{0} - u_{0} \frac{\hbar}{m} f^{j} \partial_{j}) \phi_{s} + (u^{0} f'' C_{0}{}^{r}{}_{s} - u_{0} \frac{\hbar}{m} f^{h} C_{h}{}^{r}{}_{s}) \phi_{r} .$$

The new bracket fulfills the Jacobi identity in all cases except when one and only one of the three factors belongs to $\mathcal{A}^{\mathrm{SL}}$.

3. QUANTUM THEORY

Now we consider a particle with given mass m, charge q and magnetic constant μ. Accordingly, we equip spacetime with the two total spacetime connections $\Gamma := \Gamma_{q/m}$ and $C := \Gamma_{2\mu}$ (§2.4), which carry both the given classical gravitational and electromagnetic fields. Then we formulate the quantum mechanics of our particle in the above background curved spacetime with absolute time.

3.1. Quantum Bundle

First of all we postulate the quantum bundle, "which carries the quantum kinematics". We stress that, differently from geometrical quantisation, this bundle is over spacetime.

Postulate Q1 The *scalar quantum bundle* is assumed to be a complex bundle $\pi_{\mathbf{Q}} : \boldsymbol{Q} \to \boldsymbol{E}$ over spacetime, with fibres of dimension 1 endowed with a Hermitian metric $h_{\mathbf{Q}}$. The *spin quantum bundle* is assumed to be a complex bundle $\pi_{\mathbf{S}} : \boldsymbol{S} \to \boldsymbol{E}$ over spacetime, with fibres of dimension 2 endowed with a Hermitian metric $h_{\mathbf{S}}$ and a non-singular $h_{\mathbf{S}}$-normalised 2-form $\varepsilon_{\mathbf{S}}$. ♣

In other words, $\pi_{\mathbf{Q}} : \boldsymbol{Q} \to \boldsymbol{E}$ and $\pi_{\mathbf{S}} : \boldsymbol{S} \to \boldsymbol{E}$ are bundles over specetime associated with principal bundles, whose structure groups are $U(1)$ and $SU(2)$, respectively.

We shall denote by b an $h_{\mathbf{Q}}$-normalised frame of $\boldsymbol{Q}$, and by z the corresponding chart on the fibres of $\boldsymbol{Q}$. The induced frame of $V\boldsymbol{Q} \to \boldsymbol{Q}$ will be denoted by ∂z.

We shall denote by (ζ_A), $A = 1, 2$, an $h_{\mathbf{S}}$-orthonormal frame of $\boldsymbol{S}$, and by (z^A) its dual frame. The induced frame of $V\boldsymbol{S}$ will be denoted by $(\partial_A := \partial z_A)$. The conjugate frame on $\boldsymbol{S}^{\bullet}$ will be denoted by $\bar{z}^{A^{\bullet}}$.

Now we focus our attention on the vector bundle $\mathrm{End}(\boldsymbol{S}) \to \boldsymbol{E}$ of complex linear endomorphisms of $\boldsymbol{S} \to \boldsymbol{E}$, whose fibres are equipped with the standard structure of associative algebra, given by $\phi\theta := \phi \circ \theta$, and with the induced structure of Lie algebra, given by $[\phi, \theta] := [\![\phi, \theta]\!] := \phi\theta - \theta\phi$. This bundle splits naturally into the direct sum of real subbundles as

$$\mathrm{End}(\boldsymbol{S}) = \boldsymbol{H} \underset{\mathbf{E}}{\oplus} i\boldsymbol{H} = (\langle 1\rangle \underset{\mathbf{E}}{\oplus} \boldsymbol{H}_0) \underset{\mathbf{E}}{\oplus} (\langle i1\rangle \underset{\mathbf{E}}{\oplus} i\boldsymbol{H}_0).$$

where $\boldsymbol{H}$, $\langle 1\rangle$ and $\boldsymbol{H}_0$ are the bundles of Hermitian endomorphisms, the bundle of enomorphisms generated by the identity and the bundle of traceless endomorphisms.

The bundle $\boldsymbol{H}_0$ is constituted by all endomorphisms ϕ whose matrix, in any $h_{\mathbf{S}}$-orthonormal frame of $\boldsymbol{S}$, is of the type $(\phi^A{}_B) = \begin{pmatrix} r & c \\ \bar{c} & -r \end{pmatrix}$, with $r \in \mathbb{R}$, $c \in \mathbb{C}$; then, the fibres of $\boldsymbol{H}_0$ have (real) dimension 3. Moreover, the fibres of $\boldsymbol{H}_0$ have a natural orientation and are equipped with the Euclidean metric

$$k : \boldsymbol{H}_0 \underset{\mathbf{E}}{\times} \boldsymbol{H}_0 \to \mathbb{R} : (\phi, \theta) \mapsto \tfrac{1}{2}\,\mathrm{Tr}(\phi \circ \theta)\ .$$

Hence, the bundle $\boldsymbol{H}_0 \to \boldsymbol{E}$ is associated with the principal bundle of all positively oriented k-orthonormal frames, with structure group $SO(3)$.

A positively oriented orthonormal frame is called a set of *Pauli operators*. Moreover we set $\sigma_0 := \mathbf{1}_{\mathbf{S}}$, so that (σ_α), $\alpha = 0, 1, 2, 3$, is a frame of $\boldsymbol{H}$.

Thus $\boldsymbol{H}_0$ is closed neither under the associative multiplication nor under the Lie commutator.

The metric k and the distinguished orientation yield the *cross-product* Lie algebra structure on $\boldsymbol{H}_0$, which, in terms of any set of Pauli operators, reads

$$\sigma_r \times \sigma_s = \varepsilon^p{}_{rs}\,\sigma_p\ .$$

The type fibre of this Lie algebra is $\mathfrak{su}(2)$, namely the Lie algebra of the Lie group $SU(2)$, which is usually called the *angular momentum algebra*.

The cross-product Lie algebra is related to the Lie algebra $\mathrm{End}(\boldsymbol{S})$ by the formula

$$\phi \times \theta = -\tfrac{i}{2}[\![\phi, \theta]\!]$$

which, in a set of Pauli operators, reads

$$[\![\sigma_r\,,\sigma_s]\!] = 2i\,\varepsilon^p{}_{rs}\,\sigma_p\,, \qquad \text{or} \qquad [\![-\tfrac{i}{2}\sigma_r, -\tfrac{i}{2}\sigma_s]\!] = \varepsilon^p{}_{rs}\cdot(-\tfrac{i}{2}\sigma_p)\,.$$

Then we see that $i\boldsymbol{H}_0$ is closed under the Lie bracket of $\mathrm{End}(\boldsymbol{S})$, and the map $\boldsymbol{H}_0 \to i\boldsymbol{H}_0 : \phi \mapsto -\frac{i}{2}\phi$ is a Lie algebra isomorphism.

Furthermore, the Clifford algebra bundle of $\boldsymbol{H}_0$ coincides with the real vector bundle underlying $\mathrm{End}(\boldsymbol{S})$, with the product given by ordinary composition.

Eventually, we define the *quantum spin bundle* to be the tensor product

$$\pi_{\mathbf{W}} : \boldsymbol{W} := \boldsymbol{Q} \underset{\mathbf{E}}{\otimes} \boldsymbol{S} \to \boldsymbol{E}\,.$$

The Hermitian metrics $h_{\mathbf{Q}}$ and $h_{\mathbf{S}}$, defined respectively on $\boldsymbol{Q}$ and $\boldsymbol{S}$, yield a Hermitian metric $h := h_{\mathbf{Q}} \otimes h_{\mathbf{S}}$ on $\boldsymbol{W}$.

We shall indicate by $b_A := b \otimes \zeta_A$ the orthonormal frame of $\boldsymbol{W}$ induced by a normal frame b of $\boldsymbol{Q}$ and by a orthonormal spin frame (ζ_A) of $\boldsymbol{S}$. The corresponding linear coordinates induced on $\boldsymbol{W}$ are denoted by $w^A := z \otimes z^A$, and the frame induced on $V\boldsymbol{W} \to \boldsymbol{W}$ by (∂w_A).

Quantum histories will be described as sections $\Psi : \boldsymbol{E} \to \boldsymbol{W}$. Locally:

$$\Psi = \Psi^A \otimes \zeta_A = \psi^A b \otimes \zeta := \psi^A b_A\,,$$

where $\Psi^A := \psi^A b : \boldsymbol{E} \to \boldsymbol{Q}$ is a scalar quantum history $(A = 1, 2)$, $\psi^A : \boldsymbol{E} \to \mathbb{C}$. In view of the Hilbert scalar product, it is also useful to regard a quantum section as a *quantum density*:

$$\Psi^\eta := \Psi \otimes \sqrt{\eta} : \boldsymbol{E} \to \boldsymbol{W}^\eta := \mathbb{L}^{3/2} \otimes \boldsymbol{W} \underset{\mathbf{E}}{\otimes} \sqrt{\wedge^3 V^* \boldsymbol{E}}\,.$$

3.2. Quantum Connection

Next we introduce the *quantum connection*, which is the main object of the quantum theory.

First we introduce a connection related to the scalar quantum bundle.

Let us consider the pullback bundle $\boldsymbol{Q}^\uparrow := J_1\boldsymbol{E} \times_{\mathbf{E}} \boldsymbol{Q} \to J_1\boldsymbol{E}$.

We can prove the following fact. If $\{\xi[o]\}$ is a *system of connections* of the bundle $\boldsymbol{Q} \to \boldsymbol{E}$ parametrised by the family of observers $\{o\}$, then there exists a unique connection Ч of the bundle $\boldsymbol{Q}^\uparrow \to J_1\boldsymbol{E}$, such that, for each observer o, the pullback $o^*\text{Ч}$ equals $\xi[o]$. This connection Ч is said to be *universal* and is characterised in coordinates by the condition $\text{Ч}_\lambda = \xi_\lambda$, $\text{Ч}^0_j = 0$. Conversely, a connection Ч of the bundle $\boldsymbol{Q}^\uparrow \to J_1\boldsymbol{E}$ such that $\text{Ч}^0_j = 0$ is the universal connection of a system of connections $\{\xi[o]\}$ of the bundle $\boldsymbol{Q} \to \boldsymbol{E}$.

Postulate Q2 We postulate a universal linear connection Ч on the bundle $\boldsymbol{Q}^\uparrow \to \boldsymbol{E}$, such that

$$\nabla[\text{Ч}]h_{\mathbf{Q}} = 0 \qquad R[\text{Ч}] = i\frac{m}{\hbar}\Omega_{q/m} \otimes \mathbf{1}_{\mathbf{Q}}\,,$$

where $\mathbf{1}_{\mathbf{Q}}$ is the identity of $\boldsymbol{Q}$. ♣

We can prove that the components of Ч are of the type

$$\text{Ч}_0 = -u_0\frac{H}{\hbar}\,, \qquad \text{Ч}_j = \frac{p_j}{\hbar}\,,$$

where H and p are the classical Hamiltonian and momentum associated with the frame of reference attached to the chosen chart, given a suitable gauge of the total potential a of Φ.

Next we introduce a connection on the spin quantum bundle.

Postulate Q3 We assume an orientation-preserving linear fibred isometry over $\boldsymbol{E}$,

$$\Sigma : \mathbb{L}^* \otimes V\boldsymbol{E} \to \boldsymbol{H}_0 ,$$

which will be called the *Pauli map*. ♣

If (e_r) is a positively-oriented orthonormal frame of $\mathbb{L}^* \otimes V\boldsymbol{E}$, then $(\sigma_r) := (\Sigma(e_r))$ is a set of Pauli operators. Henceforth, when dealing with Σ, we shall use the linear fibred charts on $\mathbb{L}^* \otimes V\boldsymbol{E}$ and $\boldsymbol{H}_0$ induced by a given frame (e_r) and the corresponding frame (σ_r). So, the information relative to Σ is encoded in the choice of such an adapted chart.

A Pauli map is, obviously, an isomorphism of cross-product Lie algebras (see §2.3). Moreover, we have the Lie algebra isomorphism $-\frac{i}{2}\Sigma : \mathbb{L}^* \otimes V\boldsymbol{E} \to i\boldsymbol{H}_0$.

Theorem 3.1 *There exists a unique linear connection* Б *on* $\boldsymbol{S}$ *such that*

$$\nabla[\text{Б}] h_{\mathbf{S}} = 0 \quad \nabla[\text{Б}] \varepsilon_{\mathbf{S}} = 0$$

and, for any section $v : \boldsymbol{E} \to \mathbb{L}^* \otimes V\boldsymbol{E}$,

$$\Sigma(\nabla[C]v) = \nabla[\text{Б}](\Sigma(v)) .$$

Namely, we have:

$$\text{Б}_{\lambda}{}^{A}{}_{B} = \tfrac{1}{4} \varepsilon_r{}^{sp} C_{\lambda}{}^{r}{}_{s} \sigma_p{}^{A}{}_{B} .$$

Proposition 3.1 *The curvature of* Б *is given by*

$$R[\text{Б}] = -\tfrac{i}{4}(*\Sigma)(\underline{R}[C])$$

where

$$\underline{R}[C] : \boldsymbol{E} \to \wedge^2 T^* \boldsymbol{E} \otimes \wedge^2 (\mathbb{L} \otimes V^* \boldsymbol{E})$$

is the completely covariant curvature tensor of C. *The coordinate expression of* $R[\text{Б}]$ *is*

$$R_{\lambda\mu}{}^{A}{}_{B} = \tfrac{i}{4} \varepsilon_r{}^{sp} R[C]_{\lambda\mu}{}^{r}{}_{s} \sigma_p{}^{A}{}_{B} .$$

Thus, we have an analogous role of the soldering forms $\mathbf{1_Q}$, $*\Sigma$ and of the base forms Ω, $\underline{R}[C]$ with respect to the scalar and spin quantum curvatures, respectively.

Eventually, we obtain a connection related to the quantum bundle.

In fact, the quantum connection and the spin connection yield a Hermitian linear connection $\text{Ч}^{\mathbf{W}} := \text{Ч} \otimes \text{Б}$, called the *quantum spin connection*, on the vector bundle

$$\boldsymbol{W}^{\uparrow} := J_1\boldsymbol{E} \underset{\mathbf{E}}{\times} \boldsymbol{W} \to J_1\boldsymbol{E} .$$

The components of $\text{Ч}^{\mathbf{W}}$ can be synthetically written as

$$\text{Ч}_{\lambda}{}^{A}{}_{B} = \text{ч}_{\lambda}^{\alpha} \sigma_{\alpha}{}^{A}{}_{B} = \text{ч}_{\lambda}^{0} \delta^{A}{}_{B} + \text{ч}_{\lambda}^{p} \sigma_{p}{}^{A}{}_{B} ,$$

where we have set

$$\text{ч}_{\lambda}^{0} := \text{Ч}_{\lambda} , \quad \text{ч}_{\lambda}^{h} := \text{Б}_{\lambda}^{h} .$$

The corresponding *covariant quantum differential* of a section Ψ turns out to be the section $\nabla\Psi : J_1\boldsymbol{E} \to T^*\boldsymbol{E} \otimes_{\mathbf{E}} \boldsymbol{W}$ with coordinate expression

$$\nabla\Psi = (\partial_\lambda \psi^A - i\text{Ч}_\lambda \psi^A - i\text{Б}_{\lambda}{}^{A}{}_{B} \psi^B) d^\lambda \otimes b_A .$$

Then we obtain the *time-like* and *space-like differentials*

$$\overset{\circ}{\nabla}\Psi := д \lrcorner \nabla\Psi : J_1\boldsymbol{E} \to \mathbb{T}^*\otimes\boldsymbol{W}\,, \qquad \check{\nabla}\Psi := \nabla\Psi_{|V\boldsymbol{E}} : J_1\boldsymbol{E} \to V^*\boldsymbol{E} \underset{\boldsymbol{E}}{\otimes} \boldsymbol{W}\,,$$

(we recall that $д : J_1\boldsymbol{E} \to \mathbb{T}^*\otimes T\boldsymbol{E}$ is the natural map introduced in remark 2.3) with expressions

$$\overset{\circ}{\nabla}\Psi = u^0\Big((\partial_0 + y^j_0\partial_j)\psi^A - i(Ч_0 + y^j_0 Ч_j)\psi^A - i(Б_0{}^A{}_B + y^j_0 Б_j{}^A{}_B)\psi^B\Big)\, b_A\,,$$
$$\check{\nabla}\Psi = (\partial_j\psi^A - iЧ_j\psi^A - iБ_j{}^A{}_B\psi^B)d^j\otimes b_A\,.$$

3.3. Generalised Pauli Equation

By means of the principle of projectability we exhibt a distinguished Lagrangian; then the standard Lagrangian formalism yields the quantum dynamical equations and conserved quantities.

We have the two following distinguished observer-dependent 4-forms over $\boldsymbol{E}$:

$$\overset{\circ}{\mathcal{L}}[\Psi] := \tfrac{1}{2}\Big(h(\Psi, i\overset{\circ}{\nabla}\Psi) + h(i\overset{\circ}{\nabla}\Psi, \Psi)\Big)\upsilon : \boldsymbol{E} \to \mathbb{L}^3\otimes \wedge^4 T^*\boldsymbol{E}\;;$$
$$\check{\mathcal{L}}[\Psi] := \frac{\hbar}{2m}(g^{\#}\otimes h)(\check{\nabla}\Psi, \check{\nabla}\Psi)\upsilon : \boldsymbol{E} \to \mathbb{L}^3\otimes \wedge^4 T^*\boldsymbol{E}\,,$$

where υ is the spacetime volume form (§2.3). Then we obtain a Lagrangian independent of the observers by the projectability principle. Namely:

Proposition 3.2 *The form*

$$\mathcal{L}[\Psi] := \overset{\circ}{\mathcal{L}}[\Psi] - \check{\mathcal{L}}[\Psi]$$

is the unique linear combination (up to an overall factor) of $\overset{\circ}{\mathcal{L}}$ and $\check{\mathcal{L}}$ which turns out to be independent of the observer.

Then we have the main dynamical postulate of the quantum theory:

Postulate Q4 The form $\mathcal{L}$ of proposition 3.2 is assumed to be the *quantum spin Lagrangian.* ♣

Note that adding to our Lagrangian a term proportional to the natural function

$$\frac{q}{m}h\Big(\Psi\,,\,\Sigma(B)\Psi\Big) : \boldsymbol{E} \to \mathbb{R}\,,$$

would simply amount to modifying the gyromagnetic ratio.

We have the coordinate expression:

$$\begin{aligned}\mathcal{L}[\Psi] = \tfrac{1}{2}\, h_{C^\bullet A}\Big(& i(\bar{\psi}^{C^\bullet}\partial_0\psi^A - \psi^A\partial_0\bar{\psi}^{C^\bullet}) - u_0\frac{\hbar}{m}g^{jk}\partial_j\psi^A\partial_k\bar{\psi}^{C^\bullet} \\ &+ ig^{jk}a_k(\psi^A\partial_j\bar{\psi}^{C^\bullet} - \bar{\psi}^{C^\bullet}\partial_j\psi^A) + u^0\frac{m}{\hbar}\psi^A\bar{\psi}^{C^\bullet}(2a_0 - g^{jk}a_j a_k) \\ &+ 2(Б_0{}^A{}_B - g^{jk}a_k Б_j{}^A{}_B)\psi^B\bar{\psi}^{C^\bullet} + u_0\frac{i\hbar}{m}g^{jk}Б_k{}^A{}_B(\psi^B\partial_j\bar{\psi}^{C^\bullet} - \bar{\psi}^{C^\bullet}\partial_j\psi^B) \\ &+ u_0\frac{\hbar}{m}g^{jk}Б_j{}^E{}_B Б_k{}^A{}_E\psi^B\bar{\psi}^{C^\bullet}\Big)\sqrt{|g|}\,\omega\,.\end{aligned}$$

The Lagrangian splits as:

$$\mathcal{L}[\Psi] = \mathcal{L}[\Psi^1] + \mathcal{L}[\Psi^2] + \mathcal{L}[\Psi]_{\mathsf{spin}}$$

where $\mathcal{L}[\Psi^1]$ and $\mathcal{L}[\Psi^2]$ (first two lines) are exactly the Lagrangians of the scalar wave functions Ψ^1 and Ψ^2 (see[2]). In the flat case the spin Lagrangian $\mathcal{L}[\Psi]_{\mathsf{spin}}$ reduces to the standard expression

$$\mathcal{L}[\Psi]_{\mathsf{spin}} = \tfrac{1}{2} u_0 \mu\, h(\Psi, \Sigma(B)\Psi)\sqrt{|g|}\,\omega \,.$$

Then, we obtain the following dynamical objects.

Proposition 3.3 *The Lagrangian yields, by the standard procedure, the* quantum momentum

$$\mathfrak{p} : J_1\boldsymbol{W} \to \mathbb{T}^*\otimes T\boldsymbol{E} \underset{\mathbf{E}}{\otimes} \boldsymbol{W} \;,$$

with coordinate expression

$$\mathfrak{p}[\Psi] = u^0\Big(\psi^A\partial_0 - i\frac{\hbar}{m}g^{jk}(u_0\partial_k - i\frac{m}{\hbar}a_k)\psi^A\partial_j - u_0\frac{\hbar}{4m}\varepsilon_p{}^{sr}g^{jk}C^{p}_{k\,s}\sigma_{r}{}^{A}{}_{B}\psi^B\partial_j\Big)\otimes b_A \;.$$

Proposition 3.4 *The Euler-Lagrange equations associated with the quantum Lagrangian are the generalised* Pauli equation

$$iD^o\Psi + \frac{\hbar}{2m}\check{\Delta}^o\Psi = 0 \;,$$

where

$$D^o\Psi = u^0\Big(\partial_0\psi^A + \frac{\partial_0\sqrt{|g|}}{2\sqrt{|g|}}\psi^A - iu^0\frac{m}{\hbar}a_0\psi^A - i\mathrm{B}_{0}{}^{A}{}_{B}\psi^B\Big)b_A \;,$$

$$(\check{\Delta}^o\Psi)^A = g^{jk}\Big(\delta^A_B(\partial_j - iu^0\frac{m}{\hbar}a_j) - i\mathrm{B}_{j}{}^{A}{}_{B}\Big)\Big(\delta^B_C(\partial_k - iu^0\frac{m}{\hbar}a_k) - i\mathrm{B}_{k}{}^{B}{}_{C}\Big)\psi^C \;.$$

It is remarkable that the Pauli equation can be expressed (again via the projectability principle) through the time differential and the quantum momentum as

$$\overset{\circ}{\nabla}[\Psi] + \mathrm{div}[\text{ч}^{\mathrm{S}}]\mathfrak{p}[\Psi] = 0 \;.$$

In the flat case the above equation reduces to the familiar Pauli equation

$$\begin{aligned} i\hbar\partial_0\psi^C = u_0\frac{1}{2m}g^{jk}(-i\hbar\partial_j - u^0ma_j)(-i\hbar\partial_k - u^0ma_k)\psi^C \\ - u^0ma_0\psi^C + \tfrac{1}{2}u_0\mu\hbar\tilde{B}^r\sigma_{r}{}^{C}{}_{B}\psi^B \;. \end{aligned}$$

For an electron $\mu = -e/m$ ($G = 2$), thus the last term equals $-\frac{\tilde{e}}{2m}\Sigma(B)\Psi$.

By considering the invariance of the Lagrangian with respect to the group $U(1)$, we obtain the conserved *probability current* with coordinate expression

$$\begin{aligned} j = \sqrt{|g|}\,h_{C^\bullet A}\Big[2\bar{\psi}^{C^\bullet}\psi^A\omega_0 \\ + \Big(iu_0\frac{\hbar}{m}g^{jk}(\bar{\psi}^{C^\bullet}\psi^A - \bar{\psi}^{C^\bullet}\psi^A_k) \\ - 2g^{jk}a_k\bar{\psi}^{C^\bullet}\psi^A - 2u_0\frac{\hbar}{4m}\varepsilon_p{}^{sr}g^{jk}C^{p}_{k\,s}\bar{\psi}^{C^\bullet}\sigma_{r}{}^{A}{}_{B}\psi^B\Big)\omega_j\Big] \;. \end{aligned}$$

3.4. Quantum Operators

Our geometrical scheme allows us to achieve quantum operators.

As far as scalar quantum mechanics is concerned, this subject has been discussed by Jadcyk and Modugno.[1,2] In a few words, we show that the vector fields on the quantum bundle preserving all quantum structures constitute a Lie algebra naturally isomorphic to the Lie algebra of quantisable phase functions. These vector fields yield naturally quantum operators. In a further forthcoming paper we show that this procedure can be easily extended to spin operators.

References

1. A. Jadczyk and M. Modugno, An outline of a new geometrical approach to Galilei general relativistic quantum mechanics, *in*: "Proc. XXI Int. Conf. on Differential Geometric Methods, in Theoretical Physics, Tianjin, 5-9 June 1992", C. N. Yang, M. L. Ge and X. W. Zhou (eds.), World Scientific, Singapore (1992)

2. A. Jadczyk and M. Modugno, "Galilei general relativistic quantum mechanics", Dipartimento di Matematica Applicata "G. Sansone", Firenze (1993).

MASSLESS SPINNING PARTICLES ON THE ANTI-DE SITTER SPACETIME

Stephan De Bièvre[1] and Salah Mehdi[2]

UFR de Mathématiques and
Laboratoire de Physique Théorique et Mathématique
Université Paris VII
2, place Jussieu, F-75251 Paris Cedex 05, France

Abstract

We show that unlike what happens on Minkowski spacetime, massless classical particles on the anti-de Sitter spacetime can have a true spin degree of freedom: their phase spaces are eight-dimensional and their world lines lightlike geodesics.

It is well known that the massless non-zero helicity representations of the Poincaré group do not admit a position operator.[1] This phenomenon has a classical counterpart: the corresponding classical particles are not represented by lightlike geodesics but rather by two-planes moving at the speed of light,[2,3] and interpreted as wave fronts. These particles moreover have six-dimensional phase spaces, reflecting the fact that they do not have a spin degree of freedom. On the other hand, massive particles, as well as massless particles of zero helicity, are described classically by geodesics and for them a position operator does not exist in quantum mechanics. We show here that the special status of the massless non-zero helicity particles disappears on the anti-de Sitter spacetime M_κ.

The classical and quantum description of the massive particles on M_κ as well as their behaviour when the curvature κ is taken to zero was studied in detail in Refs.4-5 (and references therein). Their classical motion follows timelike geodesics. The methods of Ref.4 are easily adapted to the zero-mass, zero-helicity case, and it is then easy to see that they move on lightlike geodesics of M_κ as expected. This leaves us with the equivalent of the zero-mass, non-zero helicity case. We turn to its study here. We will show that the phase space of these particles is eight-dimensional, so that they have a spin degree of freedom. Moreover the particles move on lightlike geodesics in M_κ.

The anti-de Sitter spacetime M_κ is the hyperboloid in $\mathbb{R}^5$ given by:

$$y \cdot y = \eta_{\mu\nu} y^\mu y^\nu = \eta^{\mu\nu} y_\mu y_\nu = (y_1)^2 + (y_2)^2 + (y_3)^2 - (y_4)^2 - (y_5)^2 = -\kappa^{-2}.$$

Here η is the standard symmetric quadratic form of signature $(+,+,+,-,-)$ on $\mathbb{R}^5$, which induces on M_κ a Lorentzian metric of signature $(+,+,+,-)$ in the usual way.

---

[1] e-mail: debievre@mathp7.jussieu.fr
[2] e-mail: mehdi@mathp7.jussieu.fr

The Einstein convention for summation over repeated indices is used throughout the paper. We choose on M_κ a time orientation by taking the vector $\dot{y}^\mu_{(0)} = \delta^\mu_4$ to be future pointing at $y^\mu_{(0)} = \delta^\mu_5 \kappa^{-1}$. The identity component of the isometry group of M_κ is $SO_0(3,2)$, which acts transitively on M_κ so that $M_\kappa \cong SO_0(3,2)/SO_0(3,1)$.

To describe the $SO_0(3,2)$-invariant one-particle systems on M_κ, we use, as in Refs.4-5, a combination of the ideas of Refs.3 and 7. We will equip the Lorentz bundle E_κ of M_κ with a $SO_0(3,2)$-invariant presymplectic form ω_κ. The kernel of ω_κ determines an integrable foliation $\mathfrak{F}_{\omega_\kappa}$ of E_κ. To each leaf $\mathfrak{L}$ of this foliation corresponds a unique motion of the particle as follows. Writing Π for the natural projection of E_κ onto M_κ, $\Pi(\mathfrak{L})$ is a world line of the particle. On the other hand, the phase space (or space of motions[3]) is obtained as the quotient of E_κ by the foliation $\mathfrak{F}_{\omega_\kappa}$ and can be naturally identified with a co-adjoint orbit of $SO_0(3,2)$. To proceed, it is convenient to identify E_κ with the space of all properly oriented η-orthonormal bases of $\mathbb{R}^5$, as follows:

$$E_\kappa = \{v = (e_1, e_2, e_3, e_4, e_5) \in \mathbb{R}^{25} \mid e_\mu \cdot e_\nu = \eta_{\mu\nu},\ \epsilon_{\alpha\beta\gamma\lambda\mu} e_1^\alpha e_2^\beta e_3^\gamma e_4^\lambda e_5^\mu = 1 \text{ and } e_5^5 e_4^4 - e_5^4 e_4^5 > 0\}$$

where $\epsilon_{\alpha\beta\gamma\lambda\mu}$ is the completely skew-symmetric tensor associated to $(\mathbb{R}^5, \eta)$ and

$$\Pi : v \in E_\kappa \mapsto y = \kappa^{-1} e_5 \in M_\kappa.$$

Clearly, (e_1, e_2, e_3, e_4) is an orthonormal bases of $T_y M_\kappa$ (the tangent space to M_κ at y) in this way. The action of $\Lambda \in SO_0(3,2)$ on E_κ is

$$\Lambda v = (\Lambda e_1, \Lambda e_2, \Lambda e_3, \Lambda e_4, \Lambda e_5). \tag{1}$$

It is transitive and free, so that E_κ is a principal homogeneous space for $SO_0(3,2)$. An explicit diffeomorphism I between $SO_0(3,2)$ and E_κ is established by picking an origin $v_{(0)}$ on E_κ. Choose $v_{(0)} = (e_{\mu(o)})$ with

$$e^\nu_{\mu(o)} = \delta^\nu_\mu$$

and write

$$I : \Lambda \in SO_0(3,2) \mapsto \Lambda v_{(0)} \in M_\kappa.$$

It is then easy to compute a basis $\{b_{\mu\nu}\}$ for the left-invariant vector fields on $E_\kappa \cong SO_0(3,2)$:

$$b_{\mu\nu} = \eta^{\mu\mu} e_\nu \cdot \frac{\partial}{\partial e_\mu} - \eta^{\nu\nu} e_\mu \cdot \frac{\partial}{\partial e_\nu}.$$

Note that the $b_{\mu\nu}$ correspond to the generators of the (pseudo-)rotations in the $\mu\nu$-plane of $\mathbb{R}^5$ and that they realize the Lie algebra $so(3,2)$ of $SO_0(3,2)$. The corresponding dual basis of left-invariant one forms is $\{\theta^{\mu\nu}\}$:

$$\theta^{\mu\nu} = e^\nu \cdot de^\mu = -e^\mu \cdot de^\nu = -\theta^{\nu\mu}$$

with

$$\theta^{\mu\nu}(b_{\mu'\nu'}) = \delta^\mu_{\mu'}\delta^\nu_{\nu'} - \delta^\mu_{\nu'}\delta^\nu_{\mu'}.$$

We are now ready to construct ω_κ as the exterior derivative of a left-invariant one form Θ_κ: $\omega_\kappa = -d\Theta_\kappa$. We will show below that the analog of the one-form used in Refs.2-3 to describe the massless non-zero helicity particle on Minkowski spacetime is

$$\Theta_\kappa = \frac{1}{\kappa}(\theta^{54} + \theta^{53}) + a\theta^{12} = \frac{1}{\kappa}(e_4 - e_3) \cdot de_5 + a e_2 \cdot de_1 \tag{2}$$

where $e^\mu = \eta^{\mu\mu} e_\mu$ and a is a non-zero real number. The case $a = 0$ corresponds to zero-helicity and will not interest us here any further. One obtains

$$\omega_\kappa = \frac{1}{\kappa} de_5 \wedge (de_4 - de_3) + a de_1 \wedge de_2.$$

A straightforward but somewhat lengthy calculation show that the kernel of ω_κ is two-dimensional and is generated by b_{12} and $b_{54} - b_{53}$. Since $[b_{12}, b_{54} - b_{53}] = 0$, it is particularly easy to compute the leaves of $\mathfrak{F}_{\omega_\kappa}$ as well as their projection on M_κ. Writing $\mathfrak{L}_{v_{(0)}}$ for the leaf through $v_{(0)}$, one finds in particular

$$\Pi(\mathfrak{L}_{v_{(0)}}) = \{y \in \mathbb{R}^5 \mid y^1 = 0 = y^2,\ y^5 = \kappa^{-1} \text{ and } y^3 = -y^4\}$$

which is a lightlike geodesic through $y_{(0)}$.

To obtain the phase space, the explicit symplectic reduction of $(E_\kappa, \omega_\kappa)$ can be constructed using the moment map J of the action (1) of $SO_0(3,2)$ as follows. First, remark that the generators of this action are the right-invariant vector fields on $E_\kappa \cong SO_0(3,2)$

$$X^{\mu\nu} = \sum_{j=1}^{5} ((e_j)^\mu . \frac{\partial}{\partial (e_j)_\nu} - (e_j)^\nu . \frac{\partial}{\partial (e_j)_\mu})$$

and that

$$L_{X_{\mu\nu}}(\omega_\kappa) = 0 = L_{X_{\mu\nu}}(\Theta_\kappa),$$

where $L_{X_{\mu\nu}}$ denotes the Lie derivative along $X_{\mu\nu}$. It follows that

$$i(X_{\mu\nu})\omega_\kappa = dJ_{\mu\nu}$$

where $J_{\mu\nu} = i(X_{\mu\nu})\Theta_\kappa$. Recalling that one can identify $so(3,2)^*$, the dual of the Lie algebra of $SO_0(3,2)$, with the space of left-invariant one forms on $SO_0(3,2)$, one defines the moment map

$$J : E_\kappa \longrightarrow so(3,2)^*, \quad v \mapsto \frac{1}{2} J_{\mu\nu}(v)\theta^{\mu\nu}$$

which is constant on the leaves of $\mathfrak{F}_{\omega_\kappa}$. It follows that $J(E_\kappa) \cong E_\kappa / \mathfrak{F}_{\omega_\kappa}$ is an eight-dimensional co-adjoint orbit of $SO_0(3,2)$. We denote it by $\mathfrak{O}_\kappa$.

It is easy to see that $\mathfrak{O}_\kappa$ does not depend on κ. Indeed, if $\{\exp(\lambda c_{34});\ \lambda \in \mathbb{R}\}$ denotes the one-parameter group of (pseudo-)rotations in the 34-plane, then

$$Ad^*(\exp(\log(\kappa) b_{34}))\Theta_1 = \Theta_\kappa.$$

So all Θ_κ belong to the same co-adjoint orbit, which shows the result. It is instructive to determine the Casimir invariants on the orbit. They are given by

$$C_1 = \frac{1}{2}\eta^{\mu\nu} J_{\mu\nu} J_{\mu\nu}$$

and

$$C_2 = A_\alpha A^\alpha$$

where $A_\alpha = \frac{1}{8}\epsilon_\alpha^{\beta\gamma\lambda\mu} J_{\beta\gamma} J_{\lambda\mu}$ is the the anti-de Sitter Pauli-Lubanski vector. Moreover a simple calculation shows that

$$J_{\mu\nu} = \frac{1}{\kappa}((e_4)_\nu (e_5)_\mu - (e_4)_\mu (e_5)_\nu) - \frac{1}{\kappa}((e_3)_\nu (e_5)_\mu - (e_3)_\mu (e_5)_\nu) + a((e_2)_\nu (e_1)_\mu - (e_2)_\mu (e_1)_\nu).$$

Thus we obtain $J_{54}(v_{(o)}) = -J_{53}(v_{(o)}) = 1/\kappa$, $J_{12}(v_{(o)}) = a$, $A^3(v_{(o)}) = a/\kappa = -A^4(v_{(o)})$, all the others $J_{\mu\nu}(v_{(o)})$ and $A^\mu(v_{(o)})$ being zero. So that

$$C_1 = a^2 \text{ and } C_2 = 0. \tag{3}$$

Note that these values are κ-independent. We will henceforth write $\mathfrak{O}_\kappa = \mathfrak{O}$.

To motivate the choice of Θ_κ in (2) we will use the well known fact that the anti-de Sitter group $SO_o(3,2)$ contracts to the Poincaré group $P^{3,1}$ and show that the orbit $\mathfrak{O}$ contracts to the orbit of the massless non-zero helicity particle of the Poincaré group. To proceed we first recall the AdS to Poincaré contraction.[5] The vector space underlying the two Lie algebra structures $\mathfrak{p}^{3,1}$ (the Lie algebra of $P^{3,1}$) and $so(3,2)$ is $V = \mathbb{R}^{10}$. We let $\{c_{\alpha\beta}\}$, with α and β in $\{1,2,3,4,5\}$, be an abstract basis of V such that:

$$[c_{\alpha\beta}, c_{\gamma\rho}] = \eta_{\alpha\gamma}c_{\beta\rho} + \eta_{\beta\rho}c_{\alpha\gamma} - \eta_{\alpha\rho}c_{\beta\gamma} - \eta_{\beta\gamma}c_{\alpha\rho}.$$

This realizes the Lie algebra $so(3,2)$. Let

$$\phi_\kappa : V \to V,\ c_{\alpha\beta} \mapsto c^\kappa_{\alpha\beta}$$

where

$$c^\kappa_{\alpha 5} = \kappa c_{\alpha 5} \text{ and } c^\kappa_{\alpha\beta} = c_{\alpha\beta} \tag{4}$$

for all α and β in $\{1,2,3,4\}$. It is clear that ϕ_κ is non-singular as long as $\kappa \neq 0$. So if we put

$$[c_{\alpha\beta}, c_{\gamma\rho}]_\kappa = \phi_\kappa^{-1}([\phi_\kappa(c_{\alpha\beta}), \phi_\kappa(c_{\gamma\rho})])$$

then $(V, [\,,\,]_\kappa)$ is a Lie algebra isomorphic to $so(3,2)$. But when κ reaches zero one obtains the Lie algebra $(V, [,]_o)$ which is no longer isomorphic to $so(3,2)$. Actually it is easy to check that $(V, [,]_o)$ is the Poincaré Lie algebra $\mathfrak{p}^{3,1}$. Note that for all α, β, γ and ρ in $\{1,2,3,4\}$:

$$\begin{aligned}
&[c_{\alpha\beta}, c_{\gamma\rho}]_0 = \eta_{\alpha\gamma}c_{\beta\rho} + \eta_{\beta\rho}c_{\alpha\gamma} - \eta_{\alpha\rho}c_{\beta\gamma} - \eta_{\beta\gamma}c_{\alpha\rho},\\
&[c_{\alpha\beta}, c_{\gamma 5}]_0 = \eta_{\alpha\gamma}c_{\beta 5} - \eta_{\beta\gamma}c_{\alpha 5},\\
&[c_{\alpha 5}, c_{\beta 5}]_0 = 0
\end{aligned}$$

are exactly the commutation relations of $\mathfrak{p}^{3,1}$. It is then obvious that the subalgebra $\sum_{\alpha,\beta=1}^4 \mathbb{R}c_{\alpha\beta}$ of $so(3,2)$ isomorphic to $so(3,1)$ is preserved in the AdS to Poincaré contraction. We now proceed as in Refs. 5 and 6. Since we are interested in the contraction of the orbit $\mathfrak{O}$ we need to consider the dual map ϕ^*_κ of ϕ_κ defined by

$$\phi^*_\kappa(f)(v) = f(\phi_\kappa^{-1}(v)) \tag{5}$$

for all f in V^* and v in V. Let us consider the family of applications $\{J^\kappa\}_{\kappa\in\mathbb{R}^*}$ defined as follows:

$$J^\kappa = (\phi^*_\kappa)^{-1} \circ J : E_\kappa \to V^*.$$

We shall write

$$J^\kappa(E_\kappa) = (\phi^*_\kappa)^{-1}(\mathfrak{O}) = (\phi^*_\kappa)^{-1}(Ad^*_{SO_o(3,2)}(\Theta_\kappa)).$$

It is then easy to see that the equations determining $J^\kappa(E_\kappa)$ are

$$\frac{1}{\kappa^2}J^{5\nu}J_{5\nu} + J^{\mu\nu}J_{\mu\nu} = a^2 \tag{6}$$

with $\mu, \nu = 1,2,3$ or 4 and $J_{54} > 0$, and

$$A^5 A_5 + \frac{1}{\kappa^2}A^\mu A_\mu = 0. \tag{7}$$

Multiplying (6) and (7) with κ^2 and taking $\kappa \to 0$ one obtains

$$J^{5\nu} J_{5\nu} = 0 \tag{8}$$

and

$$A^\mu A_\mu = 0. \tag{9}$$

These are exactly the values of the Casimirs for a massless non-zero helicity free particle on Minkowski spacetime.[2] In particular note that (9) is the Pauli-Lubanski equation related to such a system.[3] Recall that the one-form used in Ref.2 to describe the massless non-zero helicity particle on Minkowski spacetime is

$$\Theta_p = \theta^{54} + \theta^{53} + a\theta^{12}.$$

Then one sees that the co-adjoint orbit $\mathfrak{O}_p$ of $P^{3,1}$ passing through Θ_p is contained in the surface defined by (8) and (9). Moreover, we easily check that

$$\phi_\kappa^*(\Theta_p) = \Theta_\kappa.$$

We conclude that in the zero curvature limit, the surfaces defined by (6) and (7) tend to $\mathfrak{O}_p$. In this way we say that the co-adjoint orbit $\mathfrak{O}$ of the anti-de Sitter group $SO_o(3,2)$ contracts to the co-adjoint orbit $\mathfrak{O}_p$ of the Poincaré group $P^{3,1}$. Remark however that $\mathfrak{O}_p$ is only six-dimensional, whereas $\mathfrak{O}$ is eight-dimensional. This reflects the loss of the spin degree of freedom.

References

1. A. S. Wightman, On the localizability of quantum mechanical systems, *Rev. Mod. Phys.* 34:845 (1962).
2. R. Arens, Classical Lorentz invariant particles, *J. Math. Phys.* 12: 2415 (1971)
3. J-M. Souriau, "Structure des systèmes dynamiques", Dunod, Paris (1970)
4. S. De Bièvre and M. A. El Gradechi, Quantum mechanics and coherent states on the anti-de Sitter spacetime and their Poincaré contraction, *Ann. Inst. Henri Poincaré* 57:403 (1992)
5. S. De Bièvre and M. A. El Gradechi, Phase space quantum mechanics on the Anti-de Sitter spacetime and its Poincaré contraction, *Ann. Phys.* 235:1 (1994)
6. C. Cishahayo and S. De Bièvre, On the contraction of the discrete series of $SU(1,1)$, *Ann. Inst. Fourier* 43:551 (1993)
7. H. P. Künzle, Canonical dynamics of spinning particles in gravitational and electromagnetic fields, *J. Math. Phys.* 13:39 (1972)

A FAMILY OF NONLINEAR SCHRÖDINGER EQUATIONS: LINEARIZING TRANSFORMATIONS AND RESULTING STRUCTURE

H.-D. Doebner,[1,2] G. A. Goldin[3] and P. Nattermann[2]

[1] Arnold Sommerfeld Institute for Mathematical Physics
[2] Institute for Theoretical Physics
Technical University of Clausthal
D-38678 Clausthal-Zellerfeld, Germany
[3] Departments of Mathematics and Physics
Rutgers University
New Brunswick, New Jersey 08903, USA

Abstract

We examine a recently proposed family of nonlinear Schrödinger equations with respect to a group of transformations that linearize a subfamily of them. We investigate the structure of the whole family with respect to the linearizing transformations, and propose a new, invariant parameterization.

1. INTRODUCTION

Previous work[1−5] on the representation theory of an infinite-dimensional kinematical algebra on $\mathbb{R}^3$, and the corresponding infinite-dimensional group, led to a Fokker-Planck type of equation for the quantum-mechanical probability density and current,

$$\partial_t \rho = -\vec{\nabla} \cdot \vec{j} + D\Delta\rho \,, \tag{1.1}$$

and in turn to a family $\mathcal{F}_D$ of nonlinear Schrödinger equations. $\mathcal{F}_D$ is parameterized by the classification parameter D of the unitarily inequivalent group representations (the diffusion coefficient in Eq. (1.1)), and five real model parameters $D'c_1, \ldots, D'c_5$:

$$i\hbar\partial_t\psi = \left(-\frac{\hbar^2}{2m}\Delta + V(\vec{x})\right)\psi + i\frac{\hbar D}{2}\frac{\Delta\rho}{\rho}\psi + \hbar D'\left(\sum_{j=1}^{5} c_j R_j[\psi]\right)\psi \,. \tag{1.2}$$

Here D' also has the dimensions of a diffusion coefficient (so that the c_j are dimensionless), and the nonlinear functionals R_j are complex homogeneous of degree zero, defined by:

$$R_1[\psi] := \frac{\vec{\nabla}\cdot\vec{J}}{\rho}\,, \qquad R_2[\psi] := \frac{\Delta\rho}{\rho}\,, \qquad R_3[\psi] := \frac{\vec{J}^2}{\rho^2}\,,$$
$$R_4[\psi] := \frac{\vec{J}\cdot\vec{\nabla}\rho}{\rho^2}\,, \qquad R_5[\psi] := \frac{(\vec{\nabla}\rho)^2}{\rho^2}\,, \tag{1.3}$$

Quantization, Coherent States, and Complex Structures
Edited by J.-P. Antoine *et al.*, Plenum Press, New York, 1995

where $\rho := \bar{\psi}\psi$ and $\vec{J} := \text{Im}\,(\bar{\psi}\vec{\nabla}\psi) = (m/\hbar)\vec{j}$.

A subfamily of these equations, characterized by $D'c_1 = D = -D'c_4$, together with $c_2 + 2c_5 = 0$ and $c_3 = 0$, satisfies Ehrenfest's theorem in quantum mechanics,[4] and is linearizable via a nonlinear transformation.[6,7] In this short note we sketch some ideas connected with the linearizing transformations; for a detailed derivation and description with emphasis on the physical interpretation, we refer to forthcoming articles.[8]

2. LINEARIZATION

The members of the Ehrenfest subfamily $\mathcal{F}_D^{Ehr}$ can be transformed into linear Schrödinger equations by a transformation $\psi \mapsto \psi' = N(\psi)$, if the remaining unspecified model parameter $D'c_2$ satisfies

$$\frac{4m}{\hbar}D'c_2 < 1 - \frac{4m^2D^2}{\hbar^2}\,. \tag{2.1}$$

Here N depends on two real parameters $\gamma, \Lambda \in \mathbb{R}, \Lambda \neq 0$, and is given by

$$\psi' := N_{(\Lambda,\gamma)}(\psi) = \psi^{\frac{1}{2}(1+\Lambda+i\gamma)}\bar{\psi}^{\frac{1}{2}(1-\Lambda+i\gamma)} = |\psi|\, e^{i(\gamma \ln|\psi| + \Lambda \arg\psi)}\,. \tag{2.2}$$

This maps solutions of the Ehrenfest subfamily of (1.2) into solutions of the linear Schrödinger equation,

$$i\frac{\hbar}{\Lambda}\partial_t\psi' = \left(-\frac{\hbar^2}{2\Lambda^2 m}\Delta + V(\vec{x})\right)\psi'\,, \tag{2.3}$$

for the choices

$$\gamma = -\frac{2mD}{\hbar}\left(1 - \frac{4m}{\hbar}D'c_2 - \frac{4m^2D^2}{\hbar^2}\right)^{-\frac{1}{2}}, \quad \Lambda = \left(1 - \frac{4m}{\hbar}D'c_2 - \frac{4m^2D^2}{\hbar^2}\right)^{-\frac{1}{2}}. \tag{2.4}$$

However, it should be noted that for non-integer values of Λ, the map N is not actually well-defined by Eq. (2.2) for all ψ. The statements in this paper therefore depend, in some cases, on an appropriate selection of wave functions.

The transformations N are *local*, in that they depend only on the values of ψ and $\bar{\psi}$. They also respect the *projective* structure of the Hilbert space $\mathcal{H} = L^2(\mathbb{R}^3, d^3x)$; i. e. for any complex number c,

$$N_{(\Lambda,\gamma)}(c\,\psi) = |c|e^{i(\gamma \ln|c| + \Lambda \arg c)}N_{(\Lambda,\gamma)}(\psi)\,, \tag{2.5}$$

whence $(c\,\psi)'$ belongs to the same ray as ψ'. Furthermore the transformations leave the symplectic structure $\omega = \delta\psi \wedge \delta\bar{\psi}$ on the Hilbert space $\mathcal{H}$ invariant up to a factor:

$$\begin{aligned} N^*_{(\Lambda,\gamma)}\omega &= \left(\frac{\partial N_{(\Lambda,\gamma)}}{\partial\psi}\delta\psi + \frac{\partial N_{(\Lambda,\gamma)}}{\partial\bar{\psi}}\delta\bar{\psi}\right) \wedge \overline{\left(\frac{\partial N_{(\Lambda,\gamma)}}{\partial\psi}\delta\psi + \frac{\partial N_{(\Lambda,\gamma)}}{\partial\bar{\psi}}\delta\bar{\psi}\right)} \\ &= \left(\frac{\partial N_{(\Lambda,\gamma)}}{\partial\psi}\frac{\partial \bar{N}_{(\Lambda,\gamma)}}{\partial\bar{\psi}} - \frac{\partial N_{(\Lambda,\gamma)}}{\partial\bar{\psi}}\frac{\partial \bar{N}_{(\Lambda,\gamma)}}{\partial\psi}\right)\delta\psi \wedge \delta\bar{\psi} \\ &= \Lambda\omega \end{aligned} \tag{2.6}$$

This gives a Hamiltonian formulation for the Ehrenfest family $\mathcal{F}_D^{Ehr}$, as has been noted elsewhere,[4,5] and establishes a connection to the framework of Weinberg.[9]

The set $\mathcal{N} := \{N_{(\Lambda,\gamma)}\}$ of these nonlinear transformations obeys the group law of the affine group $Aff(1)$ in one dimension,

$$N_{(\Lambda_1,\gamma_1)} \circ N_{(\Lambda_2,\gamma_2)} = N_{(\Lambda_1\Lambda_2,\,\Lambda_1\gamma_2+\gamma_1)}\,. \tag{2.7}$$

A further, essential property of $N \in \mathcal{N}$ is, of course, that it leaves the probability density invariant:

$$\rho'(\vec{x},t) = \rho(\vec{x},t)\,, \tag{2.8}$$

where ρ' is the density transformed under N. In a certain sense N is a nonlinear generalization of a linear, $U(1)$-gauge transformation. We shall call $\mathcal{N}$ the set of *local projective nonlinear gauge transformations.*

3. GAUGE INVARIANCE AND REPARAMETERIZATION

The transformation of the current $\vec{J}'$ under $N \in \mathcal{N}$ is given by

$$\vec{J}' := \mathrm{Im}\left(\bar{\psi}'\vec{\nabla}\psi'\right) = \rho'\vec{\nabla}\arg\psi' = \Lambda\vec{J} + \frac{\gamma}{2}\vec{\nabla}\rho\,. \tag{3.1}$$

In order to show the invariance of the family $\mathcal{F}_D$ of equations (1.2), we rewrite $\mathcal{F}_D$ wholly in terms of densities and currents. Using the expansion of the Laplacian $\Delta\psi = \{iR_1[\psi] + (1/2)R_2[\psi] - R_3[\psi] - (1/4)R_5[\psi]\}\,\psi$, we obtain the general form,

$$i\partial_t\psi = i\sum_{j=1}^{2}\nu_j R_j[\psi]\psi + \sum_{j=1}^{5}\mu_j R_j[\psi]\psi + \mu_0 V\psi. \tag{3.2}$$

From (2.2), (2.8) and (3.1), we deduce that $\psi' = N_{(\Lambda,\gamma)}(\psi)$ again fulfills (3.2), but with primed parameters:

$$\begin{gathered}\nu_1' = \frac{\nu_1}{\Lambda}\,,\quad \nu_2' = -\frac{\gamma}{2\Lambda}\nu_1 + \nu_2\,,\\ \mu_1' = -\frac{\gamma}{\Lambda}\nu_1 + \mu_1\,,\quad \mu_2' = \frac{\gamma^2}{2\Lambda}\nu_1 - \gamma\nu_2 - \frac{\gamma}{2}\mu_1 + \Lambda\mu_2\,,\quad \mu_3' = \frac{\mu_3}{\Lambda}\\ \mu_4' = -\frac{\gamma}{\Lambda}\mu_3 + \mu_4\,,\quad \mu_5' = \frac{\gamma^2}{4\Lambda}\mu_3 - \frac{\gamma}{2}\mu_4 + \Lambda\mu_5\,,\quad \mu_0' = \Lambda\mu_0\,.\end{gathered} \tag{3.3}$$

Thus we have that the 8-parameter family $\mathcal{F}$ of Eq. (3.2) is invariant under the action of $\mathcal{N}$; i.e., under the action of the affine group $Aff(1)$. An appropriate description of $\mathcal{F}$ is by means of the *orbits* of $Aff(1)$; since there are 2 group parameters, we look next for 6 functionally independent parameters that are invariant under $Aff(1)$. These are the gauge invariants. After some calculations, we obtain gauge-invariant parameters:

$$\begin{gathered}\iota_1 = \nu_1\mu_2 - \nu_2\mu_1\,,\quad \iota_2 = \mu_1 - 2\nu_2\,,\quad \iota_3 = 1 + \mu_3/\nu_1\,,\quad \iota_4 = \mu_4 - \mu_1\mu_3/\nu_1\,,\\ \iota_5 = \nu_1(\mu_2 + 2\mu_5) - \nu_2(\mu_1 + 2\mu_4) + 2\nu_2^2\mu_3/\nu_1\,,\quad \iota_0 = \nu_1\mu_0\,.\end{gathered} \tag{3.4}$$

Now if we like we can choose ν_1 and μ_1 as our group parameters ($\nu_1 \neq 0$); this condition is fulfilled for all modifications of the linear Schrödinger equation, as ν_1 derives from the Laplacian in the Schrödinger equation. Inverting (3.4), we have

$$\begin{gathered}\nu_2 = \frac{1}{2}(\mu_1 - \iota_2)\,,\quad \mu_2 = \frac{1}{2}\nu_1^{-1}(2\iota_1 - \iota_2\mu_1 + \mu_1^2)\,,\\ \mu_3 = (\iota_3 - 1)\nu_1\,,\quad \mu_4 = \iota_4 - \mu_1 + \iota_3\mu_1 \qquad (?\\ \mu_5 = \frac{1}{2}\nu_1^{-1}\left(\iota_5 - \iota_1 + \iota_4(\mu_1 - \iota_2) + \frac{1}{2}(\mu_1^2 - \iota_2^2)(\iota_3 - 1)\right)\,,\quad \mu_0 = \nu_1^{-1}\iota_0\,.\end{gathered}$$

With this reparameterization the family of nonlinear Schrödinger equations is foliated in leaves characterized by $\iota_0, \ldots, \iota_5$, of subfamilies depending on the two group parameters ν_1, μ_1. The group of nonlinear gauge transformations $\mathcal{N}$ acts effectively on each leaf of the foliation. Because of (2.8), the time-evolving probability density for all points (ν_1, μ_1) in a given leaf is the same.

Let us identify the gauge invariants for some special leaves:

a. The linear Schrödinger equation corresponds to the values $\nu_1 = -\hbar/2m$, $\mu_2 = -\hbar/4m$, $\mu_3 = \hbar/2m$, $\mu_5 = \hbar/8m$, $\mu_0 = 1/\hbar$, and $\nu_2 = \mu_1 = \mu_4 = 0$. We then have only two nonvanishing gauge invariants:

$$\iota_1 = \frac{\hbar^2}{8m^2}, \quad \iota_0 = -\frac{1}{2m} \text{ (for } V \not\equiv 0) . \tag{3.6}$$

Note that μ_0 and ι_0 are indeterminate if $V \equiv 0$. So $\hbar$ and m (or their quotient, in the free case) are gauge-invariant quantities for the family of equations.

b. The Ehrenfest subfamily $\mathcal{F}_D^{Ehr}$ corresponds to the values $\nu_1 = -\hbar/2m$, $\nu_2 = D/2$, $\mu_1 = D$, $\mu_2 = -\hbar/4m + c_2 D'$, $\mu_3 = \hbar/2m$, $\mu_4 = -D$, $\mu_5 = \hbar/8m - c_2 D'/2$, and $\mu_0 = 1/\hbar$. Again there are just two nonzero gauge invariants: ι_0 as before, and

$$\iota_1 = \frac{\hbar^2}{8m^2} - c_2 \frac{\hbar D'}{2m} - \frac{D^2}{2} . \tag{3.7}$$

Here we rediscover the linearization of the Ehrenfest family, as when the right-hand side of (3.7) is positive, we can introduce a new constant $\hbar'$ such that a linear Schrödinger equation with $\hbar'$ replacing $\hbar$ is contained in the orbit.

c. For the more general, Galilei (Schrödinger) invariant subfamily[4,10] $\mathcal{F}_D^{Gal}$ of $\mathcal{F}_D$, characterized by the conditions $c_1 + c_4 = c_3 = 0$, we have the values: $\nu_1 = -\hbar/2m$, $\nu_2 = D/2$, $\mu_1 = c_1 D'$, $\mu_2 = -\hbar/4m + c_2 D'$, $\mu_3 = \hbar/2m$, $\mu_4 = -c_1 D'$, $\mu_5 = \hbar/8m + c_5 D'$, and $\mu_0 = 1/\hbar$. Now we get four nonvanishing gauge invariants, taking independent values: ι_0, ι_1, ι_2, and ι_5.

4. FINAL REMARK

We have noted that the transformations linearizing the Ehrenfest subfamily $\mathcal{F}_D^{Ehr}$ of a family $\mathcal{F}_D$ of nonlinear Schrödinger equations can be viewed as generalizing the usual $U(1)$-gauge transformations, and act as a gauge group $\mathit{Aff}(1)$ on the parameter space of the family. We have calculated a parameterization by means of the gauge invariants, together with the group parameters of $\mathit{Aff}(1)$.

In connection with a more physical interpretation of our foliation of $\mathcal{F}_D$, we quote a remark by Feynman and Hibbs[11]:

> *Indeed all measurements of quantum-mechanical systems could be made to reduce eventually to position and time measurements. Because of this possibility a theory formulated in terms of position measurements is complete enough in principle to describe all phenomena.* (p. 96)

If one adopts this point of view, quantum theories for which the wave functions give the same probability density in space and time are "in principle" equivalent. Hence the leaves of our foliation consist of sets of "in principle" equivalent quantum-mechanical evolution equations.

References

1. H.-D. Doebner and G. A. Goldin, On a general nonlinear Schrödinger equation admitting diffusion currents, *Phys. Lett. A* 162:397 (1992)

2. H.-D. Doebner and G. A. Goldin, Group theoretical foundations of nonlinear quantum mechanics, *in:* "Annales de Fisica, Monografias, Vol. II", p. 442, CIEMAT, Madrid (1993)

3. H.-D. Doebner and G. A. Goldin, Manifolds, general symmetries, quantization, and nonlinear quantum mechanics, *in:* "Proceedings of the First German-Polish Symposium on Particles and Fields, Rydzyna Castle, 1992", p. 115, World Scientific, Singapore (1993)

4. H.-D. Doebner and G. A. Goldin, Properties of nonlinear Schrödinger equations associated with diffeomorphism group representations, *J. Phys. A: Math. Gen.* 27:1771 (1994)

5. P. Nattermann, "Struktur und Eigenschaften einer Familie nichtlinearer Schrödingergleichungen", Diplom thesis; Technical University of Clausthal (1993)

6. P. Nattermann, Solutions of the general Doebner-Goldin equation via nonlinear transformations , *in:* "Proceedings of the XXVI Symposium on Mathematical Physics, Torun, December 7–10, 1993", p. 47, Nicolas Copernicus University Press, Torun (1994)

7. G. Auberson and P. C. Sabatier, On a class of homogemeous nonlinear Schrödinger equations, *J. Math. Phys.* 35:4028 (1994)

8. H.-D. Doebner, G. A. Goldin and P. Nattermann, work in progress, to be submitted for publication.

9. S. Weinberg, Testing quantum mechanics, *Ann. Phys. (NY)* 194:336 (1989)

10. P. Nattermann, Symmetry, local linearization, and gauge classification of the Doebner-Goldin equation, Clausthal–preprint ASI-TPA/8/95, *Rep. Math. Phys.* (to appear)

11. R. P. Feynman and A. R. Hibbs, "Quantum Mechanics and Path Integrals", McGraw-Hill, New York (1965).

MODULAR STRUCTURES IN GEOMETRIC QUANTIZATION

Gérard G. Emch

Department of Mathematics, University of Florida
Gainesville, FL 32611, USA

Abstract

The purpose of this lecture is to show how certain modular structures, borrowed from the theory of von Neumann algebras, can be exploited to extract primary representations (prequantization) and irreducible representations (quantization) from the regular representation of the symmetry group of the physical systems to be considered. The emphasis is on presenting specific examples for which the solution is exhibited explicitly.

1. REVIEW AND MOTIVATION

Depending on who is speaking, physicist or mathematician, geometric quantization is a method either to select "relevant" representations, or to construct "specific" representations of a group; the matter of which group one wishes to consider is a further question of idiosyncratic (or "classical") preferences.

The general idea of geometric quantization was discovered from a physical approach by Souriau,[1] and was formulated independently by Kirillov.[2] Kostant [3] made important contributions; see also the expository accounts by: Simms & Woodhouse,[4] Guillemin & Sternberg,[5] Abraham & Marsden,[6] Śniatycki,[7] Woodhouse.[8]

In this lecture, we want to draw attention to some, hitherto neglected, algebraic structures underlying the geometric quantization programme; to be specific, we are going to focus on the following examples: the Weyl group of the canonical commutation relations, first for the flat configuration space $\mathbb{R}^n$, and its generalisation to the simply connected manifold $\mathbb{H}^n$ of constant negative curvature; we will then show how much of what one learns from these two examples can be transferred to certain representations of groups such as $SO(2,3)$. For the moment, let us denote by: G any of these groups; $d\mu$ its left-invariant Haar measure; and $U^L(G)$ its left-regular representation on the Hilbert space $L^2(G, d\mu)$.

The Kirillov[2] method of co-adjoint orbits extracts from $U^L(G)$ unitary sub-representations $U_P(G)$; these are best understood as operating on the Hilbert space $L^2(\Omega_{\xi^*}, \omega \wedge \cdots \wedge \omega)$ where Ω_{ξ^*} is a coadjoint orbit of G, i.e. the orbit of some $\xi^* \in \mathcal{G}^*$ under the coadjoint-action of G on the dual $\mathcal{G}^*$ of its Lie algebra $\mathcal{G}$; ω is the symplectic form naturally [2] induced by the coadjoint action of G on Ω_{ξ^*}. The physical interpretation of this procedure is that the symplectic manifold (Ω_{ξ^*}, ω) models a classical phase space, e.g. in the simplest cases, the cotangent bundle T^*M of some configuration

Quantization, Coherent States, and Complex Structures
Edited by J.-P. Antoine *et al.*, Plenum Press, New York, 1995

space M. The stage on which the representations $U_P(G)$ appears is thus refered to as "prequantization". The next stage ("quantization") is to extract sub-representations of $U_P(G)$ that will now be irreducible; this is achieved by the introduction of a real (or complex, as the case may be) 'polarisation', a procedure akin to choosing to emphasise a specific CSCO ('complete system of commuting observables' à la Dirac[9]).

In the particular case of the CCR (canonical commutation relations) for the configuration space $\mathbb{R}^n$, the group G is a $(2n+1)$-dimensional Lie group W_o^{2n+1}, which corresponds to the fact that we are interested in projective representations of a group G^{2n} the generators of which are the n-components of the linear momentum and the n-components of the position. At the prequantization stage, these observables are the self-adjoint differential operators defined on (a commun dense domain of) the Hilbert space $L^2(\mathbb{R}^{2n}, d^np\, d^nq)$ by:

$$P^k = -i\hbar\partial_{q^k} \quad ; \quad Q^k = i\hbar\partial_{p^k} + q^k \quad . \tag{1.1}$$

It was noticed by Streater,[10] who attributes this remark to Segal,[11] that the same Hilbert space supports an (anti-) representation of the CCR, namely

$$\tilde{P}^k = i\hbar\partial_{q^k} + p^k \quad ; \quad \tilde{Q}^k = -i\hbar\partial_{p^k} \quad . \tag{1.2}$$

and that these two representations commute with one another; in fact,[12] these representations are primary, and they are even the commutant of one another (see next section), a phenomenon that can be traced[13] in the reproducing kernel structure of phase space quantization.

The questions we want to address in this lecture are:

(a) whether this phenomenon is a freak due to the flatness of $\mathbb{R}^n$;

(b) what are the stuctures lurking behind this phenomenon;

(c) whether these structures can be brought to bear on other systems.

From a mathematical point of view, we shall show that a clue to answering these questions is to be found in the natural isomorphism between the left- and right-regular representations of G.

2. MODULAR STRUCTURES

In his reminiscences on the seminal "rings of operators" papers,[14] Murray[15] points out that the theory acquired very early two of its major structure theorems.

The first of these is the "double commutant theorem".[16] It links topological and algebraic properties of what was later[17] to be called a von Neumann algebra, i.e. an object $\mathcal{N}$ defined as a sub-* algebra of $\mathcal{B}(\mathcal{H})$, the *-algebra of the bounded linear operators on a Hilbert space $\mathcal{H}$, with $\mathcal{N}$ satisfying one of the following two equivalent properties: (1) $\mathcal{N}$ is closed in the weak–operator topology and contains the identity in $\mathcal{B}(\mathcal{H})$; and (2) $\mathcal{N}$ is closed under the algebraic operation of taking the double commutant. For completeness, we recall[17–20] here that: (1) the closure $\mathcal{S}^w$ in weak- operator topology of a set $\mathcal{S}$ is the set of all $T \in \mathcal{B}(\mathcal{H})$ such that, given any positive ϵ and any pair of finite sequences $\{\phi_i \mid i = 1,\cdots,n\}$ and $\{\psi_i \mid i = 1,\cdots,n\}$ of vectors in $\mathcal{H}$, there is an $S \in \mathcal{S}$ such that $|<(T-S)\phi_i, \psi_i>| < \epsilon \quad \forall \quad i = 1,\cdots,n$; and (2) the commutant $\mathcal{S}'$ of $\mathcal{S}$ is the set of all $S' \in \mathcal{B}(\mathcal{H})$ such that $[S, S'] \equiv SS' - S'S = 0$; the bicommutant of $\mathcal{S}$ is the set $\mathcal{S}'' \equiv (\mathcal{S}')'$. The later characterization[18] of von Neumann algebras as (concrete) W^*-algebras, i.e. as (faithful representations of) C^*-algebras that are the dual of a Banach space), will not bear directly on what we need in this lecture.

The second of these two basic theorems was the problem von Neumann proposed to young Murray, namely let $\mathcal{N}$ be a factor, i.e. a von Neumann algebra with trivial center: $\mathcal{N} \cap \mathcal{N}' = CI$. For instance, let $\mathcal{N} = \{M \otimes I \mid M \in \mathcal{M}_m\}$ where $\mathcal{M}_m$ is the von Neumann algebra of all $m \times m$ matrices with complex entries, and I is the identity in $\mathcal{M}_n$; one has then indeed $\mathcal{N}' = \{I \otimes N \mid N \in \mathcal{M}_n\}$; and thus $\mathcal{N} \cap \mathcal{N}' = CI$. This example, we are told by Murray[15], motivated von Neumann's question, namely whether it is was true in general that every factor $\mathcal{N}$ is isomorphic to some $\mathcal{B}(\mathcal{H})$. The answer, it turned out, is negative in general, unless one adds the condition that $\mathcal{N}$ has minimal projectors, i.e. projectors $E = E^* = E^2 \in \mathcal{N}$ such that for any other projector $F = F^* = F^2 \in \mathcal{N}$ the condition $F\mathcal{H} \subseteq E\mathcal{H}$ implies that either $F = 0$ or $F = E$.

This result happens to be just what we need to appropriate in order to appreciate the phenomenon noted at the end of Section 1. Specifically, let $\mathcal{H}$ be a separable Hilbert space, and $\mathcal{S} = \{S \in \mathcal{B}(\mathcal{H}) \mid Tr\, S^*S < \infty\}$ be the space of the Hilbert-Schmidt operators, equipped with the scalar product $< S, T > = Tr\,(S^*T)$ under which it becomes a Hilbert space. We define two representations of $\mathcal{B}(\mathcal{H})$ on $\mathcal{S}$, namely

$$\mathcal{N}^L = \pi^L(\mathcal{B}(\mathcal{H})) \quad \text{with} \quad \pi^L(B)S \equiv B\,S \tag{2.1}$$

$$\mathcal{N}^R = \pi^R(\mathcal{B}(\mathcal{H})) \quad \text{with} \quad \pi^R(B)S \equiv S\,B^*. \tag{2.2}$$

Clearly

$$\mathcal{N}^L = (\mathcal{N}^R)' \quad \text{or equivalently:} \quad \mathcal{N}^R = (\mathcal{N}^L)' \tag{2.3}$$

and, with

$$J : S \in \mathcal{S} \mapsto S^* \in \mathcal{S}, \tag{2.4}$$

J is an involutive (i.e. $J^2 = I$) anti–unitary operator on $\mathcal{S}$, that implements an (anti-linear) isomophism between the von Neumann algebra $\mathcal{N}^L$ and its commutant $\mathcal{N}^R$:

$$N \in \mathcal{N}^L \mapsto JNJ \in \mathcal{N}^R. \tag{2.5}$$

This is the archetypical example of the modular structures to be now exhibited in relation with the geometric quantization programme: factors (i.e. primary representations) that are isomorphic to their commutant, the isomorphism being implemented by an involutive anti-unitary operator J, as in (2.5).

3. GEOMETRIC QUANTIZATION FOR THE CCR

The purpose of this section is to show how modular structures enter when the geometric quantization programme is applied to the canonical commutation relations. The first step is to choose the group on which this progamme will be carried out in the two steps below: prequantization and quantization.

For the flat configuration space $\mathbb{R}^n$, the familiar Weyl group W_o^{2n+1} is the undisputable natural choice; note that W_o^{2n+1} is a central extension

$$e \to S^1 \to W_o^{2n+1} \to G_o^{2n} \to e \tag{3.1}$$

of G_o^{2n} by S^1, where G_o^{2n} is the trivial (i.e. direct–product) extension

$$e \to \mathbb{R}^n \to G_o^{2n} \to \mathbb{R}^n \to e \quad , \tag{3.2}$$

corresponding to the projective representation

$$U(\mathbf{a})V(\mathbf{b}) = \exp\{i\mathbf{a} \cdot \mathbf{b}\}V(\mathbf{b})U(a) \tag{3.3}$$

associated to the Mackey system of imprimitivity

$$U(\mathbf{a})E(\Delta)U(\mathbf{a})^{-1} = E(\mathbf{a}^{-1}[\Delta]). \tag{3.4}$$

For the n-dimensional manifold $\mathcal{M}_\lambda^n$ of constant negative curvature $\kappa = -\lambda < 0$, (pictured in physicist language by the mass-shell in Minkowski space $M^{1,n}$ where the speed of light is $c = \lambda^{-\frac{1}{2}}$), we had shown[21] how to control the classical limit of the quantum systems of imprimitivity based on these manifolds. We argue elsewhere[22] (independently in quantum and classical physics) that the Weyl group W_λ^{2n+1} is the central extension

$$e \to S^1 \to W_\lambda^{2n+1} \to G_\lambda^{2n} \to e \tag{3.5}$$

of G_λ^{2n} by S^1, where G_λ^{2n} itself is an extension

$$e \to \mathbb{R}^n \to G_\lambda^{2n} \to H_\lambda^n \to e \tag{3.6}$$

of H_λ^n by $\mathbb{R}^n$. Specifically, these groups and their extensions are defined as follows (note that one recovers the respective building blocks of W_o^{2n+1} when $\lambda \mapsto 0$):

$$H_\lambda^n = \{\mathbf{a} = (a^1, a^2, \cdots, a^n) \mid a^j \in \mathbb{R},\, j = 1, \cdots, n\} \tag{3.7}$$

with product law $(\mathbf{aa}') = \mathbf{a} + \mathbf{a}[\mathbf{a}']$ where

$$\mathbf{a}[\mathbf{a}']^1 = a'^1 \quad \text{and} \quad \mathbf{a}[\mathbf{a}']^k = \exp\{2\lambda a^1\} a'^k \;\; k = 2, \cdots, n. \tag{3.8}$$

G_λ^{2n} is defined[22] by

$$G_\lambda^{2n} = \{g = (\mathbf{a}, \mathbf{b}) \mid \mathbf{a} \in H_\lambda^n, \mathbf{b} \in \mathbb{R}^n\} \tag{3.9}$$

with product law $gg' = (\mathbf{aa}', \mathbf{b} + \mathbf{a}[\mathbf{b}'])$, where

$$\mathbf{a}[\mathbf{b}]_1 = b_1 \quad \text{and} \quad \mathbf{a}[\mathbf{b}]_k = \exp\{-2\lambda a^1\}\, b_k\,,\; k = 2, \cdots, n \tag{3.10}$$

Finally, the product law

$$(g, \sigma)(g', \sigma') = (gg', \sigma\sigma'\omega(g, g')) \quad \text{in} \quad W_\lambda^{2n+1} = \{(g, \sigma) \mid g \in G_\lambda^4,\, \sigma \in S^1\} \tag{3.11}$$

is specified by the 2-cocycle

$$\omega(g, g') = a^1 b'_1 + \exp\{-2\lambda a^1\} \sum_{k=2}^{n} a^k\, b'_k\,. \tag{3.12}$$

The classical (i.e. phase space) interpretation of the groups just defined is now proposed in terms of their actions on the maximal co-adjoint orbits of W_λ^{2n+1}. For the sake of notational simplicity, we henceforth carry the discussion explicitly for the case $n = 2$ which illustrates faithfully the essential features of the general case $n \geq 2$. It is also convenient (though evidently not indispensable) to write the Lie algebra $\mathcal{W}_\lambda^5$ of the group W_λ^5 explicitly in terms of the basis $\{\xi_j \mid j = 1, \cdots, 5)\}$ given by the generators of the one-parameter subgroups corresponding to the coordinates we have chosen throughout to describe the group W_λ^5. In the dual basis, a general element of $\mathcal{W}_\lambda^*$ is written as $\xi^* = (\xi_1^*, \cdots, \xi_5^*)$. The co-adjoint action of the semi-direct product W_λ^5 on $\mathcal{W}_\lambda^*$ is:

$$\begin{pmatrix} 1 & 2\lambda\exp\{-2\lambda a^1\}\, a^2 & 0 & -2\lambda\exp\{2\lambda a^1\}\, b_2 & b_1 + 2\lambda\, a^2 b_2 \\ 0 & \exp\{-2\lambda a^1\} & 0 & 0 & b_2 \\ 0 & 0 & 1 & 0 & -a^1 \\ 0 & 0 & 0 & \exp\{2\lambda a^1\} & -a^2 \\ 0 & 0 & 0 & 0 & 1 \end{pmatrix} \tag{3.13}$$

with generators $X^5 = 0$; $X_{\xi^j} = \Lambda_{jk}\partial_{\xi^*_k}$ $j,k = 1,\cdots,4$ where Λ is the antisymmetric matrix

$$\Lambda = \begin{pmatrix} 0 & 2\lambda\,\xi^*_2 & \xi^*_5 & -2\lambda\,\xi^*_4 \\ -2\lambda\,\xi^*_2 & 0 & 0 & \xi^*_5 \\ -\,\xi^*_5 & 0 & 0 & 0 \\ 2\lambda\,\xi^*_4 & -\,\xi^*_5 & 0 & 0 \end{pmatrix} \tag{3.14}$$

Proposition 1 (Maximal Co-adjoint Orbits)

1. *The maximal coadjoint orbits* $\Omega^*_{\xi^*_5}$ *are of dimension* 4 ; *they are classified by their intersection* $(0,\cdots,0,\xi^*_5 \neq 0)$ *with the axis* $\mathbb{R}\xi^{*5}$.
2. *On each maximal coadjoint orbit* $\Omega^*_{\xi^*_5}$ *the symplectic form defined by* $\varpi_{\xi^*}(X_{\xi^j},X_{\xi^k}) =\, <\xi^*\,,\,[\xi^j\,,\,\xi^k]>$ *is given by* $\varpi_{\xi^*}(X_{\xi^j}\,,\,X_{\xi^k}) = \Lambda_{jk}$ *with* Λ *as in (3.14) ; and we have* $\varpi_{\xi^*}(\partial_{\xi^*_j}\,,\,\partial_{\xi^*_k}) = -(\Lambda^{-1})^{jk}$.
3. $\{\tilde{\xi}^*_1 = \xi^*_1 - 2\lambda\, x_2\, \xi^*_4\,;\, \tilde{\xi}^*_j = \xi^*_j\,,\, j = 2,3,4\}$ *diagonalizes the symplectic form* $\varpi_{\xi^*} = d\tilde{\xi}^*_1 \wedge d\tilde{\xi}^*_3 + d\tilde{\xi}^*_2 \wedge d\tilde{\xi}^*_4$.
4. *For* $j = 1,\cdots,4$ *the functions* $F_{\xi^j} : \xi^* \in \Omega_{\xi^*_5} \mapsto \xi^*_j \in \mathbb{R}$ *satisfy the first-order differential equations* $dF_{\xi^j}(\xi^*) = -X_{\xi^j} \rfloor \varpi_{\xi^*}$
5. *Upon assuming (without loss of generality)* $F_{\xi^5} = 0$, *we have for all* $\xi,\eta \in \mathcal{W}$: $\{F_{\xi^j}\,,\,F_{\xi^k}\} - F_{[\xi^j,\xi^k]} =\, <\xi^{*5}\,;\,[\xi\,,\,\eta]>$.

The interpretation of this result[22] is completed with the natural identifications :

$$(\xi^*_1,\xi^*_2,\xi^*_3,\xi^*_4) \in \Omega^*_1 \mapsto (p_1,p_2,-x^1,-x^2) \in T^*M^2_\lambda\,. \tag{3.15}$$

(where one has assumed, without loss of generality, that $\xi^*_5 = 1$). The group H^2_λ is then identified with the subgroup of upper-diagonal matrices in $SL(2,\mathbb{R})$ acting by the familiar fractional transformations on the base manifold $\mathcal{M}^2_\lambda$ now identified with the Poincaré half-plane $\{z = x^1 + ix^2 \in \mathbb{R}\times\mathbb{R}^+\}$ (with the metric scaled to give $\kappa = -\lambda$); the group $\mathbb{R}^2$ acts on each fibers $\{(p^1,p^2) \in \mathbb{R}^2\}$ by translation. The functions F_ξ , defined for every $\xi \in \mathcal{W}$, are Hamiltonian functions corresponding to the generators ξ of the group W^5_λ ; as such, they have an immediate physical interpretation. For instance ξ^*_1 corresponds to a component of the linear momentum; it should not be confused with the Darboux coordinate $\tilde{\xi}^*_1$ although this distinction fades out in the flat case limit $\lambda \to 0$. Note that the Hamiltonian functions $F'_\xi s$ are defined here only by their differential equation, and hence are each specified only up to a constant; this is reflected in the freedom one should have in choosing a particular extension in the cohomological equivalence class of the 2-cocycle defining the Weyl group W^5_λ . Specifically, these *constants* appear here as $<\xi^{*5}\,;\,[\xi\,,\,\eta]>$ and correspond indeed to the 2-cocycle of the *Lie algebra* extension $0 \to \mathbb{R} \to \mathcal{W}^5 \to \mathcal{G}^4 \to 0$. Whence the Poisson brackets:

$$\begin{array}{llllllll} \{x^1,x^2\} & = 0 & ; & \{p_1,x^1\} & = 1 & ; & \{p_2,x^2\} & = 1 \\ \{p_1,p_2\} & = -2\lambda p_2 & ; & \{p_1,x^2\} & = 2\lambda x^2 & ; & \{p_2,x^1\} & = 0 \end{array} \tag{3.16}$$

Having thus identified the classical system described by the Weyl group W^5_λ , we can inspect how the modular structures are borne by its prequantization. We denote by U^L (resp. U^R) the left– (resp. right–) regular representation of our Lie group W^{2n+1}_λ

on the Hilbert space $\mathcal{H} = \mathcal{L}^2(W_\lambda^{2n+1}, d\mu)$, and Δ be the modular function of W_λ^{2n+1}; i.e. with $w_o, w \in W_\lambda^{2n+1}$ and $\Psi \in \mathcal{H}$:

$$\{U^L(w_o)\Psi\}(w) = \Psi(w_o^{-1}w) \qquad \{U^R(w_o)\Psi\}(w) = \Delta(w_o)^{\frac{1}{2}}\,\Psi(ww_o) \tag{3.17}$$

which are changed one into the other by the involutive anti-unitary operator

$$\{J\Psi\}(w) = \Delta(w)^{-\frac{1}{2}}\,\Psi(w^-1)^* \tag{3.18}$$

Upon defining the von Neumann algebras

$$\mathcal{N}^L = \{U^L(w) \mid w \in W_\lambda^{2n+1}\}'' \tag{3.19}$$

and

$$\mathcal{N}^R = \{U^R(w) \mid w \in W_\lambda^{2n+1}\}'' \tag{3.20}$$

we have:[17]

$$\mathcal{N}^L = (\mathcal{N}^R)', \quad \mathcal{N}^R = (\mathcal{N}^L)', \quad J\mathcal{N}^L J = \mathcal{N}^R, \quad J\mathcal{N}^R J = \mathcal{N}^L, \tag{3.21}$$

which brings us near to the modular stuctures we are looking for, except that these regular representations are not primary, and thus these von Neumann algebras are not factors; they can however be decomposed[17] into their primary components. As W_λ^{2n+1} is a central extension by S^1, the central decomposition of $U^\natural$ (where $\natural$ stands for either R or L) is a direct sum of primary representations $U_k^\natural(W_\lambda^{2n+1})$ each of which is induced by a character χ_k of S^1. Specifically, we obtain (again with $n = 2$ for notational convenience):

Proposition 2 (Prequantization) *With $W_\lambda^5, \mathcal{H}, U^\natural, \mathcal{N}^\natural$ as above, let $\mathcal{Z} = \mathcal{N}^\natural \cap (\mathcal{N}^\natural)'$ and Z the spectrum of $\mathcal{Z}$. Then:*

1.
$$\mathcal{H} = \oplus_{k\in Z}\mathcal{H}_k\ ; \quad \mathcal{N}^\natural = \oplus_{k\in Z}\mathcal{N}_k^\natural\ ; \quad J = \oplus_{k\in Z}J_k \tag{3.22}$$
$$\mathcal{N}_k^L = (\mathcal{N}_k^R)'\ ; \quad \mathcal{N}_k^R = (\mathcal{N}_k^L)'\ ; \quad J_k\mathcal{N}_k^L J_k = \mathcal{N}_k^R\ ; \quad J_k\mathcal{N}_k^R J_k = \mathcal{N}_k^L \tag{3.23}$$

2. *With $k \in Z$, $w = (g, \sigma) \in W_\lambda^5 \cong G_\lambda \times S^1$, and $\Psi \in \mathcal{H}_k$:*
$$\{U_k^L(g,\sigma)\,\Psi\}(g') = \sigma^{-k}\exp\{-ik < \mathbf{a}, \mathbf{b}' - \mathbf{b} >\}\,\Psi(g^{-1}g'),\ \sigma \in S_1 \tag{3.24}$$

3. *For $k = 1$, the self-adjoint generators of this representation are of the form $H_j = -iD_j + \mu_j$, $j = 1, \cdots, 4$ (as $j = 5$ is now without interest) with:*
$$\begin{array}{ll} D_1 = \partial_{a^1} \ +2\lambda a^2\partial_{a^2} - 2\lambda b_2\partial_{b_2} & \mu_1 = b_1 \\ D_2 = \partial_{a^2} & \mu_2 = b_2 \\ D_3 = \partial_{b_1} & \mu_3 = 0 \\ D_4 = \partial_{b_2} & \mu_4 = 0 \end{array} \tag{3.25}$$

4. *The 2-form $d\phi$, defined from the equations $\phi(D_j) = \mu_j$, coincides with the symplectic form ϖ of Proposition 1.*

Note that U_k^L, described above as a direct summand of the left-regular representation of our Weyl group, *is* (up to a unitary equivalence which reflects the ever present ambiguity in the choice of a Kostant gauge) the representation one obtains from the usual procedure of prequantization. The generators of the primary representations obtained as direct summands of the left– and right–regular representations will reflect the particularity that one should now distinguish between right– and left–invariant vector fields due to the non-abelianness of the group $\mathcal{H}_\lambda^2$, itself a reflection of the non–zero curvature of $\mathcal{M}_\lambda^2$. In the limit $\lambda \to 0$ the above generators reduces to the P 's and Q's of the flat prequantization (1.1).

The involutive anti–unitary equivalence (3.23) [carried over in the central decomposition of (3.21)] equips then the prequantization representation of W_λ^5 with the modular structure we were looking for.

This structure now allows to describe, in a way that explicitly emphasizes the role of the complete sets of commuting observables, how the irreducible representations of ordinary quantum mechanics sit in the prequantization representation. Indeed, any maximal abelian algebra $\mathcal{A}$ in the commutant $\mathcal{N}_k^R$ of the prequantization algebra $\mathcal{N}_k^L$ gives[17] a direct integral decomposition of $\mathcal{N}_k^L$. In the specific case where one chooses $\mathcal{A} = \{U_k^R(0,\mathbf{b}) \mid \mathbf{b} \in \mathbb{R}^2\}''$ i.e. the maximal abelian algebra corresponding to the position observables, the direct integral decomposition of the prequantization representation $U_k^L(W_\lambda^5)$ into irreducible representations of W_λ^5 is labelled by the spectrum R^2 of $\mathcal{A}$, which appears here simply as the character group of R^2. To bring this into contact with the polarization one would use in this second step of the geometric quantization programme, it suffices to note that the maximal abelian subalgebra $\{U_k^\natural(0,\mathbf{b}) \mid \mathbf{b} \in \mathbb{R}^2\}'' \subset U_n^\natural(W_\lambda^5)''$ (where $\natural$ stands for either R[ight] or L[eft]) plays here the role of the Lagrangian submanifold consisting of the base manifold $M_\lambda^2 \subset T^*M_\lambda^2$, see in particular the first Poisson bracket in (3.16). Incidentally, the last Poisson bracket in (3.16) shows that in this last step, we could have used alternatively the Lagrangian submanifold $\{(p^2, x^1)\}$, the physical interpretation of which is however not quite as palatable as that of the natural choice M_λ^2 we made.

It is easy to verify that all the irreducible representations obtained in this way (from a given prequantization representation) are unitarily equivalent, a result that confirms the natural extension of Mackey's proof[23] of the von Neumann uniqueness theorem.

Proposition 3 (Quantization) *W_λ^5 admits (up to unitary equivalence) exactly one irreducible representation; this representation is given by the system of imprimitivity*

$$U(\mathbf{a})E(\Delta)U(\mathbf{a})^{-1} = E(\mathbf{a}^{-1}[\Delta]) \tag{3.26}$$

based on the algebra of Borel Δ's of $\mathcal{M}_\lambda^2$, and acting on the Hilbert space $\mathcal{L}^2(\mathcal{M}_\lambda^2, d\mu_\lambda)$, where $d\mu_\lambda$ is the measure on $\mathcal{M}_\lambda^2$ invariant under the natural action of H_λ^2.

Here again, all this generalizes from $\mathcal{M}_\lambda^2$ and W_λ^5 to $\mathcal{M}_\lambda^n$ and W_λ^{2n+1}.

4. EXTENSION TO SO(2,3)

The Weyl group of the CCR, discussed in the previous section, is an example of a Lie group for which the prequantization representation U_Ω appears as a primary direct summand U_k^L in the central decomposition of the left–regular representation U^L of the group. This is to say that the projector P_Ω that reduces U^L to U_Ω is a minimal projector P_k in the center of the von Neumann algebra $\mathcal{N}^L$ generated by U^L. Irreducible representations had then to be extracted from the prequantization representation $U_\Omega = U_k^R$, either geometrically (reduction by polarization) or algebraically (reduction by maximal abelian sub–algebras in the commutant $\mathcal{N}_k^R$ of $\mathcal{N}_k^L$).

The present section exhibits a case where instead of having $P_\Omega = P_k$, the opposite extreme occurs,[24] namely where the intersection $P_\Omega \cap P_k$ projects U^L immediately onto an *irreducible* representation of G, that appears as *the* common irreducible direct summand in both U_Ω and U_k^L. The realization that this situation could indeed occur was motivated by a recent geometric description[25] of some discrete series representations of SO(2,3).

Proposition 4 (Prequantization) *Let G be a Lie group, with left–invariant Haar measure $d\mu$; and U^L (resp. U^R) be its left–(resp. right–)regular representations on $\mathcal{H} = \mathcal{L}^2(G, d\mu)$. Let $H \subset G$ be a compact, abelian subgroup of G with Haar measure $d\mu_H$; and $\chi \in H^*$ be a character of H. Let further P_χ be the projector*

$$P_\chi = \frac{1}{\mu_H(H)} \int_H d\mu_H(h)\, \chi(h)^*\, U^R(h) \tag{4.1}$$

Then

1. *$P_\chi \mathcal{H} = \{\Psi \in \mathcal{H} \mid \Psi(gh) = \chi(h)\Psi(g) \quad \forall \quad (g,h) \in G \times H\}$*
2. *$P_\chi \in \mathcal{N}_H^R \subseteq \mathcal{N}^R = (\mathcal{N}^L)'$ and thus reduces U^L.*
3. *$U_\chi^L \equiv U^L | P_\chi \mathcal{H}$ is the prequantization representation U_Ω corresponding to the orbit $\Omega = G/H$; i.e. $P_\chi = P_\Omega$.*

Recall[26] that a UIR $V : G \to \mathcal{U}(\mathcal{H}_V)$ is said to belong to the discrete series (it is then refered to by the acronym DUIR) whenever there exists at least one pair $(\Phi, \Psi) \in \mathcal{H}_V \times \mathcal{H}_V$ with $\Phi \neq 0 \neq \Psi$) such that the function $W_{\Phi,\Psi} : g \in G \mapsto (V(g)\Psi, \Phi) \in C$ belongs to $\mathcal{H} = \mathcal{L}^2(G, d\mu)$. Note that this definition allows[27] for G to be not unimodular. The following result is then an immediate commentary[24] on Refs.27-28.

Proposition 5 (Modular Structures) *Let V be a DUIR of G on some Hilbert space $\mathcal{H}_V$, let $\mathcal{S}_V$ be the space of Hilbert–Schmidt operators on $\mathcal{H}_V$, U_V^L (resp. U_V^R) be the primary representations of G obtained by left– (resp. right–) action of V on $\mathcal{H}_V$ (see (2.1-2.2)); J_V be defined as in (2.4); U^L (resp. U^R) be the left–(resp. right–) regular representation of G on $\mathcal{H} = \mathcal{L}^2(G, d\mu)$. Then:*

1. *There exists a partial isometry $W_V : \mathcal{S}_V \mapsto \mathcal{H}$ that intertwines the modular structures $\{U_V^L, U_V^R, J_V\}$ and $\{U^L, U^R, J\}$ i.e.*

$$W_V U_V^L(g) = U^L(g) W_V, \quad \forall\, g \in G, \tag{4.2}$$

$$W_V U_V^R(g) = U^R(g) W_V, \quad \forall\, g \in G, \tag{4.3}$$

$$W_V J_V = J W_V. \tag{4.4}$$

2. *The range $\mathcal{L}_V \equiv W_V \mathcal{S}_V$ of W_V coincides with $\oplus_\mathcal{V} \tilde{\mathcal{H}}_V$, where $\mathcal{V}$ is the collection of all closed subspaces $\tilde{\mathcal{H}}_V$ satisfying: (i) $\tilde{\mathcal{H}}_V$ is stable under $U^L(G)$, and (ii) $U^L | \tilde{\mathcal{H}}_V \approx V$; moreover $\mathcal{L}_V$ is stable under $U^R(G)$.*
3. *The projector $\mathcal{P}_V : \mathcal{H} \mapsto \mathcal{L}_V$ is minimal in the center of the von Neumann algebras $\mathcal{N}^\natural$ generated by the regular representations $U^\natural$ of G on $\mathcal{H}$.*

Recall[29] that when, *in addition*: (i) G is a semi–simple Lie group, and (ii) H is maximal, i.e. is a compact Cartan subgroup of G, there exists exactly one character χ of H that appears with multiplicity 1 in a DUIR V; and, conversely, that a character χ with this property characterizes every DUIR V.

Proposition 6 (Quantization) *With the notation of Propositions 4 and 5, let H be a compact Cartan subgroup of a semi–simple Lie Group G; V be a DUIR of G; and χ be the corresponding character of H. Then :*

1. *$[P\chi, \mathcal{P}_V] = 0$ and the range $\tilde{H}_V$ of the projector $\tilde{P}_V \equiv P\chi \cap \mathcal{P}_V$ is stable under U_χ^L and U_V^L.*

2. $\tilde{V} \equiv U^L_\chi|\tilde{H}_V = U^L_V|\tilde{H}_V \approx V$ *; in particular* $\tilde{V}$ *is irreducible.*

3. Every DUIR V of G appears in this way.

Further Extensions. One of the major simplifying features of the representations we considered in this lecture was that we could restrict our attention to the discrete summands appearing in the central decomposition of the left– and right–regular representations. Nevertheless, it appears that the technique of Gelfand triplets[30] should open an avenue to dealing concretely with situations where one must consider the contribution of integral decompositions associated with the continuous part of the spectrum of the center $\mathcal{N}^L \cap \mathcal{N}^R$ of the regular representations. This avenue remains to be thoroughly traveled.

References

1. J. M. Souriau, "Structure des systèmes dynamiques", Dunod, Paris (1966)
2. A. Kirillov, "Eléments de la théorie des représentations", Mir, Moscow (1974)
3. B. Kostant, "Quantization and unitary representations", *Lecture Notes in Mathematics* 170, Springer, New York (1970); and *in* "Géometrie symplectique et Physique mathématique", CNRS, Paris (1970).
4. D.J. Simms and N.M.J. Woodhouse, "Lectures in Geometric Quantization", *Lecture Notes in Physics* 53, Springer, New York (1976)
5. V. Guillemin and S. Sternberg "Geometric Asymptotics", *Mathematical Surveys* 14, AMS, Providence (1977); "Symplectic Techniques in Physics", Cambridge Univ. Press, Cambridge (1984)
6. R. Abraham and J.E. Marsden, "Foundations of Mechanics", Benjamin, Reading (1978)
7. J. Śniatycki, "Geometric Quantization and Quantum Mechanics", App.Math.Sc. No.30, Springer, New York (1980)
8. N.M.J. Woodhouse, "Geometric Quantization", Clarendon Press, Oxford (1980)
9. P.A.M. Dirac, "The Principles of Quantum Mechanics", Clarendon, Oxford (1930)
10. R.F. Streater, Canonical quantization, *Commun. Math. Phys.* 2:354-374 (1966)
11. I.E. Segal, Quantization in nonlinear systems, *J. Math. Phys.* 1:468-488 (1960); 5:269-282 (1964); see also: *Symposia Mathematica* 14:9-117 (1974)
12. G.G. Emch, Prequantization and KMS structures, *Intern'l J. Theor. Phys. Phys.* 20:891-904 (1981); KMS structures in ggeometric quantization, *Contemporary Mathematics* 62:175-186 (1987); Geometric quantization: Regular representations and modular algebras, *in*"Group Theoretical Methods in Physics (Moscow, 1990)", V.V. Dodonov and V.I. Man'ko, eds., Springer, Heidelberg (1991)
13. S.T. Ali and G.G. Emch, Geometric quantization: Modular reduction theory and coherent states", *J. Math. Phys.* 27:2936-2943 (1986)
14. F.J. Murray and J. von Neumann, On rings of operators, *Ann. Math.(2)* 37:116-229 (1936); II, *Trans. Amer. Math. Soc.* 41:208-248 (1937)
15. F.J. Murray, The rings of operators papers, *in* "The Legacy of John von Neumann", *Proc. Symp. Pure Math.* 50:57-60 (1990)
16. J. von Neumann, Zur Algebra der Funktionaloperationen und Theorie der normalen Operatoren, *Math. Ann.* 102: 370-427 (1929)

17. J. Dixmier, "Les algèbres d'opérateurs dans l'espace hilbertien (Algèbres de von Neumann)", Gauthier-Villars, Paris (1957); 2nd ed. (1969) ; and "Les C*-algèbres et leurs représentations", Gauthier-Villars, Paris (1964)

18. S. Sakai, "C*-algebras and W*-algebras", *Ergebnisse der Mathematik und ihrer Grenzgebiete* 60, Springer, Heidelberg (1971)

19. M. Takesaki, "Theory of Operator Algebras. I", Springer, Heidelberg (1979)

20. R. V. Kadison and Ringrose, "Fundamentals of the Theory of Operator Algebras. I", Academic Press, New York (1983)

21. G. G. Emch, Quantum and classical mechanics on homogeneous Riemanian manifolds, *J. Math. Phys.* 23:1785-1791 (1982)

22. J. Bertrand, G. G. Emch and G. Rideau, The cohomology of the classical and quantum Weyl CCR in curved spaces, *Lett. Math. Phys.* (to appear)

23. G.W. Mackey, A theorem of Stone and von Neumann, *Duke Math. J.* 16:313-326 (1949)

24. S. T. Ali, A. M. El Gradechi and G. G. Emch, Modular algebras in geometric quantization, *J. Math. Phys.* 33:6237-6243 (1994)

25. A. M. El Gradechi and S. De Bièvre, Phase space quantum mechanics on the Anti-de Sitter spacetime and its Poincaré contraction, *Ann. Phys. (NY)* 235:1-35 (1994)

26. A. W. Knapp, "Representation Theory of Semisimple Lie Groups", Princeton University Press, Princeton (1986)

27. M. Duflo and C. C. Moore, On the regular representation of a nonunimodular locally compact group, *J. Funct. Analysis* 21:209-243 (1976)

28. A. Grossmann, J. Morlet and T. Paul, Transforms associated to square integrable group representations. I, *J. Math. Phys.* 26:2473-2479 (1985)

29. M. E. Taylor, "Noncommutative Harmonic Analysis", *Mathematical Surveys and Monographs* 22 (1986)

30. I. M. Gelfand and N. Ya. Vilenkin, "Generalized Functions. IV", Academic Press, New York (1968).

DIFFEOMORPHISM GROUPS AND ANYON FIELDS

Gerald A. Goldin[1] and David H. Sharp[2]

[1] Departments of Mathematics and Physics, Rutgers University
New Brunswick, NJ 08903 USA
[2] Theoretical Division, Los Alamos National Laboratory
Los Alamos, NM 87545 USA

Abstract

We make use of unitary representations of the group of diffeomorphisms of the plane to construct an explicit field theory of anyons. The resulting anyon fields satisfy q-commutators, where q is the well-known phase shift associated with a single counterclockwise exchange of a pair of anyons. Our method uses a realization of the braid group by means of paths in the plane, that transform naturally under diffeomorphisms of $\mathbf{R}^2$

1. DIFFEOMORPHISM GROUP REPRESENTATIONS AND ANYONS

The intrinsic structure of standard quantum mechanics includes representations of an infinite-dimensional group, whose infinitesimal generators are the mass density operator $\rho(\mathbf{x},t)$ and the momentum density operator $\mathbf{J}(\mathbf{x},t)$. By examining the commutation relations and other properties of ρ and $\mathbf{J}$, one determines that the corresponding group is the natural semidirect product $G = \mathcal{S}(M) \times \mathit{Diff}(M)$, where the manifold M is physical space (typically $\mathbf{R}^3$), $\mathcal{S}(M)$ is the additive group of smooth, real-valued scalar functions on M that together with all derivatives vanish rapidly at infinity (Schwartz' space), and $\mathit{Diff}(M)$ is the group of diffeomorphisms of M under composition that, together with all derivatives, become rapidly trivial at infinity.[1]

Quantum-mechanical systems are described by the continuous unitary representations (CURs) of G, or in certain cases (such as an ideal, incompressible fluid) particular subgroups of G (e.g., the volume-preserving diffeomorphisms). This fact has been established and used in our previous work to obtain a unified description of an astonishing variety of quantum systems, ranging from extended objects such as vortex configurations[2] to point particles obeying boson, fermion, and (in two space dimensions) intermediate, or "anyon", statistics.[3] The latter possibility had already been conjectured from the topology of two-particle configuration space in the plane;[4] the diffeomorphism group approach provided a rigorous prediction even without the assumed exclusion of configurations where the particle coordinates coincide. From diffeomorphism group representations there also followed many of the fundamental physical properties of anyons–the shifts in angular momentum and energy spectra, the connection with configuration space topology, the relation to charged particles circling regions of magnetic flux, and the mathematical role of the braid group.[3,5,6] Anyon statistics

Quantization, Coherent States, and Complex Structures
Edited by J.-P. Antoine *et al.*, Plenum Press, New York, 1995

find application in physics to surface phenomena, particularly the fractional quantum Hall effect.[7]

Based on the diversity of known quantum systems associated with CURs of the diffeomorphism group or its semidirect product, we believe that G can be regarded as a universal, or generic, group of local symmetries describing non-relativistic quantum theory.[6] In the formulation of quantum mechanics based on diffeomorphism group representations, canonical fields ψ and ψ^* do not play a fundamental role. While the infinitesimal generators ρ and $\mathbf{J}$ can be constructed from canonical fields in specific models (see below), is not necessary to use this fact to obtain representations of $Diff(M)$, or to establish the physical interpretations of these representations.

It is nevertheless useful to reintroduce the field operators when that is possible; for example, annihilation and creation operators provide a way to construct states with specified numbers of particles, and fields are a starting point for many computational methods. It is thus worth asking how canonical fields can be constructed, taking as a starting point a collection of CURs of G.

We show here how creation and annihilation fields can be constructed, uniquely up to unitary equivalence, as operators *intertwining a hierarchy* of representations of G. So defined, these operators create or annihilate *configurations,* where the type of object created is defined by the representations from which one starts. We take as a specific example the construction of anyon fields from diffeomorphism group representations. Then we are able to determine the algebra that the field operators satisfy. Thus we obtain q-commutation relations for anyon fields *not* as a starting assumption, nor by introducting a Chern-Simons potential into a canonical field theory, but strictly as a *consequence* of the unitary group representations that describe anyons together with the intertwining property defining the fields. In the course of doing this, we shall also make clear how elements of $Diff(\mathbf{R}^2)$ act on the braid group.

First we provide some basic facts about the infinite-dimensional Lie algebra of mass and momentum density operators, and the corresponding Lie group of unitary operators, in canonical nonrelativistic theories. One has formally,

$$\rho(\mathbf{x},t) = m\psi^*(\mathbf{x},t)\psi(\mathbf{x},t),$$

$$\mathbf{J}(\mathbf{x},t) = (\hbar/2i)\{\psi^*(\mathbf{x},t)[\nabla\psi(\mathbf{x},t)] - [\nabla\psi^*(\mathbf{x},t)]\psi(\mathbf{x},t)\}\,, \tag{1.1}$$

where the fields in (1.1) obey, at equal times t, for all $\mathbf{x},\mathbf{y}$,

$$[\psi(\mathbf{x},t),\psi(\mathbf{y},t)]_\pm = [\psi^*(\mathbf{x},t),\psi^*(\mathbf{y},t)]_\pm = 0,$$

$$[\psi(\mathbf{x},t),\psi^*(\mathbf{y},t)]_\pm = \delta(\mathbf{x}-\mathbf{y}). \tag{1.2}$$

The subscript "-" denotes the commutator, and "+" the anticommutator bracket. To interpret ρ and $\mathbf{J}$ as operator-valued distributions on the spatial manifold M, define $\rho(f) = \int_M \rho(\mathbf{x})\, f(\mathbf{x})\, d\mathbf{x}$ and $J(\mathbf{g}) = \int_M \mathbf{J}(\mathbf{x})\cdot\mathbf{g}(\mathbf{x})\, d\mathbf{x}$; where $f \in \mathcal{S}(M)$, and $\mathbf{g}$ is a vector field whose components (together with all derivatives) vanish rapidly at infinity. We shall call the set of such vector fields $Vect(M)$. We then obtain the same infinite-dimensional semidirect product Lie algebra (a nonrelativistic local current algebra) *independent* of which bracket is chosen in (1.1), namely:

$$[\rho(f_1),\, \rho(f_2)] = 0\,, \qquad [\rho(f),\, J(\mathbf{g})] = i\hbar\rho(\mathbf{g}\cdot\nabla f)\,,$$

$$[J(\mathbf{g}_1),\, J(\mathbf{g}_2)] = -i\hbar J([\mathbf{g}_1,\, \mathbf{g}_2])\,, \tag{1.3}$$

where $\mathbf{g}\cdot\nabla f = L_{\mathbf{g}}f$ is the Lie derivative, and $[\mathbf{g}_1,\, \mathbf{g}_2] = \mathbf{g}_1\cdot\nabla\mathbf{g}_2 - \mathbf{g}_2\cdot\nabla\mathbf{g}_1$ is the Lie bracket of the vector fields. The fact that the Lie algebra (1.3) is the same for fermions and bosons means that the information about the particle statistics, formerly encoded in the algebra of fields, will now be contained in the choice of a *representation* satisfying (1.3). We have shown that inequivalent representations, in spaces of dimension greater than one, describe the different statistics.[3]

It is a standard result that the C^∞ vector field $\mathbf{g}$ generates a unique one-parameter group of C^∞ diffeomorphisms, $\phi_t^{\mathbf{g}} : M \to M\ (t \in \mathbf{R})$ with $\phi_{t_1}^{\mathbf{g}} \circ \phi_{t_2}^{\mathbf{g}} = \phi_{t_1+t_2}^{\mathbf{g}}$; satisfying the ordinary differential equation $(\partial/\partial t)\phi_t^{\mathbf{g}}(\mathbf{x}) = \mathbf{g}(\phi_t(\mathbf{x}))$, together with the initial condition $\phi_{t=0}^{\mathbf{g}}(\mathbf{x}) = \mathbf{x}$. (The conditions on $\mathbf{g}$ at infinity are important to the global existence of ϕ_t.) Then, defining the unitary operators $U(f) = \exp[(i/m)\rho(f)]$ and $V(\phi_t^{\mathbf{g}}) = \exp[(it/\hbar)J(\mathbf{g})]$, we have the semidirect product group law,

$$U(f_1)V(\phi_1)U(f_2)V(\phi_2) = U(f_1 + f_2 \circ \phi_1)V(\phi_1\phi_2)\,, \tag{1.4}$$

where $\phi_1\phi_2 = \phi_2 \circ \phi_1$ is the composition of the diffeomorphisms.

The simplest representations of (1.3) and (1.4) are the N-particle Bose and Fermi representations. For specificity let $M = \mathbf{R}^2$ or $\mathbf{R}^3$ and let the wave function $\Phi_N^{s,a}$ belong to the Hilbert space $\mathcal{H}_N^{s,a}$ of symmetric (s) or antisymmetric (a) functions of N variables $(\mathbf{x}_1, \ldots \mathbf{x}_N) \in M$, square integrable with respect to Lebesgue measure μ. Then these representations are given by

$$\begin{aligned} \rho_N(f)\Phi_N^{s,a}(\mathbf{x}_1, \ldots, \mathbf{x}_N) &= m \sum_{j=1}^{N} f(\mathbf{x}_j)\, \Phi_N^{s,a}(\mathbf{x}_1, \ldots, \mathbf{x}_N)\,, \\ J_N(\mathbf{g})\Phi_N^{s,a}(\mathbf{x}_1, \ldots, \mathbf{x}_N) &= \frac{\hbar}{2i} \sum_{j=1}^{N} \{\mathbf{g}(\mathbf{x}_j) \cdot \nabla_j \Phi_N^{s,a}(\mathbf{x}_1, \ldots, \mathbf{x}_N) \\ &\quad + \nabla_j \cdot [\mathbf{g}(\mathbf{x}_j)\Phi_N^{s,a}(\mathbf{x}_1, \ldots, \mathbf{x}_N)\,]\,\}\,, \end{aligned} \tag{1.5}$$

and correspondingly

$$\begin{aligned} U_N(f)\Phi_N^{s,a}(\mathbf{x}_1, \ldots, \mathbf{x}_N) &= \exp[i \sum_{j=1}^{N} f(\mathbf{x}_j)]\, \Phi_N^{s,a}(\mathbf{x}_1, \ldots, \mathbf{x}_N)\,, \\ V_N(\phi)\Phi_N^{s,a}(\mathbf{x}_1, \ldots, \mathbf{x}_N) &= \Phi_N^{s,a}(\phi(\mathbf{x}_1), \ldots, \phi(\mathbf{x}_N))\, [\prod_{j=1}^{N} \mathcal{J}_\phi(\mathbf{x}_j)]^{\frac{1}{2}}\,, \end{aligned} \tag{1.6}$$

where $\mathcal{J}_\phi(\mathbf{x})$ denotes the Jacobian of the diffeomorphism ϕ at $\mathbf{x}$.

Note that the operators in (1.5) are self-adjoint and those in (1.6) are unitary. They preserve the particle number N, and are also manifestly symmetric with respect to exchange of particle coordinates $\mathbf{x}_j$; thus they also preserve the wave function symmetry. For $N = 0, 1, 2, \ldots$, the N-particle Bose (respectively, Fermi) representations constitute a *hierarchy*, in an obvious physical sense that we make precise below.

2. HIERARCHIES OF REPRESENTATIONS AND THEIR INTERTWINING FIELD OPERATORS

The first step in our development is to establish the conditions that have to be satisfied for creation and annihilation field operators to intertwine representations of the diffeomorphism group. This allows us to specify a well-defined sense, satisfied by the above examples, in which a collection of continuous unitary representations of the group *Diff*(M) forms a hierarchy. The representations in the hierarchy are labeled by the number N of elementary configurations; thus we establish the bracket that an intertwining field must obey with the operators in the N-configuration Hilbert space.

The required conditions follow from the structure of the commmutation relations between the fields ψ and ψ^* and the operators ρ and $\mathbf{J}$. For bosons and fermions (where we already know ρ, $\mathbf{J}$, ψ, and ψ^*), these commutation relations can be obtained by direct formal computation starting from (1.1) and (1.2). To facilitate the calculation, and in anticipation of the results of Section 3.2, we shall generalize this procedure from the outset and start directly from the q-commutation relations for the fields. These are based on the q-deformed bracket $[A,B]_q = AB - qBA$, where q is assumed to be a complex number of modulus one. When $q=1$, we recover the commutator brackets "-" in (1.2), and when $q=-1$, we have the anticommutator brackets "+". We write

$$[\psi(\mathbf{x},t),\psi(\mathbf{y},t)]_q = [\psi^*(\mathbf{x},t),\psi^*(\mathbf{y},t)]_q = 0\,,$$

$$[\psi(\mathbf{y},t),\psi^*(\mathbf{x},t)]_q = \delta(\mathbf{x}-\mathbf{y})\,. \tag{2.1}$$

Note that for the first two equations of (2.1) to be consistent when $q \neq \pm 1$, they cannot be interpreted as holding for all ordered pairs $(\mathbf{x},\mathbf{y})$, but only in a half-space H of $M \times M$. In the complementary half-space $\bar{H} = M \times M - H$, we have instead the $(1/q)$-bracket. Then the equation for $[\psi(\mathbf{x}),\, \psi(\mathbf{y})]_q$ is consistent with the equation for $[\psi^*(\mathbf{x}),\, \psi^*(\mathbf{y})]_q$, since we are assuming that $|q| = 1$. The third equation of (2.1) is written as indicated for $(\mathbf{x},\mathbf{y}) \in H$; it may be written equivalently (using $\bar{q} = 1/q$) as

$$[\psi(\mathbf{x},t),\psi^*(\mathbf{y},t)]_{1/q} = \delta(\mathbf{x}-\mathbf{y})\,. \tag{2.2}$$

Now we are ready to obtain brackets for ψ and ψ^* with ρ and $\mathbf{J}$. We shall use the algebraic identity,

$$[AB,C]_- = A[B,C]_q + q\,[A,C]_{1/q}B\,, \tag{2.3}$$

that relates the ordinary commutator to the q-commutator. Then we can calculate the commutators of the field operators with the generators of the infinite-dimensional group. We obtain, for field operators obeying (2.1) and (2.2) for any fixed value of q having modulus one,

$$[\rho(\mathbf{y},t),\psi^*(\mathbf{x},t)] = m\psi^*(\mathbf{y},t)\delta(\mathbf{x}-\mathbf{y}),$$

$$[\rho(\mathbf{y},t),\psi(\mathbf{x},t)] = -m\psi(\mathbf{y},t)\delta(\mathbf{x}-\mathbf{y}), \tag{2.4}$$

$$[\mathbf{J}(\mathbf{y},t),\psi^*(\mathbf{x},t)] = \frac{\hbar}{2i}\{\psi^*(\mathbf{y},t)\nabla_{\mathbf{y}}\delta(\mathbf{x}-\mathbf{y}) - \delta(\mathbf{x}-\mathbf{y})\nabla_{\mathbf{y}}\psi^*(\mathbf{y},t)\}\,,$$

$$[\mathbf{J}(\mathbf{y},t),\psi(\mathbf{x},t)] = -\frac{\hbar}{2i}\{\psi(\mathbf{y},t)\nabla_{\mathbf{y}}\delta(\mathbf{x}-\mathbf{y}) - \delta(\mathbf{x}-\mathbf{y})\nabla_{\mathbf{y}}\psi(\mathbf{y},t)\}. \tag{2.5}$$

The justification of these equations for all values of $\mathbf{x}$ and $\mathbf{y}$ involves performing the calculation in each half-space separately, and noting that the answer is the same.

Next we multiply (2.4) by the test functions $h(\mathbf{x})$ and $f(\mathbf{y})$, and (2.5) by the test function $h(\mathbf{x})$ and vector field $\mathbf{g}(\mathbf{y})$, and integrate over $\mathbf{x}$ and $\mathbf{y}$. We thus obtain,

$$[\rho(f),\psi^*(h)] = \psi^*(mfh)\,, \quad [\rho(f),\psi(h)] = -\psi(mfh)\,, \tag{2.6}$$

$$[J(\mathbf{g}),\psi^*(h)] = \psi^*(\frac{\hbar}{2i}\{\mathbf{g}\cdot\nabla h + \nabla\cdot(\mathbf{g}h)\})\,,$$

$$[J(\mathbf{g}),\psi(h)] = -\psi(\frac{\hbar}{2i}\{\mathbf{g}\cdot\nabla h + \nabla\cdot(\mathbf{g}h)\})\,. \tag{2.7}$$

The essential point is that we find the same *commutator* brackets independent of whether we begin the calculation with Bose fields, Fermi fields, or even fields satisfying q-commutators; that is, equations (2.6) and (2.7) are representation-independent!

It is also straightforward to verify that (1.3) together with (2.6) and (2.7) satisfy the Jacobi identity, as long as we do not include brackets of fields with each other in the identity. However, ψ and ψ^* satisfy different relations with each other in the Bose and Fermi cases, and (as we shall shortly see) in the anyon case. Only in the Bose case do we have an actual Lie algebra of fields together with densities and currents.

For the cases of bosons and fermions, we can now look again at the N-particle representations (1.5) and (1.6), and interpret ψ^* and ψ as creation and annihilation operators respectively *intertwining* these representations. In the Bose or Fermi Fock representations of the usual second-quantized nonrelativistic field operators, we have Hilbert space vectors $\Phi^{s,a} = (\Phi_N^{s,a}), N = 0, 1, 2, \ldots$, with $\Phi_N^{s,a} \in \mathcal{H}_N^{s,a}$. For bosons (s),

$$[\psi(\mathbf{x})\Phi^s]_N(\mathbf{x}_1 \ldots \mathbf{x}_N) = \sqrt{N+1}\,\Phi^s_{N+1}(\mathbf{x}_1 \ldots \mathbf{x}_N, \mathbf{x}),$$

$$[\psi^*(\mathbf{x})\Phi^s]_N(\mathbf{x}_1 \ldots \mathbf{x}_N) = \frac{1}{\sqrt{N}} \sum_{j=1}^{N} \delta(\mathbf{x} - \mathbf{x}_j)\, \Phi^s_{N-1}(\mathbf{x}_1 \ldots \hat{\mathbf{x}}_j \ldots \mathbf{x}_N), \qquad (2.8)$$

where $\hat{\mathbf{x}}_j$ means that $\mathbf{x}_j$ is to be omitted; for fermions (a),

$$[\psi(\mathbf{x})\Phi^a]_N(\mathbf{x}_1 \ldots \mathbf{x}_N) = \sqrt{N+1}\,\Phi^a_{N+1}(\mathbf{x}_1 \ldots \mathbf{x}_N, \mathbf{x}),$$

$$[\psi^*(\mathbf{x})\Phi^a]_N(\mathbf{x}_1 \ldots \mathbf{x}_N) = \frac{1}{\sqrt{N}} \sum_{j=1}^{N} (-1)^{N-j} \delta(\mathbf{x} - \mathbf{x}_j)\, \Phi^a_{N-1}(\mathbf{x}_1 \ldots \hat{\mathbf{x}}_j \ldots \mathbf{x}_N). \qquad (2.9)$$

When all but one of the N-particle components of Φ vanish, we can see from (2.8) or (2.9) that $\psi^* : \mathcal{H}_N^{s,a} \to \mathcal{H}_{N+1}^{s,a}$, while $\psi : \mathcal{H}_{N+1}^{s,a} \to \mathcal{H}_N^{s,a}$. It is straightforward to verify explicitly that both (2.8) and (2.9) satisfy (2.6) and (2.7).

Note next that the expressions mfh and $(\hbar/2i)\{\mathbf{g}\cdot\nabla h + \nabla\cdot(\mathbf{g}h)\}$, which occur on the right-hand sides of (2.6) and (2.7), are just the one-particle representations of $\rho(f)$ and $J(\mathbf{g})$ respectively, applied to h (if we regard h as an element of the Hilbert space $\mathcal{H}_1$). Then we can rewrite (2.6) and (2.7) in the form

$$[\rho(f), \psi^*(h)] = \psi^*(\rho_{N=1}(f)h), \quad [J(\mathbf{g}), \psi^*(h)] = \psi^*(J_{N=1}(\mathbf{g})h),$$

$$[\rho(f), \psi(h)] = -\psi(\rho_{N=1}(f)h), \quad [J(\mathbf{g}), \psi(h)] = -\psi((J_{N=1}(\mathbf{g})h). \qquad (2.10)$$

Finally we exponentiate $\rho(f)$ and $J(\mathbf{g})$ in (2.10), and obtain

$$U(f)\psi^*(h)U^{-1}(f) = \psi^*(U_{N=1}(f)h), \quad V(\phi)\psi^*(h)V^{-1}(\phi) = \psi^*(V_{N=1}(\phi)h),$$

$$U(f)\psi(h)U^{-1}(f) = \psi(U_{N=1}(-f)h), \quad V(\phi)\psi(h)V^{-1}(\phi) = \psi((V_{N=1}(\phi^{-1})h). \qquad (2.11)$$

When we make the dependence on N explicit in (2.11), we have

$$U_{N+1}(f)\psi^*(h) = \psi^*(U_{N=1}(f)h)U_N(f), \quad V_{N+1}(\phi)\psi^*(h) = \psi^*(V_{N=1}(\phi)h)V_N(\phi),$$

$$U_N(f)\psi(h) = \psi(U_{N=1}(-f)h)U_{N+1}(f), \quad V_N(\phi)\psi(h) = \psi((V_{N=1}(\phi^{-1})h)V_{N+1}(\phi). \quad (2.12)$$

The preceding calculations for the case of canonical fields motivate the following general perspective. For a collection of CURs of the diffeomorphism group (or its semidirect product) to form a hierarchy labeled by N, a necessary and sufficient condition is that ψ^* and ψ can be constructed obeying (2.12). Especially noteworthy is the fact that the argument of the fields ψ^* and ψ in these equations is a Hilbert space vector from the $N = 1$ space in the hierarchy. This fact *defines* the $N = 1$ space, and establishes the nature of the configuration that ψ^* creates and ψ annihilates.

We expect this general structure to occur not only for point-like configurations, but also for configurations of extended objects such as vortex filaments or ribbons. With vortex configurations, the argument of ψ^* and ψ is a one-vortex Hilbert space vector, so that the unsmeared creation and annihilation fields no longer depend on a single point in space but on a spatially extended configuration. Only the currents, in unsmeared form, have as their arguments individual points in space. In the case of quantum vortex configurations, we also have additional complications associated with the possibility of overlapping or knotted vortices. This is a topic of our current research.

In the next section we use the above results to construct explicit fields for anyons that obey (2.12), anticipating that the fields will satisfy different commutation relations from those satisfied by Bose and Fermi fields. It turns out that these are the q-commutators written above.

3. CONSTRUCTING ANYON FIELDS

In this section, we construct anyon fields from a hierarchy of continuous unitary representations of the group $\mathit{Diff}(\mathbf{R}^2)$, using the N-anyon representations.[3] We display these fields explicitly, using a convenient diagrammatic representation of the elements of the braid group. It then emerges that the fields so obtained satisfy a q-commutator algebra. We stress that the q-commutator is not put in by hand, but is one of the consequences of the diffeomorphism group approach, just as anyons themselves are a consequence of the representation theory of the diffeomorphism group.

We construct anyon fields obeying the commutation relations (2.12) in the following steps: First, we write the N-anyon representation of $\mathit{Diff}(\mathbf{R}^2)$ and its semidirect product, in the Hilbert space $\mathcal{H}_N^{eq}$ of *equivariant* wave functions, defined on the universal covering space of N-particle configuration space in $\mathbf{R}^2$. The equivariance is for a one-dimensional unitary representation (a character) of the braid group B_N. Second, we make this representation of $\mathit{Diff}(\mathbf{R}^2)$ concrete by introducing a way of labeling an element in the covering space by a set of N paths in the plane. Third, we make use of this to define ψ^* as a creation operator mapping $\mathcal{H}_N^{eq}$ to $\mathcal{H}_{N+1}^{eq}$. Finally we state the results about ψ^* and ψ that are obtained in this framework.

To write the N-anyon representation, recall that an N-particle configuration is an *unordered* set γ of N distinct points in the plane: $\gamma = \{\mathbf{x}_1 \ldots \mathbf{x}_N\} \subset \mathbf{R}^2$; the indexing of the points is arbitrary. Let the configuration space Δ_N be the set of all such configurations γ, and let μ be a normalized measure on Δ_N locally equivalent to Lebesgue measure. The fundamental group $\pi_1(\Delta_N)$ is the braid group B_N. An element $\tilde{\gamma}$ of the universal covering space $\tilde{\Delta}_N$ can be labeled by a configuration γ, together with a braid $b \in B_N$ that specifies the *sheet* in $\tilde{\Delta}_N$ to which the element belongs; we shall write $\tilde{\gamma} = (\gamma, b)$. This labeling is not unique, but conventional; the sheet associated with the identity element $e \in B_N$ may be selected arbitrarily. We denote by p the projection mapping, $p(\tilde{\gamma}) = \gamma$. The braid group also *acts* on $\tilde{\Delta}_N$; for $b' \in B_N$, we have $b'(\gamma, b) = (\gamma, bb')$. An equivariant wave function $\tilde{\Psi}$ is a complex-valued function on $\tilde{\Delta}_N$ that transforms in accordance with a character T of B_N; that is, $\tilde{\Psi}(\gamma, bb') = T(b')\tilde{\Psi}(\gamma, b)$. Because $\tilde{\Psi}$ is equivariant, the product $\overline{\tilde{\Phi}(\gamma, b)}\tilde{\Psi}(\gamma, b)$ is independent of the particular choice of b. Thus we can use the measure μ on Δ_N to define square-integrable wave functions and to introduce an inner product: $(\tilde{\Phi}, \tilde{\Psi}) = \int_{\Delta_N} \overline{\tilde{\Phi}(\gamma, b)}\tilde{\Psi}(\gamma, b)\, d\mu(\gamma)$. The result is the Hilbert space $\mathcal{H}_N^{eq}$.

As we have emphasized strongly in our earlier work,[8] these ideas are not restricted to complex-valued functions and one-dimensional representations of B_N; quantum theories based on higher-dimensional, irreducible representations are equally possible (braid parastatistics). However, we limit ourselves here to discussing the usual anyon case where, when b is the braid for a single, counterclockwise exchange of two particles, $T(b) = \exp i\theta$.

Now the action of diffeomorphisms in the base space, which is given by $\phi\gamma = \{\phi(\mathbf{x}_1), \ldots \phi(\mathbf{x}_N)\}$ for $\phi \in \mathit{Diff}(\mathbf{R}^2)$, lifts to the covering space in such a way that if $p(\tilde{\gamma}) = \gamma$, then $p(\phi\tilde{\gamma}) = \phi\gamma$. Denote by K_γ the stability subgroup of γ. Diffeomorphisms in K_γ act, as do braids, on the points $(\gamma, b) \in p^{-1}(\gamma)$ belonging to the different sheets in the covering space. There is thus a natural homomorphism from K_γ onto B_N; and T determines a CUR of K_γ in which distinct components are represented by (in general) distinct complex numbers. The N-anyon representation of the semidirect product group G may be written on $\mathcal{H}_N^{eq}$ as:

$$U_N(f)\tilde{\Psi}(\tilde{\gamma}) = \exp\left[i\langle\tilde{\gamma}, f\rangle\right]\tilde{\Psi}(\tilde{\gamma}), \quad V_N(\phi)\tilde{\Psi}(\tilde{\gamma}) = \tilde{\Psi}(\phi\tilde{\gamma})\sqrt{\frac{d\mu_\phi}{d\mu}(\gamma)}, \tag{3.1}$$

where $\langle\tilde{\gamma}, f\rangle = \sum_j f(\mathbf{x}_j)$ when $\gamma = \{\mathbf{x}_1 \ldots \mathbf{x}_N\}$, and where μ_ϕ is the transformed measure given by $d\mu_\phi(\gamma) = d\mu(\phi\gamma)$.

Next we introduce a concrete way to label points in $\tilde{\Delta}_N$ that assists in understanding the action of diffeomorphisms in this space. For $\mathbf{x} \in \mathbf{R}^2$, write $\mathbf{x}$ in Cartesian coordinates as (x^1, x^2). Choose a typical configuration $\gamma = \{\mathbf{x}_j \,|\, j = 1, \ldots, N\}$ in which (for now) all the $\mathbf{x}_j$ have distinct values of their first coordinates: i.e., $x_j^1 \neq x_k^1$ for $j \neq k$. For such a choice of γ, we consider a set Γ of N continuous, non-self-intersecting and non-mutually-intersecting paths $\{\Gamma_j \,|\, j = 1, \ldots, N\}$, coming in from infinity and terminating at the $\mathbf{x}_j$. For specificity we shall take all the Γ_j at infinity to be parallel to the x^2-axis, and to extend in the direction of the negative x^2-axis. For a fixed configuration γ, consider two such sets of paths, Γ^1 and Γ^2, terminating at γ. They are said to be homotopic if the individual paths Γ_j^1 can be continuously deformed into the paths Γ_j^2, without moving the terminal points $\mathbf{x}_j \in \gamma$, without changing the direction of the paths at infinity (though they may be translated), and of course without any of the paths intersecting each other. Denote the homotopy class containing Γ by $[\Gamma]$. An element $\tilde{\gamma}$ of the covering space, with $p(\tilde{\gamma}) = \gamma$, can now be identified with a class $[\Gamma]$ whose set of terminal points is γ.

Given a configuration γ as above, we can make a canonical choice for an element $\tilde{\gamma}$ of the covering space by letting all the Γ_j be straight half-lines parallel to the x^2-axis. We call this particular set of paths $\Gamma_0^{\{\mathbf{x}_1,\ldots,\mathbf{x}_N\}}$, or Γ_0^γ (see Fig. 1). Since the indexing of the $\mathbf{x}_j$ is to this point arbitrary, we can also label the paths and their terminal points in accordance with the order of their x^1 coordinates. Thus we have $x_1^1 < x_2^1 < \ldots < x_N^1$, with Γ_j terminating at $\mathbf{x}_j$. The homotopy class $[\Gamma_0^\gamma]$ is the element of $p^{-1}(\gamma)$ that we shall conventionally associate with the identity element in the braid group.

Now the important observation is that diffeomorphisms of the plane act not only on the configurations γ but on the sets of paths Γ, since these also lie in the plane. It is also evident that a diffeomorphism that becomes trivial at infinity respects homotopy equivalence as we have defined it, so that it actually acts on $[\Gamma]$. Thus, for fixed γ, diffeomorphisms in the stability subgroup K_γ map the classes $[\Gamma]$ of paths terminating at γ into each other.

Suppose, for specificity, that we have a fixed pair of points $\{\mathbf{x}_1, \mathbf{x}_2\}$ in the plane, and consider the canonical paths $\Gamma_0^{\{\mathbf{x}_1,\mathbf{x}_2\}}$ constructed in accordance with Fig. 1, terminating at $\{\mathbf{x}_1, \mathbf{x}_2\}$. Let ϕ be a diffeomorphism, trivial at infinity, that exchanges the points; i.e., $\mathbf{x}_2 = \phi(\mathbf{x}_1)$ and $\mathbf{x}_1 = \phi(\mathbf{x}_2)$. One way in which ϕ might act on the pair of paths $\Gamma_0^{\{\mathbf{x}_1,\mathbf{x}_2\}}$ is to map them to a pair of paths as in Fig. 2 (imagine ϕ to have support in the shaded region of the plane). Then we may regard ϕ as implementing a single *counterclockwise* exchange of $\mathbf{x}_1$ and $\mathbf{x}_2$, and associate with this diffeomorphism the corresponding generator b_{12} in the braid group. Alternatively, a diffeomorphism may implement a clockwise exchange of the points, as in Fig. 3. With such a diffeomorphism, we associate the inverse braid group generator b_{12}^{-1}. Clearly a group homomorphism is defined in this manner, from the stability subgroup K_γ of $Diff(\mathbf{R}^2)$ onto the braid group. The example generalizes in the obvious way to N-particle configurations in $\mathbf{R}^2$. We denote the homomorphism by $h_\gamma : K_\gamma \to B_N$, and write $h_\gamma(\phi) = b$ for the braid associated with ϕ.

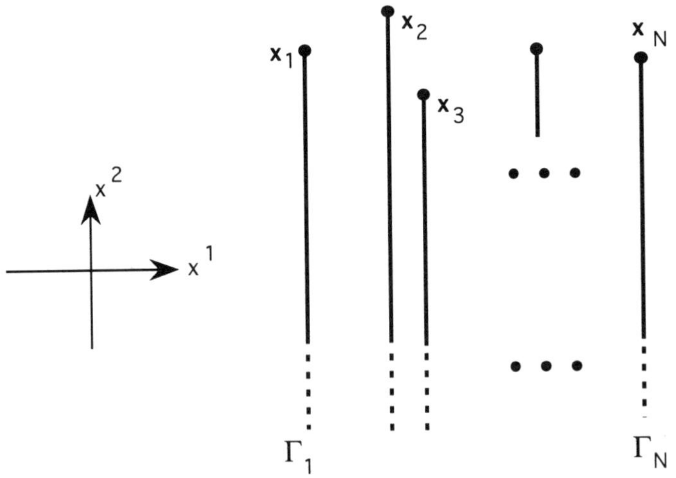

Figure 1. For $\gamma = \{\mathbf{x}_1, \ldots, \mathbf{x}_N\}$, a canonical choice Γ_0^γ of paths $\{\Gamma_j\}$ terminating at $\{\mathbf{x}_j\}$.

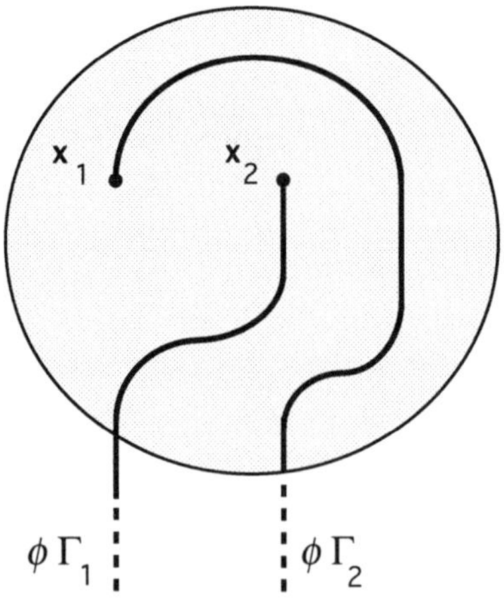

Figure 2. A diffeomorphism with support inside the indicated region moves the paths to a different homotopy class, implementing a single counterclockwise exchange of two points labeled originally as in Fig. 1.

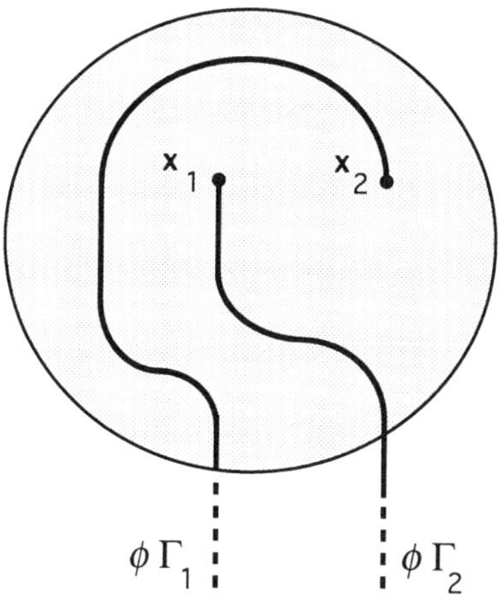

Figure 3. A diffeomorphism implementing a single clockwise exchange, as distinct from the counterclockwise exchange of Fig. 2.

The above provides a faithful mapping from B_N to the homotopy classes $[\Gamma]$. Then we can write $T(\Gamma)$ in place of $T(b)$, when b is the braid that takes $[\Gamma_0^\gamma]$ to $[\Gamma]$.

We shall next make use of this picture to define the anyon creation field ψ^*, mapping $\mathcal{H}_N^{eq}$ to $\mathcal{H}_{N+1}^{eq}$ and satisfying the desired equations (2.12). To describe the way that such an anyon field acts, we introduce one more important convention: a way to denote the procedure for adding a single anyonic particle at $\mathbf{x}$, not merely to an N-anyon configuration, but to an element of the N-anyon covering space. Doing this, of course, will *break* the homotopy equivalence, because points on *many* different sheets of the $(N+1)$-anyon covering space correspond to the introduction of a a new anyon at a point $\mathbf{x}$. We therefore need a standard way to make the choice.

Given the homotopy class $[\Gamma_0^\gamma]$, and the additional point $\mathbf{x}$, define a new set of paths $\Gamma_{\mathbf{x}}^{\{\mathbf{x}_1,\ldots,\mathbf{x}_N\}}$ by placing the point $\mathbf{x}$ in the plane among the N paths comprising Γ_0^γ, and drawing a new path Γ_{N+1} that terminates at $\mathbf{x}$, and comes in from infinity to the *right* of the N existing paths without intersecting them (see Fig. 4). The homotopy class $[\Gamma_{\mathbf{x}}^{\{\mathbf{x}_1,\ldots,\mathbf{x}_N\}}]$ is thus defined, specifying a particular element of the $(N+1)$-anyon covering space. We stress the rather subtle point that $[\Gamma_{\mathbf{x}}^{\{\mathbf{x}_1,\ldots,\mathbf{x}_N\}}]$ is defined by this procedure as a homotopy class; but in order to define it, we needed to use not merely the class $[\Gamma_0^\gamma]$, but the actual element Γ_0^γ within that class.

We now have all we need to construct the anyon creation field acting on the Hilbert space $\mathcal{H}_N^{eq}$, in close analogy with the second-quantized, nonrelativistic Bose and Fermi creation fields discussed above. Roughly speaking, we can already see from Figs. 1-4 how the q-commutator will enter. Suppose $\{\mathbf{x}_1, \mathbf{x}_2\}$ are as in Fig. 2. If we first create an anyon at $\mathbf{x}_2$, we obtain the path $\Gamma_1 = \Gamma_0^{\{\mathbf{x}_2\}}$, which is a straight line parallel to the x^2-axis, terminating at $\mathbf{x}_2$. Creating the next anyon at $\mathbf{x}_1$ gives us the paths in the class $[\Gamma_{\mathbf{x}_1}^{\{\mathbf{x}_2\}}]$. Such a pair of paths is depicted in Fig. 2, corresponding to the braid group generator. On the other hand, if we first create the anyon at $\mathbf{x}_1$, we consider the path $\Gamma_1 = \Gamma_0^{\{\mathbf{x}_1\}}$. Creating the next anyon at $\mathbf{x}_2$ gives us the paths in the class $[\Gamma_{\mathbf{x}_2}^{\{\mathbf{x}_1\}}]$, which for this example is just the class $\Gamma_0^{\{\mathbf{x}_1,\mathbf{x}_2\}}$ associated with the identity element of the braid group. There will thus be a relative phase $q = T(b_{12})$ occurring in the two products of creation operators, where T is the one-dimensional unitary representation of the braid group characterizing the anyons in the hierarchy.

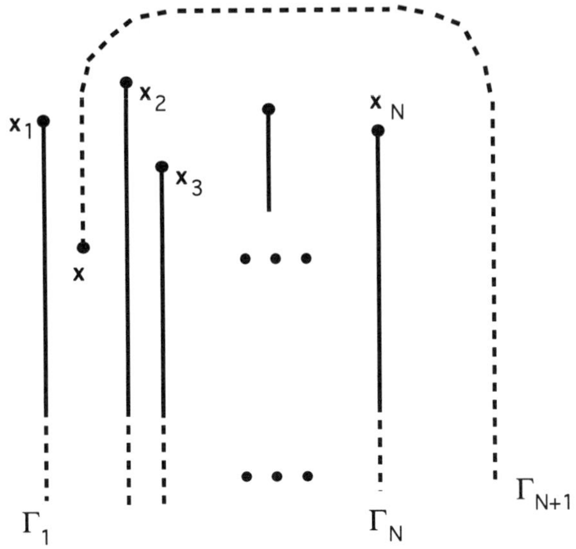

Figure 4. An anyonic particle is created at $\mathbf{x}$, defining the element $\Gamma_{\mathbf{x}}^{\{\mathbf{x}_1,\ldots,\mathbf{x}_N\}}$ of the $(N+1)$-anyon covering space.

More precisely, consider an equivariant wave function $\tilde{\Psi}_N$ in $\mathcal{H}_N^{eq}$. We write $\tilde{\Psi}_N = \tilde{\Psi}_N(\gamma, \Gamma)$, where $\gamma = \{\mathbf{x}_1, \ldots, \mathbf{x}_N\}$ and the paths in Γ terminate at the points in γ. For fixed γ it is convenient to regard $\tilde{\Psi}_N$ as defined for the *individual* sets of paths Γ, but constant on the equivalence classes $[\,\Gamma\,]$. The equivariance of $\tilde{\Psi}_N$ is put in by requiring that if Γ is obtained from Γ_0^γ by the braid $b = h_\gamma(\phi)$, then

$$\tilde{\Psi}_N(\gamma, \Gamma) = T(b)\tilde{\Psi}_N(\gamma, \Gamma_0^\gamma)\,, \tag{3.2}$$

or with an alternative notation, $\tilde{\Psi}_N(\gamma, \Gamma) = T(\Gamma)\tilde{\Psi}_N(\gamma, \Gamma_0^\gamma)$. That is, specifying $\tilde{\Psi}_N$ for (almost) all values on a single sheet in the covering space $\tilde{\Delta}_N$, defines its values on any other sheet. We see now that the condition we imposed earlier in defining Γ_0^γ, that all the $\mathbf{x}_j$ have distinct values of their first coordinates, can be regarded merely as the omission of an arbitrary boundary (having measure zero) associated with crossing from one sheet to the next in $\tilde{\Delta}_N$.

Now we write, in analogy with Eqs. (2.8) and (2.9), the anyon annihilation and creation fields. Let $\tilde{\Psi}$ denote the sequence $(\tilde{\Psi}_N), N = 0, 1, 2, \ldots$, with $\tilde{\Psi}_N \in \mathcal{H}_N^{eq}$, and $(\tilde{\Psi}, \tilde{\Psi}) = \sum_N (\tilde{\Psi}_N, \tilde{\Psi}_N) < \infty$. We define

$$[\psi(\mathbf{x})\tilde{\Psi}]_N(\{\mathbf{x}_1, \ldots, \mathbf{x}_N\}, \Gamma_0^{\{\mathbf{x}_1,\ldots,\mathbf{x}_N\}}) = \tilde{\Psi}_{N+1}(\{\mathbf{x}_1, \ldots, \mathbf{x}_N, \mathbf{x}\}, \Gamma_{\mathbf{x}}^{\{\mathbf{x}_1,\ldots,\mathbf{x}_N\}})\,,$$

$$[\psi^*(\mathbf{x})\tilde{\Psi}]_N(\{\mathbf{x}_1, \ldots, \mathbf{x}_N\}, \Gamma_0^{\{\mathbf{x}_1,\ldots,\mathbf{x}_N\}}) =$$

$$\sum_{j=1}^N \delta(\mathbf{x} - \mathbf{x}_j)\, \tilde{\Psi}_{N-1}(\{\mathbf{x}_1, \ldots, \hat{\mathbf{x}}_j, \ldots, \mathbf{x}_N\}, \Gamma_0^{\{\mathbf{x}_1,\ldots,\hat{\mathbf{x}}_j,\ldots,\mathbf{x}_N\}})\, T^*(\Gamma_{\mathbf{x}}^{\{\mathbf{x}_1,\ldots,\hat{\mathbf{x}}_j,\ldots,\mathbf{x}_N\}}), \tag{3.3}$$

where as before $\hat{\mathbf{x}}_j$ means that the point $\mathbf{x}_j$ is omitted. This defines $[\psi(\mathbf{x})\tilde{\Psi}]_N$ and $[\psi^*(\mathbf{x})\tilde{\Psi}]_N$ on one sheet. We extend the definitions in (3.3) to the other sheets of $\tilde{\Delta}_N$ (in general, infinitely many of them) using the equivariance property (3.2).

Comparing (3.3) with (2.8) and (2.9), note that the factors $\sqrt{N+1}$ and $1/\sqrt{N}$ no longer appear. This is because, in the case of anyons, the inner product is defined with respect to integration in the base space Δ_N (or, equivalently, over just one sheet of $\tilde{\Delta}_N$). It *must* be defined so, as the number of sheets in the covering space is not $N!$ any longer, but is now infinite.

When all but one of the N-particle components of Ψ vanish, we see from (3.3) that $\psi : \mathcal{H}^{eq}_{N+1} \to \mathcal{H}^{eq}_N$, and $\psi^* : \mathcal{H}^{eq}_N \to \mathcal{H}^{eq}_{N+1}$. In fact, ψ and ψ^* defined in this way are just the intertwining fields obeying Eqs. (2.12).

Finally, we are in a position to determine, by straightforward calculation from (3.3), what commutation relations the anyon fields we have obtained satisfy with each other. The answer is just the q-commutation relations given by Eqs. (2.1), where q is the phase specified by the representation T of the braid group generator. Furthermore, we recover the operators $\rho(\mathbf{x})$ and $\mathbf{J}(\mathbf{x})$ in terms of the anyon fields as the desired expressions given by (1.1), with $\int(\tilde{\Psi}_N, \rho(\mathbf{x})\tilde{\Psi}_N)\, d^2x = Nm$.

4. CONCLUSION

To sum up we have proposed a way, beginning with a collection of diffeomorphism group representations, to classify them into hierarchies based on the existence of intertwining field operators. Our method works not only for the N-particle Bose and Fermi representations, $N = 0, 1, 2, \ldots$ (by which it is motivated), but for the N-anyon representations of *Diff*$(\mathbf{R}^2)$ that we previously obtained. Anyons with distinct values of the phase characterizing the intermediate statistics belong naturally to different hierarchies.

Then we obtain q-commutation relations for the resulting anyon fields as a *consequence* of our prescription. Assuming little more than the fundamental role played by CURs of the diffeomorphism group, we thus arrive by entirely natural means at a framework for treating the many-anyon system. Our approach provides an alternative to beginning with the introduction of fields obeying noncanonical commutation relations, or to beginning with particles obeying canonical fields and introducing a Chern-Simons potential to describe the anyons.

We expect this method to generalize to still other hierarchies of diffeomorphism group representations, such as those describing extended objects like quantized vortex loops and filaments.

References

1. For more detailed reviews of background material, see G. A. Goldin and D. H. Sharp, Diffeomorphism groups and local symmetries: Some applications in quantum physics, *in*: "Symmetries in Science III", B. Gruber, F. Iachello, eds., Plenum, New York, p. 181 (1989); G. A. Goldin, Predicting anyons: The origins of fractional statistics in two-dimensional space, *in*: "Symmetries in Science V", B. Gruber, L. C. Biedenharn, H.-D. Doebner, eds., Plenum, New York, p. 259 (1991); and G. A. Goldin and D. H. Sharp, The diffeomorphism group approach to anyons, *Int. J. Mod. Phys.* B5:2625 (1991)

2. G. A. Goldin, R. Menikoff, and D. H. Sharp, Diffeomorphism groups and quantized vortex filaments, *Phys. Rev. Lett.* 58: 2162 (1987); Quantum vortex configurations in three dimensions, *Phys. Rev. Lett.* 67:3499 (1991)

3. G. A. Goldin, R. Menikoff, and D. H. Sharp, Particle statistics from induced representations of a local current group, *J. Math. Phys.* 21:650 (1980); Representations of a local current algebra in non-simply connected space and the Aharonov-Bohm effect, *J. Math. Phys.* 22:1664 (1981)

4. J. M. Leinaas and J. Myrheim, On the theory of identical particles, *Nuovo Cimento* 37B:1 (1977)

5. G. A. Goldin and D. H. Sharp, Rotation generators in two-dimensional space and particles obeying unusual statistics, *Phys. Rev.* D28:830 (1983)

6. G. A. Goldin, R. Menikoff, and D. H. Sharp, Diffeomorphism groups, gauge groups, and quantum theory, *Phys. Rev. Lett.* 51:2246 (1983)

7. R. B. Laughlin, *Phys. Rev. Lett.* 50:1395 (1983); B. I. Halperin, Statistics of quasiparticles and the hierarchy of fractional quantized Hall states, *Phys. Rev. Lett.* 52: 1583; R. B. Laughlin, Fractional statistics in the quantized Hall effect, *in*: "Two Dimensional Strongly Correlated Electronic Systems", Zi-Zhao Gan, Zhao-Bin Su, eds., Gordon and Breach, London (1989) F. Wilczek, "Fractional Statistics and Anyon Superconductivity", Singapore: World Scientific (1990)

8. G. A. Goldin, R. Menikoff, and D. H. Sharp, Comment on 'General theory for quantum statistics in two dimensions', *Phys. Rev. Lett.* 54:603 (1985)

ON A FULL QUANTIZATION OF THE TORUS

Mark J. Gotay

Department of Mathematics
University of Hawai'i
Honolulu, HI 96822
United States of America

Abstract

I exhibit a prequantization of the torus which is actually a "full" quantization in the sense that a certain minimal complete set of classical observables is irreducibly represented. Thus in this instance there is no Groenewold-Van Hove obstruction to quantization.

1. INTRODUCTION

The prequantization procedure produces a faithful representation of the entire Poisson algebra of a quantizable symplectic manifold.[1] In general, these prequantization representations are flawed physically; for instance, the prequantization of $\mathbf{R}^{2n}$ with its standard symplectic structure is not unitarily equivalent to the Schrödinger representation. One usually remedies this by imposing an irreducibility requirement. But there is seemingly a price to be paid for irreducibility: one can no longer quantize *all* classical observables, but rather only proper subalgebras thereof.

This "obstruction" to quantization has been known since the 1940s. In a series of papers, Groenewold and later Van Hove showed that it is impossible to quantize the entire Poisson algebra of polynomials on $\mathbf{R}^{2n}$ in such a way that the Heisenberg $\mathrm{h}(2n)$ subalgebra of inhomogeneous linear polynomials is irreducibly represented.[2-4] Some of the maximal subalgebras of polynomials that *can* be consistently quantized subject to this irreducibility requirement are the inhomogeneous quadratic polynomials, and polynomials which are at most affine in the momenta or the configurations. See Refs. 5-7 for further discussions of this example. Recently, a similar phenomenon was observed for S^2. In this case it was shown that the maximal subalgebra of the Poisson algebra of spherical harmonics that can be consistently quantized while irreducibly representing the $\mathrm{u}(2)$ subalgebra of spherical harmonics of degree at most one is just this $\mathrm{u}(2)$ subalgebra itself.[8]

Based on these results as well as general quantization theory, it would seem reasonable to conjecture that a "no-go" theorem must always hold, to wit:

It is impossible to quantize the entire Poisson algebra of any symplectic manifold subject to an irreducibility requirement.

To make this conjecture precise, I introduce some terminology. Let (M,ω) be a connected symplectic manifold, and let $\mathcal{O}$ be a Poisson subalgebra of $C^\infty(M)$.

Definition 1. *A* prequantization *of $\mathcal{O}$ is a linear map $\mathcal{Q}$ from $\mathcal{O}$ to an algebra of (essentially) self-adjoint operators*[a] *on a Hilbert space $\mathcal{H}$ such that*

$$(i)\ \ \mathcal{Q}(\{f,g\}) = 2\pi i\,[\mathcal{Q}(f),\mathcal{Q}(g)]\,,$$

where $\{\,,\,\}$ denotes the Poisson bracket and $[\ ,\ ]$ the commutator.[b] *If $\mathcal{O}$ contains the constant function 1, then also*

$$(ii)\ \ \mathcal{Q}(1) = I.$$

As the nomenclature suggests, it appears necessary in practice to supplement these conditions, often by requiring that a certain subset $\mathcal{F}$ of observables be represented irreducibly. In favorable circumstances one can take $\mathcal{F}$ to consist of the components of a momentum map associated to a transitive Lie symmetry group. In the nonhomogeneous case, the corresponding notion is a "complete set of observables." This is a subset $\mathcal{F}$ of $C^\infty(M)$ with the property that the Hamiltonian vector fields of elements of $\mathcal{F}$ span the tangent spaces to M at every point.[c] This means in particular that $\{f,g\} = 0$ for all $f \in \mathcal{F}$ implies that $g = \text{const}$. I say that $\mathcal{F}$ is *minimal* if, modulo the constants, the linear span of $\mathcal{F}$ contains no proper complete subspace.

Similarly, a set of operators $\mathcal{A}$ on $\mathcal{H}$ is *irreducible* provided the only (bounded) operator which commutes with each $A \in \mathcal{A}$ is a multiple of the identity. Since irreducibility is the quantum analogue of completeness, I make

Definition 2. *Let $\mathcal{F} \subset \mathcal{O}$ be a complete set of observables. A prequantization $\mathcal{Q}$ of $\mathcal{O}$ is a* quantization *of the pair $(\mathcal{O},\mathcal{F})$ provided*

(iii) the corresponding operators $\{\mathcal{Q}(f)\,|\,f \in \mathcal{F}\}$ form an irreducible set.

If $\mathcal{F}$ is minimal and one can take $\mathcal{O} = C^\infty(M)$, the quantization is full.[d]

Thus there do not exist full quantizations of either $(\mathbf{R}^{2n}, \mathrm{h}(2n))$ or $(S^2, \mathrm{u}(2))$. According to the conjecture above, it should be impossible to fully quantize *any* symplectic manifold. But here I show that this is *false*: there exists a full quantization of the torus T^2. The existence of this full quantization, which can be obtained as a particular Kostant-Souriau prequantization, is surprising, especially given the Groenewold-Van Hove obstruction to a full quantization of its covering $\mathbf{R}^2$. The dichotomy stems from the fact that, unlike on $\mathbf{R}^2$, any prequantum bundle on T^2 is nontrivial. This twisting forces the wave functions to be *quasi*-periodic, and these boundary conditions effectively override the obstruction in the case of interest. That the prequantization I consider may indeed provide a full quantization is signalled by the observation that it has, in a sense, the Schrödinger representation of the Heisenberg algebra built in. It turns out that this representation of the Heisenberg algebra is well known in solid state physics, where it goes under the name "*kq*-representation".[10]

---

[a] For the most part technical difficulties with unbounded operators will be ignored, as they are not essential for what follows.

[b] I use units in which Planck's constant $h = 1$.

[c] This definition of a complete set of observables is slightly stronger than that given in Ref. 1.

[d] Without some sort of minimality condition on $\mathcal{F}$, it is often possible to find full quantizations; indeed, it may happen that a prequantization representation is itself irreducible.[4,9]

2. PREQUANTIZATION

It is convenient to realize the torus T^2 as $\mathbf{R}^2/\mathbf{Z}^2$ with symplectic form

$$\omega = N dx \wedge dy,$$

where N is a nonzero integer. Then one can identify $C^\infty(T^2)$ with the space of doubly periodic functions f on the plane:

$$f(x+m, y+n) = f(x,y), \qquad m,n \in \mathbf{Z}.$$

For the prequantization of the torus, I follow Ref. 1, §2.3. (See also Ref. 11 for a gauge-theoretic treatment.) Let L_N be a prequantum bundle over T^2 with Chern class N.[e] To explicitly construct it,[f] introduce the sets

$$U_- = \{(x,y) \in (-\delta, 1-\delta) \times [0,1]\} \quad \text{and} \quad U_+ = \{(x,y) \in (\delta, 1+\delta) \times [0,1]\}$$

for $\delta > 0$ small. Identifying $y = 0$ with $y = 1$ gives a pair of cylinders, and then identifying x with $x+1$ for $x \in (-\delta, \delta)$ pastes the cylinders together into a torus. On $U_\pm$ define the connection potentials

$$\theta_\pm = Nx\, dy.$$

On the overlap $U_+ \cap U_-$,

$$\theta_- - \theta_+ = \begin{cases} 0, & x \in (\delta, 1-\delta) \\ -N\, dy, & x \in (-\delta, \delta) \end{cases}$$

whence the transition functions are

$$c(x,y) = \begin{cases} 1, & x \in (\delta, 1-\delta) \\ e^{-2\pi i N y}, & x \in (-\delta, \delta). \end{cases}$$

The space $\Gamma(L_N)$ of smooth sections of L_N may thus be identified with the space of smooth 'quasi-periodic' complex-valued functions ϕ on the plane:

$$\phi(x+m, y+n) = e^{2\pi i N m y}\phi(x,y), \qquad m,n \in \mathbf{Z}. \tag{2.1}$$

The corresponding prequantum Hilbert space $\mathcal{H}_N$ consists of those $\phi(x,y)$ satisfying (2.1) which are square integrable over $[0,1)^2$.

The prequantization map $\mathcal{Q}_N$ is

$$\mathcal{Q}_N(f) = \frac{1}{2\pi i N}\left(\frac{\partial f}{\partial x}\left(\frac{\partial}{\partial y} - 2\pi i N x\right) - \frac{\partial f}{\partial y}\frac{\partial}{\partial x}\right) + f. \tag{2.2}$$

---

[e] Up to equivalence, there is exactly one line bundle per value of N, but each L_N carries a two-parameter family of inequivalent connections. Below I fix a particular connection.

[f] The associated principal circle bundle over T^2 may be realized as $\Gamma_N \backslash H_N(2)$, where $H_N(2)$ is the (polarized) Heisenberg group consisting of all matrices of the form

$$\begin{pmatrix} 1 & a & -c/N \\ 0 & 1 & b \\ 0 & 0 & 1 \end{pmatrix}$$

with $a, b, c \in \mathbf{R}$, and Γ_N is the subgroup of $H_N(2)$ consisting of those matrices for which a, b and c are integers, cf. Ref. 12, §2.

These operators are essentially self-adjoint on $\Gamma(L_N)$.

Now define, for each $N \neq 0$, the minimal complete sets[g]

$$\mathcal{F}_N = \left\{1, e^{\pm 2\pi i N x}, e^{\pm 2\pi i N y}\right\}.$$

Then (2.2) gives

$$\mathcal{Q}_N\left(e^{\pm 2\pi i N x}\right) = e^{\pm 2\pi i N x}\left(1 \mp 2\pi i N x \pm \frac{\partial}{\partial y}\right) \tag{2.3}$$

$$\mathcal{Q}_N\left(e^{\pm 2\pi i N y}\right) = e^{\pm 2\pi i N y}\left(1 \mp \frac{\partial}{\partial x}\right). \tag{2.4}$$

Note that $\mathcal{Q}_N\left(e^{-2\pi i N x}\right) = \mathcal{Q}_N\left(e^{2\pi i N x}\right)^*$ and $\mathcal{Q}_N\left(e^{-2\pi i N y}\right) = \mathcal{Q}_N\left(e^{2\pi i N y}\right)^*$.

3. THE GO THEOREM

The key observation is that the operators

$$\hat{X} = \frac{1}{2\pi i}\left(\frac{\partial}{\partial y} - 2\pi i N x\right), \quad \hat{Y} = -\frac{1}{2\pi i}\frac{\partial}{\partial x}, \quad \hat{Z} = -\frac{N}{2\pi i}, \tag{3.1}$$

the first two of which appear in (2.2), provide a representation of the Heisenberg algebra

$$\mathrm{h}(2) = \left\{X, Y, Z \in \mathbf{R}^3 \,|\, [X,Y] = Z,\ [X,Z] = [Y,Z] = 0\right\}$$

on $\mathcal{H}_N$. For $|N| = 1$ this is equivalent to the Schrödinger representation, as I now show.

For the moment fix $N = 1$. ($N = -1$ will work as well.) The analysis is simplified by applying the Weil-Brezin-Zak transform W [cf. Refs. 6, §1.10, and 10] to the data in the previous section. Let $\mathcal{S}(\mathbf{R})$ be the Schwartz space. Define a linear map $W : \mathcal{S}(\mathbf{R}) \to \Gamma(L_1)$ by

$$(W\psi)(x,y) = \sum_{k \in \mathbf{Z}} \psi(x+k)e^{-2\pi i k y}$$

with inverse

$$(W^{-1}\phi)(x) = \int_0^1 \phi(x,y)\,dy.$$

That W extends to a unitary map of $L^2(\mathbf{R})$ onto $\mathcal{H}_1$ is readily checked, cf. Thm. 1.109 in Ref. 6. Under this transform, $\hat{X}$ and $\hat{Y}$ go over to the operators $-x$ and $-\frac{1}{2\pi i}\frac{d}{dx}$ on $L^2(\mathbf{R})$, respectively. Thus the $N = 1$ prequantization carries the Schrödinger representation of the Heisenberg algebra.

This leads one to suspect that $\mathcal{Q}_1$ might yield a full quantization of the torus. Indeed this is the case:

Theorem 1 (Go Theorem). *The prequantization $\mathcal{Q}_1$ gives a full quantization of $(C^\infty(T^2), \mathcal{F}_1)$.*

[g] Although in principle I consider only real-valued observables, here it is convenient to use complex notation.

Proof: Since $\mathcal{Q}_1$ is a prequantization, conditions (i) and (ii) are automatically satisfied. Thus it is only necessary to verify (iii), i.e.,

$$\left\{\mathcal{Q}_1(e^{\pm 2\pi i x}), \mathcal{Q}_1(e^{\pm 2\pi i y})\right\}$$

forms an irreducible set. Applying W to (2.3) and (2.4), one has, as operators on $\mathcal{S}(\mathbf{R})$,

$$\begin{aligned}\left(\mathcal{Q}_1(e^{\pm 2\pi i x})\psi\right)(x) &= e^{\pm 2\pi i x}(1 \mp 2\pi i x)\psi(x)\\ \left(\mathcal{Q}_1(e^{\pm 2\pi i y})\psi\right)(x) &= \left(1 \mp \frac{d}{dx}\right)\psi(x \pm 1).\end{aligned}$$

Suppose that T is a bounded linear operator on $L^2(\mathbf{R})$ which commutes with both $A := \mathcal{Q}_1(e^{2\pi i x})$, $B := \mathcal{Q}_1(e^{2\pi i y})$ and their adjoints. Then T must commute with $A^*A = 1 + 4\pi^2 x^2$ and $B^*B = 1 - d^2/dx^2$. So $[T, x^2] = 0$ and $[T, d^2/dx^2] = 0$, whence T must commute with $[d^2/dx^2, x^2] = 4x\, d/dx + 2$. Thus T commutes with the generators of the symplectic algebra $\mathrm{sp}(2,\mathbf{R})$ in the metaplectic representation μ. Since $\mathcal{S}(\mathbf{R})$ contains a dense subspace of analytic vectors for μ (viz. the Hermite functions, cf. Ref. 6, §4.3), it follows that $[T, \mu(\mathcal{M})] = 0$ for all $\mathcal{M}$ in some neighborhood of the identity in the metaplectic group $\mathrm{Mp}(2,\mathbf{R})$ and hence, as this group is connected, for all $\mathcal{M}$ in $\mathrm{Mp}(2,\mathbf{R})$.

Although the metaplectic representation is reducible, the subrepresentations μ_e and μ_o on each invariant summand of $L^2(\mathbf{R}) = L^2_e(\mathbf{R}) \oplus L^2_o(\mathbf{R})$ of even and odd functions are irreducible [cf. Ref. 6, §4.4]. Writing $T = P_eT + P_oT$, where P_e and P_o are the even and odd projectors, one has

$$[P_eT, \mu(\mathcal{M})] = P_e[T, \mu(\mathcal{M})] + [P_e, \mu(\mathcal{M})]T = 0 \tag{3.2}$$

for any $\mathcal{M} \in \mathrm{Mp}(2,\mathbf{R})$. It then follows from the irreducibility of the subrepresentation μ_e that $P_eT = k_eP_e + RP_o$ for some constant k_e and some operator $R : L^2_o(\mathbf{R}) \to L^2_e(\mathbf{R})$. Substituting this expression into (3.2) yields $[RP_o, \mu(\mathcal{M})] = 0$, and Schur's Lemma then implies that R is either an isomorphism or is zero. But R cannot be an isomorphism as the representations μ_e and μ_o are inequivalent [cf. Ref. 6, Thm. 4.56]. Thus $P_eT = k_eP_e$. Similarly $P_oT = k_oP_o$, whence $T = k_eP_e + k_oP_o$.

But now a short calculation shows that T commutes with

$$A - A^* = 2i(\sin 2\pi x - 2\pi x \cos 2\pi x)$$

only if $k_e = k_o$. Thus T is a multiple of the identity, and so $\{A, A^*, B, B^*\}$ is an irreducible set, as was to be shown. □

This proof is not valid for $|N| > 1$. The problem is that the map $W : L^2(\mathbf{R}) \to \mathcal{H}_N$, which in order to maintain (2.1) now takes the form

$$(W\psi)(x,y) = \sum_{k\in\mathbf{Z}} \psi(x+k)e^{-2\pi i N k y},$$

is no longer onto; indeed, from this formula it is apparent that the image of W consists of those functions in $\mathcal{H}_N$ which have period $1/N$ in y. Denote the subspace of all such functions by $\mathcal{P}_N$. Since the operators $\hat{X}$ and $\hat{Y}$ preserve periodicity, $\Gamma(L_N) \cap \mathcal{P}_N$ is invariant and it follows that the representation (3.1) of $\mathrm{h}(2)$ on $\mathcal{H}_N$ is no longer irreducible. While these facts do not *a priori* preclude the existence of a full quantization for $|N| > 1$, they do yield the following.

Proposition 2. *For* $|N| > 1$, $\mathcal{Q}_N$ *does* not *represent* $\mathcal{F}_N$ *irreducibly.*

Proof: It suffices to observe that the operators (2.3) and (2.4) commute with the orthogonal projector $\mathcal{H}_N \to \mathcal{P}_N$. □

Thus $\mathcal{Q}_N$ provides a full quantization of $(C^\infty(T^2), \mathcal{F}_N)$ iff $|N| = 1$.

4. DISCUSSION

The $|N| = 1$ quantization of the torus is curious in several respects. First, it does not mesh well with what one intuitively expects on the basis of geometric quantization theory.[1] For one thing, it is not necessary to introduce a polarization in order to obtain an acceptable quantization. (Since the prequantum wave functions are quasi-periodic – or, more crudely, since $\mathcal{H}_1 \approx L^2(\mathbf{R})$ – they are in effect "already polarized.") And if one does polarize T^2, then one is guaranteed to be able to consistently quantize only a substantially smaller set of observables – namely, those whose Hamiltonian vector fields preserve the polarization – but not necessarily any larger subalgebra. On the other hand, it is known that the torus admits a strict deformation quantization,[13] so perhaps it is not entirely unexpected that it admits a full quantization as well.

A second point concerns the role of the Heisenberg group H(2) in this example, especially as compared to $\mathbf{R}^2$. In both cases, it acts transitively (and factors through a translation group.) On $\mathbf{R}^2$, there is a momentum mapping for this action, and it is natural to insist that quantization provide an irreducible representation of h(2). But on T^2, there is no momentum map; the torus is "classically anomalous" in this regard.[14] So it does not make sense to require that quantization produce a representation of h(2) in this case. Nonetheless, there is the representation (3.1) of the Heisenberg algebra on $\mathcal{H}_N$ which, for $|N| = 1$, is irreducible. The reason one can obtain a full quantization while irreducibly representing the Heisenberg algebra in this instance is because $\hat{X}$ and $\hat{Y}$ are not the quantizations of any observables; in other words, the representation (3.1) does not arise via quantization. Thus in this sense the quantization is also anomalous. On $\mathbf{R}^2$ this is not the case: demanding that $\mathcal{Q}$ be h(2)-equivariant is too stringent a requirement to allow the quantization of every observable in the Schrödinger representation.

As an aside, it is interesting to observe that if x and y *were* globally defined observables, then from (2.2) one would have

$$\mathcal{Q}(x) = \frac{1}{2\pi i N}\frac{\partial}{\partial y} \quad \text{and} \quad \mathcal{Q}(y) = -\frac{1}{2\pi i N}\left(\frac{\partial}{\partial x} - 2\pi i N y\right). \tag{4.1}$$

Of course the operators $\partial/\partial y$ and $\partial/\partial x - 2\pi i N y$ are not defined on $\Gamma(L_N)$,[h] although they do make sense on $\Gamma(L_{-N})$ with the connection given by the potentials $\overline{\theta}_\pm = Ny\,dx$. In fact, one has

$$\frac{\partial}{\partial y} = \overline{\nabla}_{\frac{\partial}{\partial y}} \quad \text{and} \quad \frac{\partial}{\partial x} - 2\pi i N y = \overline{\nabla}_{\frac{\partial}{\partial x}}$$

where $\overline{\nabla}$ is the aforesaid connection on L_{-N}. This explains the remark of Asorey [Ref. 14, §4], to the effect that 'the quantum generators of translation symmetries are

[h] In Ref. 15 such operators are called "bad."

given by multiples of $-i\nabla_j$ for $j = 1, 2$.' Furthermore, if one were on $\mathbf{R}^2$ rather than T^2, the operators (4.1) would be well defined and would commute with $\hat{X}$ and $\hat{Y}$, indicating that the representation (3.1) of h(2) is reducible. So it is apparent how the nontriviality of the prequantum bundles removes this obstruction to the irreducibility of this representation of the Heisenberg algebra.

It is perhaps also surprising that the irreducibility of the representation (3.1) of h(2) on $\mathcal{H}_1$ apparently does not, in and by itself, imply that $\mathcal{Q}_1$ is a full quantization of $(C^\infty(T^2), \mathcal{F}_1)$. Indeed, that the prequantum Hilbert space carries the Schrödinger representation only seems to be a portent; the proof of the Go Theorem devolves instead upon the properties of the metaplectic representation. I do not know if this theorem can be proved directly using the irreducibility of h(2). More generally, I wonder to what extent the fullness of the prequantization $\mathcal{Q}_N$, for a given minimal complete set $\mathcal{F}$, is correlated with the irreducibility of the h(2) representation? For instance, is it possible to obtain a full quantization using $\mathcal{Q}_N$ for $|N| > 1$? Or perhaps one could prove a no-go result in this context, thereby strengthening Proposition 2?

Thirdly, the requirement that quantization irreducibly represent a complete set $\mathcal{F}$ typically leads to "von Neumann rules" for elements of $\mathcal{F}$.[8] Roughly speaking, these rules govern the extent to which $\mathcal{Q}$ preserves the multiplicative structure of the Poisson algebra. In particular, they determine how $\mathcal{Q}(f^2)$ is related to $\mathcal{Q}(f)^2$ for $f \in \mathcal{F}$. For $\mathbf{R}^{2n}$ and S^2 this relation is relatively 'tight'. But for T^2 both $\mathcal{Q}_N(f^2)$ and $\mathcal{Q}_N(f)^2$ are completely determined for any observable f and any N by the simple fact that $\mathcal{Q}_N$ is a prequantization; irreducibility is irrelevant. Moreover, one sees from (2.2) that $\mathcal{Q}_N(f^2)$ is a first order differential operator whereas $\mathcal{Q}_N(f)^2$ is of second order. This indicates that, in general, one cannot expect quantization to respect the classical multiplicative structure to any significant degree.

Finally, a salient feature of this example is that the Poisson subalgebra $\wp(\mathcal{F}_1)$ generated by the minimal complete set $\mathcal{F}_1$ – the trigonometric polynomials – is infinite dimensional, and is in fact dense in $C^\infty(T^2)$. This does not happen for either $\mathbf{R}^{2n}$ or S^2, in which cases the relevant subalgebras are finite dimensional. Since the irreducibility of $\mathcal{F}_1$ under $\mathcal{Q}_1$ is tantamount to that of $\wp(\mathcal{F}_1)$, it follows that the irreducibility requirement on T^2 is substantially weaker than the corresponding requirements on either $\mathbf{R}^{2n}$ or S^2. This, along with the *a posteriori* observation that $\mathcal{Q}_1$ irreducibly represents $C^\infty(T^2)$,[i] is likely the underlying reason why $\mathcal{Q}_1$ provides a full quantization of $(C^\infty(T^2), \mathcal{F}_1)$.[j] On the other hand, there are prequantizations which do not represent $C^\infty(T^2)$ irreducibly, for instance, the prequantization of Avez.[9,15] Consequently such a prequantization cannot yield a full quantization of $(C^\infty(T^2), \mathcal{F})$ for any subset $\mathcal{F}$. Notice, however, that Avez' prequantization is not a Kostant-Souriau prequantization.

An alternate approach to the quantization of the torus, which touches upon some of the issues discussed here, and which contains applications to the quantum Hall effect, is given in the recent paper by Aldaya et al.[16]

ACKNOWLEDGMENTS

I thank V. Aldaya, G. Emch, G. Goldin, J. Guerrero, A. Hurst, M. Karasev and J. Tolar for enlightening discussions on no-go theorems, and H. Grundling and G. Tuynman for their helpful comments on this paper. I would also like to express my

[i] This follows immediately from Theorem 1.

[j] I am indebted to G. Tuynman for clarifying this point.

appreciation to both the organizers of the XIII Workshop on Geometric Methods in Physics and the University of Warsaw, who provided a lively and congenial atmosphere for working on this problem. Finally, I am most grateful to Jean-Pierre Antoine for his boundless patience! This research was partially supported by NSF grant DMS-9222241.

References

1. A. A. Kirillov, Geometric quantization, *in*: "Dynamical Systems IV: Symplectic Geometry and Its Applications", V.I. Arnol'd and S.P. Novikov (eds.), Encyclopaedia Math. Sci. **IV**, Springer, New York 137-172 (1990)

2. H. J. Groenewold, On the principles of elementary quantum mechanics, *Physics* 12:405-460 (1946)

3. L. Van Hove, Sur le problème des relations entre les transformations unitaires de la mécanique quantique et les transformations canoniques de la mécanique classique, *Acad. Roy. Belgique Bull. Cl. Sci.* (5) 37:610-620 (1951)

4. L. Van Hove, Sur certaines représentations unitaires d'un groupe infini de transformations, *Mém. Acad. Roy. Belgique Cl. Sci.* 26:61-102 (1951)

5. P. Chernoff, Mathematical obstructions to quantization, *Hadronic J.* 4:879-898 (1981)

6. G. B. Folland, "Harmonic Analysis in Phase Space", Ann. Math. Studies 122, Princeton Univ. Press, Princeton (1989)

7. A. Joseph, Derivations of Lie brackets and canonical quantization, *Comm. Math. Phys.* 17:210-232 (1970)

8. M. J. Gotay, H. Grundling and C. A. Hurst, A Groenewold-Van Hove theorem for S^2, dg-ga/9502008. To appear in *Trans. Amer. Math. Soc.* (1995)

9. P. Chernoff, Irreducible representations of infinite-dimensional transformation groups and Lie algebras, I (1994). To appear in *J. Funct. Anal.*

10. J. Zak, Dynamics of electrons in solids in external fields, *Phys. Rev.* 168:686-695 (1968)

11. N. S. Manton, The Schwinger model and its axial anomaly, *Ann. Phys.* 159:220-251 (1985)

12. M. Fernández, M. J. Gotay and A. Gray, Compact parallelizable four dimensional symplectic and complex manifolds, *Proc. Amer. Math. Soc.* 103:1209-1212 (1988)

13. M. A. Rieffel, Deformation quantization of Heisenberg manifolds, *Commun. Math. Phys.* 122:531-562 (1989)

14. M. Asorey, Classical and quantum anomalies in the quantum Hall effect, *in*: "Proc. on the Fall Workshop on Differential Geometry and Its Applications", X. Gràcia, M.C. Muñoz and N. Román (eds.), CPET, Barcelona 75-80 (1994)

15. A. Avez, Symplectic group, quantum mechanics and Anosov's systems, *in*: "Dynamical Systems and Microphysics", A. Blaquière et al. (eds.), Springer, New York, 301-324 (1980)

16. V. Aldaya, M. Calixto and J. Guerrero, Algebraic quantization, good operators and fractional quantum numbers, hep-th/9507016 (1995).

DIFFERENTIAL FORMS ON THE SKYRMION BUNDLE

Christian Gross

Fachbereich Mathematik
Technische Hochschule Darmstadt
Germany

1. INTRODUCTION

During the last years many physicists seem to have rediscovered the SKYRME model[1] in theoretical nuclear physics as an effective field theory related to quantum chromodynamics (QCD) by its underlying symmetry. Yet most of the articles deal with the ungauged, purely hadronic case treating interactions between baryons and mesons and do not cover interactions between these particles and electromagnetic fields, although especially for the latter case, the SKYRME model reveals some interesting features. Grand unification theory in its present form implies that magnetic monopoles M are able to catalyze baryon-number-violating processes like

$$\mathrm{M} + p^{+} \longrightarrow \mathrm{M} + e^{+} + \text{ pions},$$

and these processes can conveniently be described within the SKYRME model.[2]

The correct settings for the purpose of treating interactions with electromagnetic fields are that of a skyrmion bundle (and a lepton bundle) associated with a principal U_1 bundle and a MAXWELL connection on it. The possibility of describing baryonic processes by means of the mesonic fields alone, is essentially based on the topological properties of the unitary groups SU_m. While their stable homotopy groups and their DE RHAM cohomology including the generators ω_{2n+1} of $H^{2n+1}(\mathrm{SU}_m)$ are well known, little is noted on the topology of the skyrmion bundle. In order to treat baryon-number-violating processes, one needs an analog of the (normalized) differential form ω_3, which counts the number of baryons described by a certain mesonic field configuration.

Also ω_5, which serves as a base for the anomalous action, has to be generalized to the bundle case. To this end, we have examined the cohomology of the skyrmion bundle in general.[3] For the purpose of recovering (global) differential forms on the bundle from those on the fiber SU_m, we have used spectral sequences. In a second step we then gauged these differential forms by adapting them to the given MAXWELL connection.

2. THE UNGAUGED SKYRME MODEL

In the SKYRME model, the meson fields π^a on space-time M generate differentiable functions $U: M \to \mathrm{SU}_{N_F}$ defined by (N_F denotes the number of flavors in QCD)

$$U = \exp(i \sum_{a=1}^{N_F^2-1} \pi^a \lambda_a) \qquad \text{with} \quad \lambda_a = (\lambda_a)^\dagger \in \mathbb{C}^{N_F \times N_F},\ \mathrm{Tr}(\lambda_a) = 0.$$

The vacuum is represented by the unit matrix $\mathbb{1} \in \mathrm{SU}_{N_F}$. Requiring $\pi^a(r) \to 0$ and thus $U(r) \to \mathbb{1}$ for $r \to \infty$ one can compactify euclidian space $\mathbb{R}^3$, resp., space-time $\mathbb{R}^4$, so that the meson fields constitute functions $U: \mathbb{R}_{(t)} \times \mathbb{S}^3 \to \mathrm{SU}_{N_F}$, resp., $U: \mathbb{S}^4 \to \mathrm{SU}_{N_F}$.

Let $L := U^{-1}\,dU = U^\dagger\,dU$ and $R := (dU)U^{-1} = (dU)U^\dagger \in \mathcal{A}_1(\mathrm{U}_m, \mathbb{C}^{m\times m})$ denote the left, resp., right invariant currents: $\mathbb{C}^{m\times m}$-valued 1-forms that are invariant under multiplication with constant elements of U_m from the left, resp., from the right and obey $L(\mathcal{X})(\mathbb{1}) = R(\mathcal{X})(\mathbb{1}) = X$ for all vector fields $\mathcal{X} \in \mathcal{D}^1(\mathrm{U}_m)$ with $\mathcal{X}(\mathbb{1}) = X \in \mathfrak{u}_m$. For any constant $Q \in \mathbb{C}^{m\times m}$, we define λ_k^Q and $\rho_k^Q \in \mathcal{A}_k(\mathrm{U}_m, \mathbb{C})$ by

$$\lambda_k^Q := \mathrm{Tr}(QL^k) := \mathrm{Tr}(Q\underbrace{L\wedge\cdots\wedge L}_{k}), \qquad \rho_k^Q := \mathrm{Tr}(QR^k) := \mathrm{Tr}(Q\underbrace{R\wedge\cdots\wedge R}_{k}),$$

These are left, resp., right invariant complex-valued k-forms on U_m; for $Q = \mathbb{1}$ we have

$$\omega_k := \lambda_k^{\mathbb{1}} = \rho_k^{\mathbb{1}} = \mathrm{Tr}(L^k) = \mathrm{Tr}(R^k) \in \mathcal{A}_k(\mathrm{U}_m, \mathbb{C}),$$

which are invariant under all multiplications. Obviously $\omega_{2l} = 0$. The forms ω_{2l+1} are closed since the MAURER-CARTAN identities $dL = -L \wedge L$, $dR = R \wedge R$ yield

$$\begin{aligned} dL^{2l+1} &= -L^{2l+2}, & dR^{2l+1} &= R^{2l+2}, & dL^{2l+2} &= dR^{2l+2} = 0, & (2.1)\\ d(UL^{2l}) &= UL^{2l+1}, & d(L^{2l}U^\dagger) &= -L^{2l+1}U^\dagger, & d(UL^{2l+1}) &= d(L^{2l+1}U^\dagger) = 0. & (2.2)\end{aligned}$$

Moreover, ω_{2l+1} generate the DE RHAM cohomology $H^*(\mathrm{SU}_m, \mathbb{C})$, resp., $H^*(\mathrm{U}_m, \mathbb{C})$.

In the SKYRME model, baryons appear as topological soliton solutions of the meson fields. Their number B can be computed by an integration over the space manifold:

$$B(U) = -\int_{\mathbb{S}^3} \frac{1}{24\pi^2} U^\star\omega_3 = -\int_{\mathbb{R}^3} \frac{1}{24\pi^2} \sum_{i,j,k=1}^{3} \mathrm{Tr}(L_i L_j L_k)\, dx^i \wedge dx^j \wedge dx^k. \qquad (2.3)$$

Compactification of space-time is crucial: normally there is no guarantee that the integral in (2.3) is an integer, but for spheres we have the following Index theorem:[4]

Theorem 2.1 *For every map $U: \mathbb{S}^{2n-1} \to \mathrm{U}_m$ the integral*

$$n(U) = \int_{\mathbb{S}^{2n-1}} \left(\frac{i}{2\pi}\right)^n \frac{(n-1)!}{(2n-1)!} U^\star\omega_{2n-1} \quad \textit{is an integer.}$$

The assignment $[U] \mapsto n(U): \pi_{2n-1}(\mathrm{U}_m) \to \mathbb{Z}$ is an isomorphism for $m \geq n$.

We are thus able to identify $\left(\frac{i}{2\pi}\right)^n \frac{(n-1)!}{(2n-1)!}\,\omega_{2n-1}$ with the generators of the integer valued cohomology of the unitary groups. At any time the meson fields respresent elements of the homotopy groups $\pi_3(\mathrm{U}_{N_F}) \cong \mathbb{Z}$ for $N_F \geq 2$; the integer characterizing the homotopy class is a topological invariant, the "topological charge" $B(U)$.

The vacuum map represents the zero element, and so $B(U \equiv \mathbb{1}) = 0$. For proton and neutron we have $B = 1$, for their antiparticles $B = -1$. Annihilation of proton and antiproton corresponds to the "addition" of their maps within the homotopy group and generates a mesonic field of topological charge $B = 0$.

The meson fields obey the field equations derived as EULER-LAGRANGE equations from a lagrangian $\mathcal{L}(U, dU)$ by variation of the action integral $\mathbf{\Gamma}(U) = \int_{\mathbb{S}^4} \mathcal{L}\, dV$. The latter splits into two parts: one of them (N_C denotes the number of colors in QCD),

$$\mathbf{\Gamma}_{AN}(U) = \lambda \int_{D^5} (U')^\star \omega_5 \qquad \text{with} \qquad \lambda = \frac{i\, N_C}{240\pi^2}, \tag{2.4}$$

describes the anomalous processes of QCD: one uses $\pi_4(\mathrm{SU}_{N_F}) = 0$ and extends U to a differentiable map $U': D^5 \to \mathrm{SU}_3$ from a five-dimensional disc D^5 whose boundary ∂D^5 is space-time $\mathbb{S}^4$. The topological quantization of the coupling constant λ in (2.4) is again a consequence of Theorem 2.1, and of the requirement that for any extension U' the result has to be unique.[5]

3. THE MAXWELL CONNECTION ON $P(M, G_{em})$

It is well known that MAXWELL's equations can conveniently be presented by use of differential forms on a principal fiber bundle $P(M, G_{em})$ with $G_{em} \cong \mathrm{U}_1 \cong \mathbb{S}^1$. For any fiber bundle $B(M, F, G)$ with bundle space B, base space M, fiber F and LIE group G, let $\pi: B \to M$ denote the projection onto the base and $\{(U_\alpha, \psi_\alpha)\}_{\alpha\in A}$ a bundle atlas, where $\mathfrak{U} = \{U_\alpha\}_{\alpha\in A}$ is an open cover of M and $\psi_\alpha: \pi^{-1}(U_\alpha) \to U_\alpha \times F: b \mapsto (\pi(b), \pi_\alpha(b))$ define local projections $\pi_\alpha: \pi^{-1}(U_\alpha) \to F$ onto the fiber. For the left action of G on F, we write $L_g(f) = \tau_f(g)$, where $L_g: F \to F$ and $\tau_f: G \to F$ are differentiable for all $g \in G$ and $f \in F$. For all $\alpha, \beta \in A$ with $U_{\alpha\beta} := U_\alpha \cap U_\beta \neq \emptyset$, $g_{\alpha\beta}: U_{\alpha\beta} \to G$ means the C^∞-map defined by $g_{\alpha\beta} := \pi_\alpha|_{\pi^{-1}(U_{\alpha\beta})} \circ (\pi_\beta|_{\pi^{-1}(U_{\alpha\beta})})^{-1}: F \to F$.

Recall that any connection Γ on a principal bundle is uniquely defined by a connection 1-form $\omega^\Gamma \in \mathcal{A}_1(P, \mathfrak{g})$. For the 1-dimensional LIE algebra $\mathfrak{g} = L(G_{em})$, the curvature 2-form $\Omega^\Gamma \in \mathcal{A}_2(P, \mathfrak{g})$ reads $\Omega^\Gamma = d\omega^\Gamma$. If $\sigma_{\alpha,f}: U_\alpha \to \pi^{-1}(U_\alpha): x \mapsto \psi_\alpha^{-1}(x, f)$ denote local sections for all $\alpha \in A$ and $f \in F$, then ω^Γ and Ω^Γ are related to the electromagnetic gauge potentials A^α and the electromagnetic gauge fields F^α by

$$A^\alpha = \sigma^\star_{\alpha,0}(\omega^\Gamma|_{\pi^{-1}(U_\alpha)}) \in \mathcal{A}_1(U_\alpha, \mathfrak{g}), \qquad F^\alpha = \sigma^\star_{\alpha,0}(\Omega^\Gamma|_{\pi^{-1}(U_\alpha)}) \in \mathcal{A}_2(U_\alpha, \mathfrak{g})$$

(we write the group operation in G_{em} additively, so 0 is the neutral element). We have[6]

Theorem 3.1 *If Γ is a connection on $P(M, G_{em})$ and $\{(U_\alpha, \psi_\alpha)\}_{\alpha\in A}$ is a bundle atlas for P, then for all $\alpha, \beta \in A$ with $U_{\alpha\beta} := U_\alpha \cap U_\beta \neq \emptyset$ and for all $x \in U_{\alpha\beta}$:*

$$\begin{aligned} F^\alpha &= dA^\alpha, & dF^\alpha &= 0, \\ A^\alpha|_{U_{\alpha\beta}} = A^\beta|_{U_{\alpha\beta}} + dg_{\beta\alpha} &= A^\beta|_{U_{\alpha\beta}} - dg_{\alpha\beta}, & F^\alpha|_{U_{\alpha\beta}} &= F^\beta|_{U_{\alpha\beta}} \end{aligned} \tag{3.1}$$

Vice versa, if for a bundle atlas $\{(U_\alpha, \psi_\alpha)\}_{\alpha\in A}$ on the principal bundle $P(M, G_{em})$ a family $\{A^\alpha \in \mathcal{A}_1(U_\alpha, \mathfrak{g})\}_{\alpha\in A}$ is given such that (3.1) holds, then there exists one unique connection Γ on $P(M, G_{em})$ such that $A^\alpha = \sigma^\star_{\alpha,0}(\omega^\Gamma|_{\pi^{-1}(U_\alpha)})$ for all $\alpha \in A$.

So the F^α constitute a global $F \in \mathcal{A}_2(M, \mathfrak{g})$ and the homogeneous MAXWELL equations simply take the form $dF = 0$, resp., $d\Omega^\Gamma = 0$.

We are interested in the case of a single (anti-)monopole of magnetic charge $m \cdot g_D$ (at first let $m \in \mathbb{R}$), which rests in the origin of our space manifold. Then space-time $M \cong \mathbb{R}_{(t)} \times \mathbb{R}^+_{(r)} \times \mathbb{S}^2_{(\theta,\phi)}$ and $P(\mathbb{R}_{(t)} \times \mathbb{R}^+_{(r)} \times \mathbb{S}^2, G_{\mathrm{em}}) \cong P(\mathbb{S}^2, G_{\mathrm{em}}) \times \mathbb{R}_{(t)} \times \mathbb{R}^+_{(r)}$, where $P(\mathbb{S}^2, G_{\mathrm{em}}) \cong P(\mathbb{S}^2, \mathbb{S}^1)$ is the only topological interesting part. For these sphere bundles the following classification theorem holds:[7]

Theorem 3.2 *If G is arcwise connected, then the set of equivalence classes of bundles over $\mathbb{S}^n$ with group G is in 1–1 correspondence with $\pi_{n-1}(G)$.*

So there is a countable number of different principal bundles $P_m(\mathbb{S}^2, \mathbb{S}^1)$ with $m \in \mathbb{Z}$, thus the magnetic charge $m \cdot g_D$ is quantized, and we can put $2eg_D = 1$.

For all $P_m(\mathbb{S}^2, \mathbb{S}^1)$ local trivializations exist over the northern and the southern hemisphere U_+, resp., U_-. The transition function $g_{-+} = -g_{+-}$ is

$$g_{-+}: U_+ \cap U_- \to \mathbb{S}^1: (\theta, \phi) \mapsto m\phi,$$

and so for $p_m \in P_m$ with $\pi(p_m) \in U_+ \cap U_-$ and $p_m \cong (\pi(p_m), \varphi_m^{(+)}) \cong (\pi(p_m), \varphi_m^{(-)})$ we have $\varphi_m^{(-)} \equiv \varphi_m^{(+)} + m\phi \pmod{2\pi}$. Equation (3.1) thus yields

$$A^{(+)} = A^{(-)} + 2mg_D\, d\phi, \qquad F^{(+)} = F^{(-)}.$$

Every connection Γ on $P(M,G)$ induces horizontal and vertical projections h, v of vector fields and differential forms on every associated bundle $B(M,F,G) = P \times_G F$. For G_{em}-invariant differential forms χ on F, i. e. $L_g^\star \chi = \chi$ for all $g \in G_{em}$, we obtain:

Theorem 3.3 *Let Γ be a connection on $P(M, G_{em})$ and $B(M, F, G_{em})$ an associated bundle. For any G_{em}-invariant $\chi \in \mathcal{A}_n(F,V)$ define $\nu \in \mathcal{A}_{n-1}(F,V)$ by*

$$\nu_f(F_f^{(1)}, \ldots, F_f^{(n-1)}) := n \cdot \chi_f(F_f^{(1)}, \ldots, F_f^{(n-1)}, d\tau_f(E)) \quad \text{for all } f \in F,\ F^{(i)} \in \mathcal{D}^1(F),$$

where $\mathfrak{g} = E\mathbb{R}$. For any $U_\alpha \in \mathfrak{U}$ denote $\chi^\alpha := \pi_\alpha^\star \chi$, $\nu^\alpha := \pi_\alpha^\star \nu$. Then on all $U_{\alpha\beta} \neq \emptyset$

$$\begin{aligned} \chi^\alpha &= \chi^\beta + \frac{1}{E} dg_{\alpha\beta} \wedge \nu, \qquad \chi^\alpha v = \chi^\alpha + \frac{1}{E} A^\alpha \wedge \nu = \chi^\beta + \frac{1}{E} A^\beta \wedge \nu = \chi^\beta v, \\ \nu^\alpha &= \nu^\alpha v = \nu^\beta = \nu^\beta v. \end{aligned}$$

Thus χv and ν define global G_{em}-invariant, vertical V-valued forms on B.

4. THE SKYRMION BUNDLE $B(M, \mathrm{SU}_n, G_{em})$

Next for the skyrmions: instead of considering maps $U: M \to \mathrm{SU}_n$ we now think of the meson fields as of global sections in a bundle $B(M, \mathrm{SU}_n, G_{em})$ associated to $P(M, G_{em})$.[8] The left action of G_{em} on SU_n is given by the inner automorphisms

$$L_g(U) = \tau_U(g) = e^{-iegQ} U e^{+iegQ},$$

which do not effect the vacuum being diagonal symmetry operations. Q is the hermitian $n \times n$-matrix containing the quark charges in units of e (again $n = 2$, resp., 3)

$$Q = \begin{pmatrix} \frac{2}{3} & 0 \\ 0 & -\frac{1}{3} \end{pmatrix}, \qquad \text{resp.,} \qquad Q = \begin{pmatrix} \frac{2}{3} & 0 & 0 \\ 0 & -\frac{1}{3} & 0 \\ 0 & 0 & -\frac{1}{3} \end{pmatrix}$$

Thus the transition functions are

$$U^\alpha(x) = e^{-ieg_{\alpha\beta}(x)Q}\, U^\beta(x)\, e^{+ieg_{\alpha\beta}(x)Q}, \tag{4.1}$$

$$\text{resp.,} \quad U^{(-)}(x) = e^{-im\phi Q}\, U^{(+)}(x)\, e^{+im\phi Q} \tag{4.2}$$

for the skyrmion bundle $B_m(M, \mathrm{SU}_n, G_{em}) \cong B_m(\mathbb{S}^2, \mathrm{SU}_n, G_{em}) \times \mathbb{R}_{(t)} \times \mathbb{R}^+_{(r)}$ associated with P_m. So not only vacuum $U \equiv \mathbb{1}$ is a global section but every $U(x) = e^{i\chi(x)Q}$ with a differentiable map $\chi: M \to \mathbb{S}^1$. From (4.1) we get $d\tau_U(X) = -ieX\,[Q, U]$ for all $X \in \mathfrak{g}$. Since ω_{2l+1}, ρ_l^Q and λ_l^Q are G_{em}-invariant, Theorem 3.3 yields:

Lemma 4.1 *$\omega_{2l+1}v$, $\rho_l^Q v$ and $\lambda_l^Q v$ for $l \in \mathbb{N}_0$ are global forms on B and we have:*

$$\begin{aligned}
\omega_{2l+1}^{\alpha}v &= \omega_{2l+1}^{\alpha} - (2l+1)ie\, A^{\alpha} \wedge (\rho_{2l}^Q - \lambda_{2l}^Q),\\
(\rho_{2l}^Q)^{\alpha}v &= (\rho_{2l}^Q)^{\alpha} - ie\, A^{\alpha} \wedge \sum_{j=1}^{2l} (-1)^j\ \mathrm{Tr}(QUL^{j-1}QL^{2l-j}U^{\dagger})^{\alpha},\\
(\lambda_{2l}^Q)^{\alpha}v &= (\lambda_{2l}^Q)^{\alpha} - ie\, A^{\alpha} \wedge \sum_{j=1}^{2l} (-1)^j\ \mathrm{Tr}(QUL^{j-1}QL^{2l-j}U^{\dagger})^{\alpha},\\
(\rho_{2l+1}^Q)^{\alpha}v &= (\rho_{2l+1}^Q)^{\alpha} - ie\, A^{\alpha} \wedge \sum_{j=1}^{2l+1} \mathrm{Tr}(QR^{j-1}QR^{2l+1-j} - QL^{j-1}U^{\dagger}QUL^{2l+1-j})^{\alpha},\\
(\lambda_{2l+1}^Q)^{\alpha}v &= (\lambda_{2l+1}^Q)^{\alpha} - ie\, A^{\alpha} \wedge \sum_{j=1}^{2l+1} \mathrm{Tr}(QUL^{j-1}QL^{2l+1-j}U^{\dagger} - QL^{j-1}QL^{2l+1-j})^{\alpha},\\
(\rho_{2l}^Q - \lambda_{2l}^Q)^{\alpha}v &= (\rho_{2l}^Q - \lambda_{2l}^Q)^{\alpha}, \qquad\qquad (\rho_1^Q + \lambda_1^Q)^{\alpha}v = (\rho_1^Q + \lambda_1^Q)^{\alpha},\\
(\rho_3^Q + \lambda_3^Q)^{\alpha}v &= (\rho_3^Q + \lambda_3^Q)^{\alpha} - 2ie\, A^{\alpha} \wedge \mathrm{Tr}\left[Q^2(R^2 - L^2) + Q\, dU^{\dagger} \wedge Q\, dU\right]^{\alpha},\\
(\rho_{2l+1}^Q + \lambda_{2l+1}^Q)^{\alpha}v &= (\rho_{2l+1}^Q + \lambda_{2l+1}^Q)^{\alpha} - 2ie\, A^{\alpha} \wedge \sum_{j=1}^{l} \mathrm{Tr}(QUL^{2j-1}QL^{2l-2j+1}U^{\dagger})^{\alpha}\\
&\quad - ie\, A^{\alpha} \wedge \sum_{j=0}^{l} \mathrm{Tr}(QR^{2j}QR^{2l-2j} - QL^{2j}QL^{2l-2j})^{\alpha}.
\end{aligned}$$

Analogous relations — with $dg_{\alpha\beta}$ instead of A^{α} — hold for the transformation rules, proving that $\rho_{2l}^Q - \lambda_{2l}^Q$ and $\rho_1^Q + \lambda_1^Q$ are global forms on B.

For calculations we need the action integral and the topological charge for the skyrmion bundle. Both consist of forms on B, whose pullbacks by the mesonic sections $U: M \to B$ are integrated over space-time, resp., the space manifold only. Yet, for the anomalous action and the topological charge, a difficulty arises because we have to integrate over forms on the fiber. Whereas every form on M can be lifted onto B by use of the pullback $\pi^{\star}$, there is no such mean for the forms ω_3, resp., $\omega_5 \in \mathcal{A}(\mathrm{SU}_n, \mathbb{C})$. One can introduce generalized forms by "trial-and-error" and afterwards check gauge invariance and closedness,[5,2] but this has to be done for every form separately and says nothing about exactness of these forms. Therefore we chose a different way:

5. SPECTRAL SEQUENCES FOR THE SKYRMION BUNDLE

For a countable ordered good cover $\mathfrak{U} = \{U_{\alpha}\}_{\alpha \in A}$ of M (i. e. all finite intersections $U_{\alpha_0 \cdots \alpha_p} := U_{\alpha_0} \cap \cdots \cap U_{\alpha_p}$, $p \in \mathbb{N}_0$ are diffeomorphic to $\mathbb{R}^n$), let $C(\pi^{-1}\mathfrak{U}, \mathcal{A})$ denote the Čech-de-Rham complex[9]

$$C(\pi^{-1}\mathfrak{U}, \mathcal{A}) := \bigoplus_{p,q \in \mathbb{N}_0} C^p(\pi^{-1}\mathfrak{U}, \mathcal{A}_q), \quad \text{where} \quad C^p(\pi^{-1}\mathfrak{U}, \mathcal{A}_q) := \prod_{\alpha_0 < \cdots < \alpha_p} \mathcal{A}_q(\pi^{-1}(U_{\alpha_0 \cdots \alpha_p})).$$

We have two commuting differential operators: on the vertical lines we have d, and on the horizontal lines we have δ defined by ("$\hat{\ }$" denotes omission)

$$(\delta\omega)_{\alpha_0 \cdots \alpha_{p+1}} := \sum_{j=0}^{p+1} (-1)^j \omega_{\alpha_0 \cdots \widehat{\alpha_j} \cdots \alpha_{p+1}}|_{U_{\alpha_0 \cdots \alpha_{p+1}}} \quad \forall \omega = \prod_{\alpha_0 < \cdots < \alpha_p} \omega_{\alpha_0 \cdots \alpha_p} \in \prod_{\alpha_0 < \cdots < \alpha_p} \mathcal{A}(U_{\alpha_0 \cdots \alpha_p}).$$

$D := D' + D''$, with $D' := \delta$ and $D'' := (-1)^p d$, is the differential operator for the single (graded) complex with $C(\pi^{-1}\mathfrak{U}, \mathcal{A})^n = \bigoplus_{p+q=n} C^p(\pi^{-1}\mathfrak{U}, \mathcal{A}_q)$. The Mayer-Vietoris principle guarantees that the restriction map $r: \mathcal{A}(B) \to \prod_{\alpha} \mathcal{A}(\pi^{-1}(U_{\alpha})) \subseteq C(\pi^{-1}\mathfrak{U}, \mathcal{A})$ induces an isomorphism $r^{\star}: H^{*}(B) \to H_D^{*}(C(\pi^{-1}\mathfrak{U}, \mathcal{A}))$, $H^n(B) \to H_D^n(C(\pi^{-1}\mathfrak{U}, \mathcal{A}))$.

The inverse map that collates together the components of an element in the Čech-de-Rham complex into a global form on B is less intuitive. For any partition of unity $\{\rho_\alpha\}_{\alpha\in A}$ subordinate to $\mathfrak{U}$ define $K: C^p(\pi^{-1}\mathfrak{U}, \mathcal{A}_q) \to C^{p-1}(\pi^{-1}\mathfrak{U}, \mathcal{A}_q)$ by

$$(K\omega)_{\alpha_0\cdots\alpha_{p-1}} := \sum_{\alpha\in A} \rho_\alpha\, \omega_{\alpha\alpha_0\cdots\alpha_{p-1}}, \tag{5.1}$$

then we have $K\delta + \delta K = \mathrm{id}$ and the following Collating formula:[9]

Theorem 5.1 *Let $\alpha = \sum_{i=0}^n \alpha_i \in C(\pi^{-1}\mathfrak{U}, \mathcal{A})^n$ with $\alpha_i \in C^i(\pi^{-1}\mathfrak{U}, \mathcal{A}_{n-i})$ and $D\alpha = \beta = \sum_{i=0}^{n+1} \beta_i$ with $\beta_i \in C^i(\pi^{-1}\mathfrak{U}, \mathcal{A}_{n+1-i})$, and define K by (5.1). Then*

$$f(\alpha) := \sum_{i=0}^{n} (-D''K)^i \alpha_i - \sum_{i=0}^{n} K(-D''K)^i \beta_{i+1} \in C^0(\pi^{-1}\mathfrak{U}, \mathcal{A}_n)$$

is a global form on B (resp., the restriction of such a form to the sets $\pi^{-1}(U_\alpha)$). The induced maps $f^\star$ and $r^\star$ on the cohomology level are inverse isomorphisms.

For any double complex the sequence $K_p := \bigoplus_{i\geq p}\bigoplus_{q\geq 0} K^{i,q}$, $p \in \mathbb{N}_0$ is a filtration by the columns of K with associated graded complex

$$GK = \bigoplus_{p\in\mathbb{N}_0} K_p/K_{p+1} = \bigoplus_{p\in\mathbb{N}_0} \left[\left(\bigoplus_{q\geq 0} K^{p,q}\right) + K_{p+1}\right].$$

Obviously the induced differential operator on $GC(\pi^{-1}\mathfrak{U}, \mathcal{A})$ is just $(-1)^p d$.

Let $\{E_r, D_r\}_{r\in\mathbb{N}_0}$ denote the spectral sequence for the Čech-de-Rham complex: $E_0 = GC(\pi^{-1}\mathfrak{U}, \mathcal{A})$ and $E_{r+1} = H^*_{D_r}(E_r)$, where $D_r: E_r^{p,q} \to E_r^{p+r,q-r+1}$ is the differential operator induced by D on E_r. If E_R becomes stationary, i. e. $E_r = E_{r+1}$ for all $r \geq R$, we denote E_R by E_∞ and say that the spectral sequence converges to some filtered complex H if $E_\infty \cong GH$. $\beta \in C(\pi^{-1}\mathfrak{U}, \mathcal{A})$ "lives to" E_r iff it represents a cohomology class $[\beta]_r \in E_r$, i. e. if β is D_i-closed in $E_0, \ldots, E_{r-1}$. Then β is d-closed and we get a "zig-zag" $\Xi = \xi_0 + \ldots + \xi_{r-1}$ of elements $\xi_i \in C(\pi^{-1}\mathfrak{U}, \mathcal{A})$ with $\xi_0 := \beta$ and

$$D'\xi_i = \delta\xi_i = -D''\xi_{i+1}, \qquad i = 0, \ldots, r-2 \tag{5.2}$$

(cf. Figure 1). Since $D_r[\beta]_r = [\delta\xi_{r-1}]_r = [\chi]_r$, D_r is given by δ at the end of the zig-zag.

Now Leray's theorem[9] states that $\{E_r, D_r\}_{r\in\mathbb{N}_0}$ converges to $H^*(B)$ and that

$$E_2^{p,q} = H^p(M, H^q(F)) \cong H^p(M) \otimes H^q(F) \oplus \mathrm{Tor}\,[H^{p+1}(M), H^q(F)],$$

if $H^*(F)$ is finitely generated and M is simply connected. Finally, closed forms on F transform into closed forms on B as follows: for any $\omega \in H^q(F)$, there exists $\psi \in H^q(B)$, such that ω is the restriction of ψ, iff ω lives to E_{q+2}.

Let us compute the zigzag for ω_{2l+1} in a skyrmion bundle, where the atlas is given over a good cover $\mathfrak{U}$. Using the local trivializations we inject ω_{2l+1} into $C^0(\pi^{-1}\mathfrak{U}, \mathcal{A}_{2l+1})$, so ξ_0 in Figure 1 is given by $(\xi_0)_\alpha = \omega_{2l+1}^\alpha$. By Lemma 4.1, (2.1) and (2.2) we find

$$(\xi_1)_{\alpha\beta} =: ie\; dg_{\beta\alpha} \wedge (\chi_{2l-1}^1)^{\alpha/\beta}, \qquad (\xi_2)_{\alpha\beta\gamma} =: (ie)^2\; dg_{\gamma\beta} \wedge dg_{\beta\alpha} \wedge (\chi_{2l-3}^2)^{\alpha/\beta/\gamma}$$

($\alpha/\beta/\gamma$ indicates that one may use any trivialization), and for $\chi^j_{2(l-j)+1} \in \mathcal{A}_{2(l-j)+1}(\mathrm{SU}_n)$:

$$\begin{aligned}
\chi_{2l-1}^1 &= (2l+1)(\rho_{2l-1}^Q + \lambda_{2l-1}^Q),\\
\chi_{2l-3}^2 &= (2l+1)\Big[2(\rho_{2l-3}^{Q^2} + \lambda_{2l-3}^{Q^2}) + \sum\nolimits_{j=1}^{l-2} \mathrm{Tr}(QR^{2j-1}QR^{2l-2j-2} + QL^{2j-1}QL^{2l-2j-2})\\
&\qquad + \sum\nolimits_{j=1}^{l-1} \mathrm{Tr}(QUL^{2j-1}QL^{2l-2j-2}U^\dagger + QUL^{2j-2}QL^{2l-2j-1}U^\dagger)\Big].
\end{aligned}$$

For our special case, Figure 2 shows a good cover of $\mathbb{S}^2$ consisting of four open sets, where U_0 covers the northern hemisphere and $U_1 \cup U_2 \cup U_3$ covers the southern hemisphere. Obviously $dg_{\gamma\beta} \wedge dg_{\beta\alpha} = 0$ for all combinations of α, β and γ. This proves:

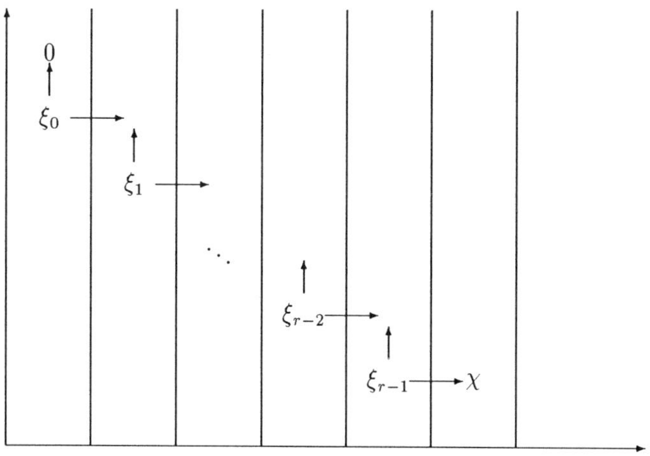

Figure 1: Illustration of the Differential operator D_r: $D_r[\xi_0]_r = [\delta\xi_{r-1}]_r = [\chi]_r$

Figure 2: A good cover on $\mathbb{S}^2$

Theorem 5.2 *The cohomology of the skyrmion bundle $B_m(M, \mathrm{SU}_n, G_{em})$ is independent of the monopole charge mg_D, but isomorphic to the cohomology of $M \times \mathrm{SU}_n$:*

$$H^k(B_m(M, \mathrm{SU}_n, G_{em}), \mathbb{Z}) \cong \bigoplus_{p+q=k} H^p(M, \mathbb{Z}) \otimes H^q(\mathrm{SU}_n, \mathbb{Z}), \qquad k \in \mathbb{N}_0.$$

The same holds for all skyrmion bundles of manifolds, where a good cover $\mathfrak{U} = \{U_\alpha\}_{\alpha \in A}$ exists such that $dg_{\gamma\beta} \wedge dg_{\beta\alpha} = 0$ for all $\alpha, \beta, \gamma \in A$.

Finally, since $\chi_1^1 = \rho_1^Q + \lambda_1^Q$ and $\chi_1^2 = 10(\rho_1^{Q^2} + \lambda_1^{Q^2}) + 5\,\mathrm{Tr}(Q\;dU\;QU^\dagger - QUQ\;dU^\dagger)$ are global forms, we have proved that $D_2(\omega_3) = D_3(\omega_5) = 0$ for *every* skyrmion bundle, so ω_3 and ω_5 live to E_∞ and we have

$$H^3(B(M, \mathrm{SU}_n, G_{em}), \mathbb{Z}) \cong \mathbb{Z} \qquad \text{and} \qquad H^5(B(M, \mathrm{SU}_n, G_{em}), \mathbb{Z}) \cong \mathbb{Z}.$$

6. COHOMOLOGY AND CONNECTION ON $B(M, \mathrm{SU}_n, G_{em})$

Now Theorem 5.1 exhibits representatives for the cohomology groups. Yet we would like to obtain forms that are anyhow adapted to Γ. To this end, observe that, given any partition of unity $\{\rho_\alpha\}_{\alpha \in A}$ and gauge potentials A^α, $\{A^\alpha - \sum_{\gamma \in A} \rho_\gamma\, dg_{\gamma\alpha}\}_{\alpha \in A}$

defines a global $\mathfrak{g}$-valued 1-form on M. Suppose $\Xi^0 = \xi_0^0 + \cdots + \xi_r^0$ is a zigzag for a G_{em}-invariant $\omega_q \in H^q(\mathrm{SU}_n)$, where every ξ_j^0 is of the type

$$(\xi_j^0)_{\alpha_0\cdots\alpha_j} = (ie)^j\, dg_{\alpha_j\alpha_{j-1}} \wedge \cdots \wedge dg_{\alpha_1\alpha_0} \wedge (\chi_{q-2j}^j)^{\alpha_0/\cdots/\alpha_j}$$

with G_{em}-invariant $\chi_{q-2j}^j \in \mathcal{A}_{q-2j}(\mathrm{SU}_n)$ and $d(\chi_{q-2(j+1)}^{j+1}) = \nu_{q-2j}^j$ from Theorem 3.3. χ_{q-2r}^r be global such that $\delta\xi_r = D_{r+1}[\omega_q]_{r+1} = D\Xi^0 = 0$; we put $\chi_{q-2j}^j = 0$ for $j > r$. If

$$(\xi_k^j)_{\alpha_j\cdots\alpha_{j+k}} := (ie)^k\, dg_{\alpha_{j+k}\alpha_{j+k-1}} \wedge \cdots \wedge dg_{\alpha_{j+1}\alpha_j} \wedge (\chi_{q-2(j+k)}^{j+k})^{\alpha_j/\cdots/\alpha_{j+k}},$$

then $\Xi^j = \xi_0^j + \cdots + \xi_{r-j}^j$ is a zigzag for χ_{q-2j}^j, as induction on k shows. Define inductively

$$\widetilde{\chi}_{q-2r}^r := \chi_{q-2r}^r, \quad \widetilde{\chi}_{q-2j}^j := (\chi_{q-2j}^j)^{\alpha_j} + ie\, d[\sum_{\alpha_{j+1}\in A} \rho_{\alpha_{j+1}}\, dg_{\alpha_{j+1}\alpha_j} \wedge \widetilde{\chi}_{q-2(j+1)}^{j+1}], \quad \text{and}$$

$$\widehat{\chi}_{q-2j}^j := [(\chi_{q-2j}^j)^{\alpha_j} - ie\, A^{\alpha_j} \wedge d\widetilde{\chi}_{q-2(j+1)}^{j+1}] + ie\, dA^{\alpha_j} \wedge \widetilde{\chi}_{q-2(j+1)}^{j+1} = \widetilde{\chi}_{q-2j}^j + d\beta^{j+1},$$

where $\beta^j := (A^{\alpha_{j-1}} - \sum_{\alpha_j} \rho_{\alpha_j} dg_{\alpha_j\alpha_{j-1}}) \wedge \widetilde{\chi}_{q-2j}^j$. Obviously $d\widehat{\chi}_{q-2j}^j = d\widetilde{\chi}_{q-2j}^j$. By Theorem 5.1 all $\widetilde{\chi}_{q-2j}^j$ are global; since all β^j are global, too, the same holds for all $\widehat{\chi}_{q-2j}^j$. Now once again induction shows that

$$(\omega_q^A)^\alpha := \sum_{j=0}^r (ie\, F^\alpha)^j \wedge (\chi_{q-2j}^j)^\alpha v \tag{6.1}$$

is global within the cohomology class of $\widetilde{\chi}_q^0 = f(\Xi^0)$. This proves:

Theorem 6.1 *$\{(\omega_q^A)^\alpha\}_{\alpha\in A}$ defines a G_{em}-invariant global $\omega_q^A \in H^q(B(M, \mathrm{SU}_n, G_{em}))$ whose restriction to the fiber is $\omega_q = i_x^\star \omega_q^A$ for all $x \in M$.*

Our explicit zigzags for ω_{2l+1} yield the following corollary to Theorem 6.1:

Corollary 6.2 *Gauge invariant generalizations of ω_3 and ω_5 adapted to Γ and generating cohomology groups isomorphic to $\mathbb{Z}$, are*

$$\begin{aligned}
(\omega_3^A)^\alpha &= \omega_3 v + ie\, F \wedge \chi_1^1 v \;=\; [\omega_3^\alpha - 3ie\, A^\alpha \wedge (\rho_2^Q - \lambda_2^Q)] + 3ie\, F \wedge (\rho_1^Q + \lambda_1^Q),\\
(\omega_5^A)^\alpha &= \omega_5 v + ie\, F \wedge \chi_3^1 v + (ie)^2 F \wedge F \wedge \chi_1^2 v \;=\; [\omega_5^\alpha - 5ie\, A^\alpha \wedge (\rho_4^Q - \lambda_4^Q)]\\
&\quad + 5ie\, F \wedge \{(\rho_3^Q + \lambda_3^Q)^\alpha - 2ie\, A^\alpha \wedge \mathrm{Tr}[Q^2(R^2 - L^2) + Q\, dU^\dagger \wedge Q\, dU]^\alpha\}\\
&\quad + 5(ie)^2\, F \wedge F \wedge [2(\rho_1^{Q^2} + \lambda_1^{Q^2})^\alpha + \mathrm{Tr}(Q\, dU\, QU^\dagger - QUQ\, dU^\dagger)^\alpha].
\end{aligned}$$

These forms coincide with the ones in the literature.[5,11] The integral over $U^\star\omega_3^A$ gives the topological charge, and the integral over $U^\star\omega_5^A$ is the anomalous action for the skyrmion bundle. Nevertheless these forms are not unique in the sense that they are the only possible generalizations of type (6.1). An additional term

$$r\,(ie)^l F^l \wedge d\,\mathrm{Tr}(QU^\dagger QU), \qquad r \in \mathbb{R},$$

may be added to ω_{2l+1}^A, and this is still of the given type, because $d\,\mathrm{Tr}(QU^\dagger QU)$ is global, G_{em}-invariant and vertical.[11] One could even add any $F^l \wedge d\alpha$ with $\alpha = L_g^\star \alpha \in \mathcal{A}_0(\mathrm{SU}_n)$. In order to exclude these, one needs further physical requirements like parity invariance, equality of the numbers of F's and Q's, etc.

These forms now allow for the treatment of the monopole-induced proton decay within the skyrmion bundle. In fact, although we have proven that ω_3^A is a correct closed differential form for the topological charge, and although $H^3(B, \mathbb{Z}) \cong \mathbb{Z}$, the number of baryons $B^A(U)$ is not topologically conserved any more, whenever magnetic monopoles are present. This is due to the fact, that in contrast to the ungauged SKYRME model, the Index theorem 2.1 does not apply any more. There is no possibility to compactify space to an $\mathbb{S}^3$, so the topological charge can vanish through the monopole singularities.[2]

References

1. T. H. R. Skyrme, A unified field theory of mesons and baryons, *Nucl. Phys.* 31:556 (1962)
2. C. E. Callan and E. Witten, Monopole catalysis of skyrmion decay, *Nucl. Phys. B* 239:161 (1984)
3. C. Gross, Topology of the skyrmion bundle, *J. Math. Phys.* (1995) (to appear)
4. R. Bott, R. Seeley, Some remarks on the paper of Callias, *Commun. math. Phys.* 62:235 (1978)
5. E. Witten, Global aspects of current algebra, *Nucl. Phys. B* 223:422 (1983)
6. S. Kobayashi and K. Nomizu, "Foundations of Differential Geometry", Vol. I, John Wiley & Sons, New York (1963)
7. N. Steenrod, "The Topology of Fibre Bundles", Princeton Univ. Press, Princ., New Jersey (1951)
8. M. Chemtob, Cross section of monopole-induced skyrmion decay, *Phys. Rev. D* 39:2013 (1989)
9. R. Bott and L. W. Tu, "Differential Forms in Algebraic Topology", Springer, New York (1982)
10. E. H. Spanier, "Algebraic Topology", McGraw-Hill, New York (1966)
11. Ö. Kaymakcalan, S. Rajeev and J. Schechter, Non-abelian anomaly and vector-meson decays, *Phys. Rev. D* 30:594 (1984).

EXPLICITLY COVARIANT ALGEBRAIC REPRESENTATIONS FOR TRANSITIONAL CURRENTS OF SPIN-1/2 PARTICLES

M. I. Krivoruchenko

Institute for Theoretical and Experimental Physics
ITEP, 117259, Moscow, Russia

Abstract

In non-relativistic quantum mechanics and in relativistic quantum theory, there exist explicitly covariant algebraic representations for transitional currents of massive spin-1/2 particles. We show that in the massless case such representations exist only for transitional currents which are non-diagonal in helicity. The diagonal currents have algebraic representations which are not explicitly covariant.

1. INTRODUCTION

In non-relativistic quantum mechanics, the transitional currents $\chi^{+}(\boldsymbol{\zeta}')\chi(\boldsymbol{\zeta})$ or $\chi^{+}(\boldsymbol{\zeta}')\boldsymbol{\sigma}\chi(\boldsymbol{\zeta})$, where $\boldsymbol{\sigma}$ are the Pauli matrices and $\chi(\boldsymbol{\zeta}'), \chi(\boldsymbol{\zeta})$ are spinors describing particles with polarizations $\boldsymbol{\zeta}', \boldsymbol{\zeta}$, determine the amplitudes of processes involving spin-1/2 particles. The cross-sections are usually computed by squaring the absolute values of the amplitudes. The spinors are then removed with the use of the technique of projection operators.[1] In this way one obtains for the cross-sections algebraic expressions which depend on the vectors $\boldsymbol{\zeta}'$, $\boldsymbol{\zeta}$ in an explicitly covariant form with respect to the three-dimensional rotations.

An analogous prescription is used in the relativistic quantum theory. Cross-sections and decay probabilities contain currents defined through the Dirac bispinors. The explicit dependence on momenta and particle polarization vectors remains hidden in the currents of this type. Just as in non-relativistic quantum mechanics, the bispinors are removed in a covariant way from the squares of absolute values of the matrix elements.[2–3]

In the case of a complicated tensor structure, it seems easier to deal with the algebraic expressions for the amplitudes rather than with transition probabilities. For example, the Compton effect amplitude is a contraction of a rank-two tensor with polarization vectors of two photons. The cross-section of this process is quadratic in the amplitude and requires already a rank-four tensor for its determination.

The explicitly covariant algebraic representations for transitional currents can be constructed in non-relativistic quantum mechanics and in the relativistic quantum theory of massive particles.[4–6]

In the next section, we discuss these representations and apply them for the analysis of the problem of electron scattering in an external Coulomb field. In Sect. 3, we discuss

Quantization, Coherent States, and Complex Structures
Edited by J.-P. Antoine *et al.*, Plenum Press, New York, 1995

the massless case and demonstrate that explicitly covariant algebraic representations exist only for spin-flip transitional currents. The currents that are diagonal in the helicity have algebraic representations which are not explicitly covariant.

2. TRANSITIONAL CURRENTS IN QUANTUM MECHANICS AND IN THE RELATIVISTIC QUANTUM THEORY. THE MOTT CROSS-SECTION

In non-relativistic quantum mechanics, the wave function of a spin-1/2 particle, polarized in direction of $\boldsymbol{\zeta}$, is described by a spinor satisfying the equations

$$\hat{\zeta}\chi(\boldsymbol{\zeta}) = \chi(\boldsymbol{\zeta})\,, \quad \chi^{+}(\boldsymbol{\zeta})\chi(\boldsymbol{\zeta}) = 1\,, \tag{2.1}$$

where $\hat{\zeta} = \boldsymbol{\zeta}\cdot\boldsymbol{\sigma}\,,\ \ \boldsymbol{\zeta}^2 = 1$, and $\boldsymbol{\sigma}$ are the Pauli matrices.

It is possible to check directly the correctness of the following equation:

$$\chi_\alpha(\boldsymbol{\zeta})\chi^{+}_{\beta}(\boldsymbol{\zeta}) = (\frac{1+\hat{\zeta}}{2})_{\alpha\beta}. \tag{2.2}$$

Acting on the spinors $\chi(\boldsymbol{\zeta})$ by the projection operator $P(\boldsymbol{\zeta}') = (1+\hat{\zeta}')/2$, we obtain an arbitrary normalized spinor $\chi(\boldsymbol{\zeta}')$. Having normalized $\chi(\boldsymbol{\zeta}')$ to unity, we get

$$\chi(\boldsymbol{\zeta}') \doteq \frac{2}{\sqrt{2+2\boldsymbol{\zeta}\cdot\boldsymbol{\zeta}'}}\frac{1+\hat{\zeta}'}{2}\chi(\boldsymbol{\zeta})\,. \tag{2.3}$$

The sign $\doteq$ is used to express the fact that two values (in this case two spinors) are equal up to a phase factor. Substituting Eq.(2.3) into Eq.(2.2), we get

$$\chi_\alpha(\boldsymbol{\zeta})\chi^{+}_{\beta}(\boldsymbol{\zeta}') \doteq \frac{2}{\sqrt{2+2\boldsymbol{\zeta}\cdot\boldsymbol{\zeta}'}}(\frac{1+\hat{\zeta}}{2}\frac{1+\hat{\zeta}'}{2})_{\alpha\beta}\,. \tag{2.4}$$

We see that the nontrivial dependence on the vectors $\boldsymbol{\zeta}$, $\boldsymbol{\zeta}'$ has an explicitly covariant form with respect to the rotation group $O(3)$.

The transitional currents can be represented as follows

$$\begin{aligned} \chi^{+}(\boldsymbol{\zeta}')\chi(\boldsymbol{\zeta}) &\doteq \sqrt{\frac{1+\boldsymbol{\zeta}\cdot\boldsymbol{\zeta}'}{2}}\,, \\ \chi^{+}(\boldsymbol{\zeta}')\boldsymbol{\sigma}\chi(\boldsymbol{\zeta}) &\doteq \frac{\boldsymbol{\zeta}+\boldsymbol{\zeta}'+i\boldsymbol{\zeta}\times\boldsymbol{\zeta}'}{\sqrt{2+2\boldsymbol{\zeta}\cdot\boldsymbol{\zeta}'}}\,. \end{aligned} \tag{2.5}$$

These currents constitute an exhaustive set of transitional currents in the non-relativistic theory, since the matrices 1 and $\boldsymbol{\sigma}$ form an exhaustive set of two-dimensional Hermitian matrices.

The above arguments allow a straightforward relativistic extension to the case of massive spinor particles. A full list of the currents is given in Refs. 5,6. For instance, the relativistic vector current has the form

$$\begin{aligned} \bar{u}(p',s')\gamma_\mu u(p,s) &= A(p,s,p',s')(m(p+p')_\mu(1-s\cdot s') \\ &\quad + i\varepsilon_{\mu\alpha\beta\gamma}p'_\alpha p_\beta(s'+s)_\gamma + mp'\cdot ss'_\mu + mp\cdot s's_\mu), \end{aligned}$$

where the bispinors $u(p,s)$ and $u(p',s')$ describe particles with momenta p_μ and p'_μ and polarizations s_μ and s'_μ, such that $p^2 = m^2$, $s^2 = -1$, $p\cdot s = 0$. The normalization conditions for the bispinors are of the form $\bar{u}(p,s)u(p,s) = 1$, the gamma matrices are defined as in Ref. 3, $\varepsilon_{0123} = +1$. The normalization coefficient $A(p,s,p',s')$ is given by

$$A(p,s,p',s') \doteq \frac{1}{2m}[(m^2 + p \cdot p')(1 - s \cdot s') + p \cdot s' p' \cdot s]^{-1/2} \ . \tag{2.6}$$

To illustrate the new method of eliminating bispinors, we consider the problem of electron scattering in a Coulomb field. Let $p = (E, \mathbf{p}), p' = (E, \mathbf{p}')$ be the momenta of the electron in the initial and final states, $\mathbf{n} = \mathbf{p}/|\mathbf{p}|, \mathbf{n}' = \mathbf{p}'/|\mathbf{p}|$. The polarization vectors corresponding to definite helicities of the electron are given by $s = \sigma(|\mathbf{p}|, \mathbf{n}E)/m, s' = \sigma'(|\mathbf{p}|, \mathbf{n}'E)/m$, where $\sigma, \sigma' = \pm 1$. We get then $p \cdot p' = E^2 - p^2 \mathbf{n}\cdot\mathbf{n}', \mathbf{s}\cdot\mathbf{s}' = \sigma\sigma'(p^2 - E^2 \mathbf{n}\cdot\mathbf{n}')/m^2, p\cdot s' = \sigma'|\mathbf{p}|E(1-\mathbf{n}\cdot\mathbf{n}')/m, p'\cdot s = \sigma|\mathbf{p}|E(1-\mathbf{n}\cdot\mathbf{n}')/m$. Substituting these expressions into the normalization coefficient (2.6) and into the zeroth component of the vector current, we obtain

$$\bar{u}(p',s')\gamma_0 u(p,s) \doteq \begin{cases} \frac{E}{m}\sqrt{\frac{1+\mathbf{n}\cdot\mathbf{n}'}{2}}, & \sigma' = \sigma \\ \sqrt{\frac{1-\mathbf{n}\cdot\mathbf{n}'}{2}}, & \sigma' = -\sigma \end{cases} \tag{2.7}$$

It is known that the cross-section for the backscattering is suppressed by an additional power of $(m/E)^2$. For $\mathbf{n}' = -\mathbf{n}$, the helicity preserving current vanishes identically as may be seen from (2.7). This can be interpreted as conservation of the projection of the total angular momentum on the direction of the electron velocity. The spin-flip current does not vanish, but has an additional power of (m/E) in comparison with the non-flip current component. The cross-section therefore contains an additional factor $(m/E)^2$. In the case of forward scattering, instead, the helicity conserving component of the current contributes to the cross-section. This is also a consequence of the conservation of the projection of the total angular momentum on the direction of the electron velocity.

Introducing the Fourier transform of the Coulomb potential and restoring the kinematical coefficients, we reproduce the well-known Mott cross-section:

$$\frac{d\sigma}{d\Omega} = \frac{4(\alpha Z)^2 m^2}{(\mathbf{p}' - \mathbf{p})^4}\left\{\left(\frac{E}{m}\right)^2 \frac{1 + \mathbf{n}\cdot\mathbf{n}'}{2} + \frac{1 - \mathbf{n}\cdot\mathbf{n}'}{2}\right\}. \tag{2.8}$$

The first term in the bracket describes the scattering with conservation of the electron helicity, while the second one describes the spin-flip electron scattering. In the non-relativistic case, the cross section (2.8) coincides with the Rutherford cross section.

3. TRANSITIONAL CURRENTS IN THE MASSLESS CASE

Dirac bispinors describing massless particles and antiparticles with fixed helicity have algebraically the same form, so we shall not distinguish between them in the sequel.

The bispinors corresponding to the left-handed (L) and right-handed (R) helicity particles are defined as solutions of the following equations

$$\begin{aligned} \hat{k}u_R(k) &= \hat{k}u_L(k) &&= 0, \\ \frac{1-\gamma_5}{2}u_R(k) &= \frac{1+\gamma_5}{2}u_L(k) &&= 0, \\ \bar{u}_R(k)\gamma_\mu u_R(k) &= \bar{u}_L(k)\gamma_\mu u_L(k) &&= 2k_\mu. \end{aligned} \tag{3.1}$$

Here $k^2 = 0$. The normalization condition is taken in such a form because of the fact that in the massless case the scalar current vanishes identically: $\bar{u}_R(k)u_R(k) = \bar{u}_L(k)u_L(k) = 0$.

Now, we do not have at hand projection operators on the states with definite k. Unlike in the massive case, the right-hand sides of the equalities

$$\begin{aligned} u_{L\alpha}(k)\bar{u}_{L\beta}(k) &= (\hat{k}\frac{1+\gamma_5}{2})_{\alpha\beta} \\ u_{R\alpha}(k)\bar{u}_{R\beta}(k) &= (\hat{k}\frac{1-\gamma_5}{2})_{\alpha\beta} \end{aligned} \tag{3.2}$$

vanish when squared. Nevertheless, the following relations hold:

$$\begin{aligned} u_L(k') = C\hat{k}'u_R(k), \\ u_R(k') = C\hat{k}'u_L(k), \end{aligned} \tag{3.3}$$

where the normalization coefficient

$$C \doteq \frac{1}{\sqrt{2k' \cdot k}} \tag{3.4}$$

becomes singular in the limit $k' \to k$. The bispinors in the left-hand side of Eq.(3.3) obey (3.1) identically if the bispinors in the right-hand side obey (3.1). The coefficient (3.4) can be found from the relations (3.2). The tensor products of bispinors with different helicity take the form

$$\begin{aligned} u_{L\alpha}(k)\bar{u}_{R\beta}(k') &\doteq \frac{1}{\sqrt{2k \cdot k'}}(\hat{k}\hat{k}'\frac{1-\gamma_5}{2})_{\alpha\beta}, \\ u_{R\alpha}(k)\bar{u}_{L\beta}(k') &\doteq \frac{1}{\sqrt{2k \cdot k'}}(\hat{k}\hat{k}'\frac{1-\gamma_5}{2})_{\alpha\beta}. \end{aligned} \tag{3.5}$$

With the help of these expressions, one can find explicitly covariant representations for transitional currents nondiagonal in helicity. First of all, it is obvious that the currents containing an odd number of gamma matrices vanish identically, since the right-hand sides of Eqs.(3.5) have an even number of gamma matrices, whereas the trace of an odd number of gamma matrices equals zero. We list the expressions for the nonvanishing currents:

$$\begin{aligned} \bar{u}_R(k')u_L(k) &\doteq \bar{u}_L(k')u_R(k) \doteq \sqrt{2k \cdot k'}, \\ \bar{u}_R(k')\sigma_{\mu\nu}u_L(k) &\doteq \sqrt{\frac{2}{k \cdot k'}}(k'_\mu k_\nu - k'_\nu k_\mu + i\varepsilon_{\mu\nu\tau\sigma}k'_\tau k_\sigma), \\ \bar{u}_L(k')\sigma_{\mu\nu}u_R(k) &\doteq \sqrt{\frac{2}{k \cdot k'}}(k'_\mu k_\nu - k'_\nu k_\mu - i\varepsilon_{\mu\nu\tau\sigma}k'_\tau k_\sigma). \end{aligned} \tag{3.6}$$

Consider now the currents which are diagonal in helicity. Since in the non-relativistic theory, in the relativistic theory of massive particles, and in the massless case considered above, the currents have explicitly covariant representations, it is unexpected that the currents diagonal in helicity do not have such a representation.

Eqs.(3.3) relate bispinors with different helicity. The only possibility to express, say, a left bispinor with momentum k' in terms of a left bispinor with momentum k is to introduce an auxiliary vector h into the right-hand side of Eq.(3.3). As a result we obtain

$$\begin{aligned} u_L(k') &= C'\hat{k}'\hat{h}u_L(k), \\ u_R(k') &= C'\hat{k}'\hat{h}u_R(k). \end{aligned} \tag{3.7}$$

The normalization coefficient equals

$$C' \doteq \frac{1}{\sqrt{4k' \cdot hk \cdot h - 2k' \cdot kh^2}} \, . \tag{3.8}$$

The tensor products of different helicity bispinors can be written now in the following form

$$\begin{aligned} u_{L\alpha}(k)\bar{u}_{L\beta}(k') &\doteq \frac{1}{\sqrt{4k' \cdot hk \cdot h - 2k' \cdot kh^2}}(\hat{k}\hat{h}\hat{k}'\frac{1+\gamma_5}{2})_{\alpha\beta} \, , \\ u_{R\alpha}(k)\bar{u}_{R\beta}(k') &\doteq \frac{1}{\sqrt{4k' \cdot hk \cdot h - 2k' \cdot kh^2}}(\hat{k}\hat{h}\hat{k}'\frac{1+\gamma_5}{2})_{\alpha\beta} \, . \end{aligned} \tag{3.9}$$

The right-hand side depends on h, whereas the left-hand side does not. The only possible explanation is that the h-dependence of the right-hand side is such that a variation of the vector h affects only the phase factor. In this sense, the vector h fixing a certain frame does not break the covariance.

Let us consider a well-known example. Average over directions a vector $\mathbf{n}(\mathbf{na})$ where $\mathbf{n}$ is a unit vector and $\mathbf{a}$ an additional (constant) vector. We have at hand only the vector $\mathbf{a}$, thus the result must be proportional to $\mathbf{a}$. The result can be reconstructed unambiguously:

$$\int \frac{d\Omega}{4\pi}\mathbf{n}(\mathbf{na}) = \frac{1}{3}\mathbf{a} \, . \tag{3.10}$$

It follows from (3.9) that, for quantities defined up to a phase factor, this reasoning is *not* conclusive. In general there exist auxiliary vectors – apart from the ones already involved – by means of which the solution is expressed. These vectors break the initial symmetry (rotational, for example). However, their variation affects only the phase factor.

We are able now to construct explicit algebraic expressions for the vector currents. They contain the auxiliary vector h, so they are not explicitly covariant. Multiplying the relations (3.9) by γ_μ and calculating the trace we find[6] :

$$\begin{aligned} \bar{u}_L(k')\gamma_\mu u_L(k) &\doteq \frac{2(k'_\mu h \cdot k + k_\mu h \cdot k' - h_\mu k' \cdot k + i\varepsilon_{\mu\nu\tau\sigma}k'_\nu k_\tau h_\sigma)}{\sqrt{4k' \cdot hk \cdot h - 2k' \cdot kh^2}} \, , \\ \bar{u}_R(k')\gamma_\mu u_R(k) &\doteq \frac{2(k'_\mu h \cdot k - k_\mu h \cdot k' - h_\mu k' \cdot k + i\varepsilon_{\mu\nu\tau\sigma}k'_\nu k_\tau h_\sigma)}{\sqrt{4k' \cdot hk \cdot h - 2k' \cdot kh^2}} \, . \end{aligned} \tag{3.11}$$

We thus conclude that the tensor algebra tools are not sufficient to represent the left-hand side of Eq.(3.11) in algebraic form through the vectors k' and k only. Any explicit expression which does not refer to some third vector in principle does not exist.

There are no other currents with an odd number of gamma matrices. As for the currents with an even number of gamma matrices, they vanish identically. It immediately follows also that the right-hand sides of Eqs. (3.9) can be written as convolutions of currents (3.11) with the matrices $\gamma_\mu(1 \pm \gamma_5)$. That is why the phase factors in the right-hand side of Eqs.(3.9) and (3.11) change in the same manner under a variation of the vector h.

Let us find the components of the spin current (2.5) for $\boldsymbol{\zeta}' = -\boldsymbol{\zeta}$. Since the normalization coefficient goes to infinity at $\boldsymbol{\zeta}' = -\boldsymbol{\zeta}$, whereas all the vector components in the denominator vanish, it is necessary to consider the limit $\boldsymbol{\zeta}' = -\boldsymbol{\zeta} + \boldsymbol{\Delta}, \boldsymbol{\Delta} \to 0$ in a plane perpendicular to z. The result must be independent of the direction of $\boldsymbol{\Delta}$. We get

$$j(-\boldsymbol{\zeta}, \boldsymbol{\zeta}) = \lim_{\boldsymbol{\Delta} \to 0} \frac{\boldsymbol{\Delta} + i\boldsymbol{\zeta} \times \boldsymbol{\Delta}}{|\boldsymbol{\Delta}|} \, . \tag{3.12}$$

We choose the vectors to be $\boldsymbol{\zeta} = (0,0,1)$, $\boldsymbol{\Delta} = (|\boldsymbol{\Delta}|,0,0)$, in which case $j(-\boldsymbol{\zeta},\boldsymbol{\zeta}) = (1,i,0)$. If $\boldsymbol{\Delta} = (0,|\boldsymbol{\Delta}|,0)$, then $j(-\boldsymbol{\zeta},\boldsymbol{\zeta}) = (-i,1,0)$. It is clear now that the arbitrariness in the choice of $\boldsymbol{\Delta}$ is removed by the phase ambiguity of the currents. Therefore the limit exists and the result does not depend on the direction of $\boldsymbol{\Delta}$.

A similar uncertainty is present in the currents (3.6). Let us consider the component $\bar{u}_L(k')\sigma_{0\alpha}u_R(k)$ at $k'_0 = k_0$, $\mathbf{k}' \to \mathbf{k}$. It is not difficult to verify that in such a case $\bar{u}_L(k')\sigma_{0\alpha}u_R(k) = 2j_\alpha(\boldsymbol{\zeta}',\boldsymbol{\zeta})$, where $\boldsymbol{\zeta}' = -\mathbf{k}'/|\mathbf{k}'|, \boldsymbol{\zeta} = \mathbf{k}/|\mathbf{k}|$. However, we have already showed that the non-relativistic spin current is not singular and well defined.

4. CONCLUSION

For massless particles, explicitly covariant algebraic representations exist only for the currents nondiagonal in helicity (see Eq.(3.6)). The diagonal currents have algebraic representations which are not explicitly covariant (see (3.11)), because of their dependence on an auxiliary vector absent from the initial conditions. Its variation, however, affects only an unobservable common phase factor.

Considering massless particles as an example, we have observed the breakdown of the well-known principle according to which the solution of a problem in the presence of a symmetry should contain only vectors entering the statement of the problem. This principle does not work if the required formula is defined up to a phase factor.

References

1. L. D. Landau, E. M. Lifshitz, "Quantum Mechanics", Nauka, Moscow (1989)
2. V. B. Berestetsky, E. M. Lifshitz , L. P. Pitaevsky, "Quantum Electrodynamics", Nauka, Moscow (1980)
3. J. D. Bjorken, S. D. Drell, "Relativistic Quantum Mechanics", Mc Graw-Hill, New York (1964)
4. M. I. Krivoruchenko, Explicit analytical expression for transitional current $J_\mu = u(p',s')\gamma_\mu u(p,s)$, *Sov. J. Nucl. Phys.* 47:1153 (1988)
5. M. I. Krivoruchenko, I.V. Kudrya, Explicit analytical expressions for transitional currents of spin-1/2 particles, *Nuovo Cim. B* 108:15 (1993)
6. M. I. Krivoruchenko, Transitional currents for spin-1/2 particles, *Uspekhi Fiz. Nauk* 164:643 (1994).

THE QUANTUM $SU(2,2)$-HARMONIC OSCILLATOR

Wojciech Mulak

Institute of Theoretical Physics
University of Wrocław
Pl. Maxa Borna 9, 50-204 Wrocław
Poland

Abstract

We consider the classical $SU(2,2)$-harmonic oscillator, with phase space $\mathcal{A}(2,2) = SU(2,2)/S(U(2) \times U(2))$, and quantize it by the coherent state method. The quantum Hamiltonian is the Toeplitz operator corresponding to the square of the distance in the $SU(2,2)$-invariant Kähler metric on the phase space. Its spectrum is computed, and found to depend on the choice of representation of $SU(2,2)$.

1. INTRODUCTION

The $SU(2,2)$-harmonic oscillator is the generalization of the model harmonic oscillator on a flat phase space. In our case the phase space $\mathcal{A}(2,2) = SU(2,2)/S(U(2)\times U(2)) \simeq SO(4,2)/SO(4) \times SO(2)$ is the eight dimensional conformal domain, on which canonical coordinates $(x^\mu, p^\mu), \mu = 0, ..., 3$ can be globally introduced.

Spaces of this type are well known as Cartan classical domains. They appear in physics and mathematics and have been considered by many authors. The complex geometry of these spaces and, in particular, its applications in conformal theories has been investigated in the work of Coquereaux and Jadczyk (see Ref.1 and references therein). As the phase space of scalar massive conformal particle it has been considered by A. Odzijewicz.[2,3]

The geometry of $\mathcal{A}(2,2)$ is related to the space-time geometry. The Shilov boundary of $\mathcal{A}(2,2)$ is the compactified Minkowski space-time, endowed with the conformal structure of signature $(+,-,-,-)$. The compactification is obtained by adding a light cone at infinity to the usual Minkowski space-time.

As it is suggested in Ref.4, the conformal domain can be considered as a replacement of space-time on the microscale. This interpretation is based on Born's idea of a reciprocity symmetry between space-time and energy-momentum space. The reciprocity symmetry can be reformulated as the symmetry of the conformal domain. As a consequence, these spaces are not distinguished on the microscale. Minkowski space is interpreted as the very-high-mass, or very-high-energy-momentum-transfer limit of the conformal domain.

The $SU(2,2)$-harmonic oscillator is a one-body system. It is obtained from a two-body interacting system by introducing "center of the mass" coordinates. The interaction is $SU(2,2)$-invariant. The covariant harmonic oscillators are used in quark

Quantization, Coherent States, and Complex Structures
Edited by J.-P. Antoine *et al.*, Plenum Press, New York, 1995

models. In these models the interaction between quarks are given by the harmonic oscillator potential. The model of a relativistic hadron consisting of two quarks interacting in that way can be found in Ref.5. These models have been considered by many authors (see references in Ref.5). It is tempting to interpret our model along similar lines.

The quantization with the Berezin-Weyl calculus[6] yields the quantum Hamiltonian as a Toeplitz operator.[1] This scheme of quantization involves a system of Perelomov generalized coherent states for $SU(2,2)$.[7,8] The representation spaces of the quantized theory are Hilbert spaces of holomorphic functions on the domain, corresponding to the members of the discrete series of unitary irreducible representations of $SU(2,2)$. These representation spaces have their counterparts in Minkowski space-time as spaces of distributional boundary values.[9]

Applying the quantization procedure for different representations, we obtain different spectra for the quantum Hamiltonian. In contrast to the geometric quantization, this quantization method does not contain the prequantization stage. For all representations the quantum Hamiltonian has discrete and degenerate spectrum. The $SU(1,1)$-harmonic oscillator has been considered in Ref.10.

We use the S–parametrization of $\mathcal{A}(2,2)$ introduced in Ref.1. This parametrization provides the description of the geometry of $\mathcal{A}(2,2)$ in terms of its symmetries.

2. THE SU(2,2)-HARMONIC OSCILLATOR

The classical Hamiltonian of the $SU(2,2)$-harmonic oscillator in the S-parametrization[1] of $\mathcal{A}(2,2)$ is given by the function:

$$H = \frac{1}{4}\mathrm{Tr}(\ln^2(S_0 S)). \tag{2.1}$$

Up to a multiplicative constant, this function is the square of the distance from the origin S_0 of $\mathcal{A}(2,2)$, calculated with the $SU(2,2)$-invariant Kähler metric on $\mathcal{A}(2,2)$. The function (2.1) generalizes the Hamiltonian of the harmonic oscillator on a flat phase space. In the flat case, the Hamiltonian of the harmonic oscillator can be obtained in this way from the Kähler metric on the phase space $\Gamma = \boldsymbol{C}^N$.

The Hamiltonian $H = H(S)$ is invariant under the action of the isotropy group at the origin S_0. The classical dynamics of the oscillator is determined by the Hamiltonian equation:

$$i_{X_H}\omega = dH, \tag{2.2}$$

where X_H denotes the Hamiltonian vector field corresponding to H.

In order to find the solution of the above equation, we postulate that the solution has the form:

$$S(t) = e^{-\frac{t}{2}X} S e^{\frac{t}{2}X}, \tag{2.3}$$

where X is an element of the Lie algebra of the isotropy group at the origin S_0. Then we obtain from (2.2) the equation:

$$i\ln(S_0 S) = [S, X]. \tag{2.4}$$

Under the assumption that $\det([S, S_0]) \neq 0$, the solution of (2.4) is:

$$X(S) = iS_0\ln(S_0 S)[S, S_0]^{-1} = iS_0\ln(S_0 S)S_0 S(\mathbf{1} - S_0 S S_0 S)^{-1}\,. \tag{2.5}$$

In the case $\det([S,S_0])=0$, the solutions of (2.4) can be found by observing that equation (2.4) is invariant under the action of the isotropy group at S_0.

$\mathcal{A}(2,2)$ can be realized as the bounded complex domain

$$\mathbf{1}-ZZ^{+}>0, \tag{2.6}$$

where the points of $\mathcal{A}(2,2)$ are parametrized by $Z\in M_2(\boldsymbol{C})$. Let us introduce the following coordinates on $\mathcal{A}(2,2)$ given in Ref.9:

$$Z=\begin{vmatrix} z_{11} & z_{12} \\ z_{21} & z_{22} \end{vmatrix}=u_1\begin{vmatrix} \lambda_1 & 0 \\ 0 & \lambda_2 \end{vmatrix}u_2 \tag{2.7}$$

$$u_1=e^{i\phi_1\sigma_1}e^{i\theta_1\sigma_3},\qquad u_2=e^{i\theta_2\sigma_3}e^{i\phi_2\sigma_1},$$

$$\lambda_1=r_+e^{i\alpha},\qquad \lambda_2=r_-e^{i\beta},$$

where the Pauli matrices are: $\sigma_1=\begin{pmatrix} 1 & 0 \\ 0 & -1 \end{pmatrix}$, $\sigma_3=\begin{pmatrix} 0 & -i \\ i & 0 \end{pmatrix}$. The conditions for the coordinates are:

$$\begin{aligned} 0 \ &\leq\ r_+,\, r_-\ <\ 1 \\ 0 \ &\leq\ \theta_1,\, \theta_2\ \leq\ \frac{\pi}{2} \\ 0 \ &\leq\ \alpha,\, \beta,\, \phi_1,\, \phi_2\ \leq\ 2\pi \end{aligned} \tag{2.8}$$

Let us express the function (2.1) in the variables (2.8). Using the relation between parametrizations, we can write:

$$Y\equiv\frac{S_0S-\mathbf{1}}{S_0S+\mathbf{1}}=\frac{1}{2}\begin{vmatrix} Z+Z^{+} & i(Z-Z^{+}) \\ i(Z-Z^{+}) & -(Z+Z^{+}) \end{vmatrix} \tag{2.9}$$

Then we have:

$$H=\frac{1}{4}\mathrm{Tr}\left(\ln^2(S_0S)\right)=\frac{1}{4}\mathrm{Tr}\left(\ln^2\frac{\mathbf{1}+Y}{\mathbf{1}-Y}\right) \tag{2.10}$$

The Hamiltonian (2.1) is invariant under the action of the isotropy group at the origin S_0, thus it is invariant under the transformation $Z\rightarrow UZV^{+}=\begin{vmatrix} r_+ & 0 \\ 0 & r_- \end{vmatrix}$, where U and V are the unitary matrices given by the singular value decomposition for Z. Using this invariance we obtain:

$$H=\frac{1}{2}\left(\ln^2\frac{1+r_+}{1-r_+}+\ln^2\frac{1+r_-}{1-r_-}\right). \tag{2.11}$$

The canonical variables can be introduced by:

$$\mathcal{Z}=\left(x^{\mu}+i\hbar\frac{p^{\mu}}{p^2}\right)\sigma_{\mu}, \tag{2.12}$$

where the unbounded parametrization of $\mathcal{A}(2,2)$ by $\mathcal{Z}\in M_2(\boldsymbol{C})$ is used. In this parametrization the condition (2.6) reads:

$$-i(\mathcal{Z}-\mathcal{Z}^{+})>0. \tag{2.13}$$

The relation between the parametrizations (2.6) and (2.13) is given by the Cayley transformation:

$$Z=\frac{\mathbf{1}+i\mathcal{Z}}{\mathbf{1}-i\mathcal{Z}} \tag{2.14}$$

3. COHERENT STATES FOR SU(2,2)

Let us consider the discrete series of unitary irreducible representations of $SU(2,2)$, which are realized in spaces of holomorphic functions on $\mathcal{A}(2,2)$, namely the representation of the series d_0 in Graev's classification.[9] The members of the series d_0 are labeled by the integer number $n = 4, 5, \ldots$ and two spin labels j_1, j_2. In our case $j_1 = j_2 = 0$.

Let $|dZ|$ denotes the Euclidean measure on $\mathcal{A}(2,2)$. Let $d\mu_n$ denotes the normalized measure given by:

$$d\mu_n(Z) \;=\; N_n \left[\det(\mathbf{1} - ZZ^+)\right]^{n-4} |dZ| \,, \qquad n = 4, 5, \ldots, \tag{3.1}$$

where the normalization constant $N_n = \pi^{-4}(n-3)(n-2)^2(n-1)$ ensures that $\int d\mu_n = 1$. The space of functions on $\mathcal{A}(2,2)$:

$$\mathcal{F}_n = \left\{ f \text{ holomorphic } : ||f||_n^2 \;=\; \int |f(Z)|^2 d\mu_n(Z) \;<\; \infty \right\} \tag{3.2}$$

is a Hilbert space with the scalar product:

$$\langle f|g\rangle = \int \overline{f(Z)} g(Z)\, d\mu_n(Z) \,, \; f, g \in \mathcal{F}_n. \tag{3.3}$$

The transformation:

$$\left(\overset{(n)}{T}(g) f\right)(Z) \;=\; [\det(CZ + D)]^{-n} \, f((AZ + B)(CZ + D)^{-1}) \,, \tag{3.4}$$

$$f \in \mathcal{F}_n \,, \quad g^{-1} = \begin{vmatrix} A & B \\ C & D \end{vmatrix} \in SU(2,2) \,,$$

defines the unitary irreducible representation of $SU(2,2)$ in $\mathcal{F}_n$. The system of coherent states of type $(\overset{(n)}{T}, |\Psi_0> = 1)$ is obtained by the action of the representation (3.4):

$$\overset{(n)}{T} |\Psi_0> = [\det(CZ + D)]^{-n} \,, \qquad 1 = |\Psi_0> \in \mathcal{F}_n. \tag{3.5}$$

The states obtained by (3.5) can be parametrized by points of $\mathcal{A}(2,2)$:

$$|\zeta> = \frac{[\det(\mathbf{1} - \zeta^+\zeta)]^{n/2}}{[\det(\mathbf{1} - \zeta^+ Z)]^n} \,, \qquad \mathbf{1} - \zeta\zeta^+ \;>\; 0. \tag{3.6}$$

The family $\{|\zeta> : \mathbf{1} - \zeta\zeta^+ \;>\; 0\}$ forms a system of generalized coherent states for $SU(2,2)$.[8] It yields a resolution of unity:

$$N_n \int |\zeta> < \zeta| d\mu(\zeta) \;=\; \mathbf{1}_{\mathcal{F}_n} \,, \tag{3.7}$$

where $d\mu(\zeta) = [\det(\mathbf{1} - \zeta\zeta^+)]^{-4} |d\zeta|$ is the $SU(2,2)$-invariant measure on $\mathcal{A}(2,2)$. Every state $|\Psi> \in \mathcal{F}_n$ has the continuous representation:

$$|\Psi> \longmapsto < \zeta|\Psi> = C_\Psi(\zeta) \;=\; \left[\det(\mathbf{1} - \zeta^+\zeta)\right]^{n/2} \Psi(\zeta). \tag{3.8}$$

The representation (3.8) has the property:

$$C_\Psi(\zeta') = \int K(\zeta', \zeta) C_\Psi(\zeta) d\mu(\zeta) \,, \tag{3.9}$$

where

$$K(\zeta', \zeta) \;=\; N_n < \zeta'|\zeta> = N_n \frac{[\det(\mathbf{1} - \zeta'^+\zeta')]^{n/2} \, [\det(\mathbf{1} - \zeta^+\zeta)]^{n/2}}{[\det(\mathbf{1} - \zeta^+\zeta')]^n} \tag{3.10}$$

is the reproducing kernel:

$$K(\zeta, \zeta'') \;=\; \int K(\zeta, \zeta') K(\zeta', \zeta'') \, d\mu(\zeta'). \tag{3.11}$$

4. THE QUANTUM SU(2,2)-HARMONIC OSCILLATOR

The classical system on $\mathcal{A}(2,2)$ may be quantized with the Berezin-Weyl calculus. This scheme of quantization involves a system of coherent states.

For every representation $\overset{(n)}{T}$ of $SU(2,2)$, $n = 4,5,\ldots$ we obtain different quantizations in the representation spaces $\mathcal{F}_n$. The operator corresponding to the classical observable is a Toeplitz operator constructed by using the generalized Bergman projection.

Let $L^2(d\mu_n)$ denotes the Hilbert space of the measurable and square integrable functions on $\mathcal{A}(2,2)$ with respect to the measure $d\mu_n$. The generalized Bergman projection[6]

$$P_B : L^2(d\mu_n) \longrightarrow \mathcal{F}_n\,,\, P_B^+ = P_B = P_B^2 \tag{4.1}$$

is given by:

$$(P_B\, f)(Z) \;=\; \int L_n(Z,\zeta) f(\zeta,\zeta^+) d\mu_n(\zeta)\,, \quad f \in L^2(d\mu_n)\,, \tag{4.2}$$

where $L_n(\zeta',\zeta) = [\det(\mathbf{1}-\zeta^+\zeta')]^{-n}$ is the generalized Bergman kernel. The quantization associates[1] to each function $f \in L^2(d\mu_n)$ an operator $\hat{f}$ in $\mathcal{F}_n$:

$$f \longmapsto \hat{f} = N_n \int f(\zeta,\zeta^+)\,|\,\zeta><\zeta|\; d\mu(\zeta). \tag{4.3}$$

Acting with $\hat{f}$ on $|\Psi> \in \mathcal{F}_n$ we have:

$$\hat{f}\,|\,\Psi\; >= \;P_B(f\cdot\Psi). \tag{4.4}$$

Then the operator (4.3):

$$\hat{f} = P_B \circ f \circ P_B \tag{4.5}$$

is the Toeplitz operator corresponding to the function f.

Let us describe an orthonormal basis in $\mathcal{F}_n$.[9] The basis consists of the functions:

$$\triangle^{jm}_{q_1q_2}(Z) \;=\; (\mathcal{N}^{jm})^{-1}(\det Z)^m D^j_{q_1q_2}(Z)\,, \tag{4.6}$$

$$m \;=\; 0,1,2,\ldots \quad 2j \;=\; 0,1,2,\ldots \quad -j \;\le\; q_1\,,q_2 \;\le\; j$$

where the function $D^j_{q_1q_2}$ is the extension of the polynomial well known from the $SU(2)$ representation theory:

$$D^j_{q_1q_2}(Z) \;=\; \left[\frac{(j+q_1)!(j-q_1)!}{(j+q_2)!(j-q_2)!}\right]^{1/2} \sum_{s=\max(0,q_1+q_2)}^{s=\min(j+q_1,j+q_2)} \binom{j+q_2}{s} \times \tag{4.7}$$

$$\times \binom{j-q_2}{s-q_1-q_2} z_{11}^s z_{12}^{j+q_1-s} z_{21}^{j+q_2-s} z_{22}^{s-q_1-q_2}$$

and the normalization constant is given by:

$$\left(\mathcal{N}^{jm}\right)^2 = (n-1)(n-2)^2(n-3)\frac{(n-3)!(n-4)!(m+2j+1)!m!}{(2j+1)(m+n-2)!(m+2j+n-1)!}. \tag{4.8}$$

The orthonormality of (4.6) reads:

$$\left\langle \triangle^{jm}_{q_1q_2} \,|\, \triangle^{j'm'}_{q_1'q_2'} \right\rangle \;=\; \delta_{j',j}\delta_{m',m}\delta_{q_1',q_1}\delta_{q_2',q_2} \tag{4.9}$$

By the quantization (4.3) of the $SU(2,2)$-harmonic oscillator, we obtain the quantum Hamiltonian:

$$\hat{H} = P_B \circ H \circ P_B\,, \quad H \in L^2(d\mu_n) \tag{4.10}$$

In order to find the spectrum of the operator (4.10), let us compute the matrix element:

$$\left\langle \Delta^{j'm'}_{q'_1 q'_2} \mid \hat{H} \Delta^{jm}_{q_1 q_2} \right\rangle = \left\langle \Delta^{j'm'}_{q'_1 q'_2} \mid H \Delta^{jm}_{q_1 q_2} \right\rangle, \tag{4.11}$$

using the coordinates (2.7). After some calculations we obtain:

$$\left\langle \Delta^{j'm'}_{q'_1 q'_2} \mid \hat{H} \Delta^{jm}_{q_1 q_2} \right\rangle = \delta_{j',j}\delta_{m',m}\delta_{q'_1,q_1}\delta_{q'_2,q_2} \overset{(n)}{E}{}^{jm}_{q_1 q_2} \tag{4.12}$$

$$\overset{(n)}{E}{}^{jm}_{q_1 q_2} = \left\langle \Delta^{jm}_{q_1 q_2} \mid \hat{H} \Delta^{jm}_{q_1 q_2} \right\rangle$$

Then the operator $\hat{H}$ is diagonal in the basis (4.6), and its eigenvalues are $\overset{(n)}{E}{}^{jm}_{q_1 q_2}$:

$$\hat{H} \Delta^{jm}_{q_1 q_2} = \overset{(n)}{E}{}^{jm}_{q_1 q_2} \Delta^{jm}_{q_1 q_2}. \tag{4.13}$$

The eigenvalues are given by the integral:

$$\overset{(n)}{E}{}^{jm}_{q_1 q_2} = \alpha_{j,m,n} \sum_{q=-j}^{q=j} \sum_{i=0}^{n-4} \sum_{l=0}^{n-4} (-1)^{i+l} \binom{n-4}{i} \binom{n-4}{l} \times \tag{4.14}$$

$$\times \int_0^1 dr_+ \int_0^1 dr_- \left(\ln^2 \frac{1+r_+}{1-r_+} + \ln^2 \frac{1+r_-}{1-r_-} \right) r_+^{2(j+m+q+i)+1} r_-^{2(j+m-q+l)+1} \left(r_+^2 - r_-^2\right)^2,$$

where

$$\alpha_{j,m,n} = \frac{(m+n-2)!(m+2j+n-1)!}{(2j+1)(n-3)!(n-4)!m!(m+2j+1)!}.$$

Let us define, for $N = 0, 1, 2, \ldots$,

$$S_1(N) \equiv \frac{1}{N+1} \sum_{a=0}^{N} \frac{1}{2a+1}$$

$$S_2(N) \equiv \begin{cases} \dfrac{1}{N+1} \displaystyle\sum_{b=1}^{N} \sum_{a=b}^{N} \frac{1}{2a+1} \cdot \frac{1}{2b} & , \quad N = 1, 2, \ldots \\ 0 & , \quad N = 0 \end{cases} \tag{4.15}$$

$$S(N) \equiv S_1(N) \ln 2 + S_2(N).$$

The integral (4.14) can be computed using the formula:

$$\int_0^1 \ln^2 \frac{1+r}{1-r} \cdot r^{2N+1} dr = 4S(N) \tag{4.16}$$

The eigenvalues are given by the formula:

$$\begin{aligned} \overset{(n)}{E}{}^{jm}_{q_1 q_2} = {} & \frac{4}{(2j+1)(n-3)!} \sum_{i=0}^{n-4} (-1)^i \binom{n-4}{i} \times \\ & \times \left\{ \frac{(m+2j+n-1)!}{(m+2j+1)!} [(m+n-2)S(2j+m+2+i) \right. \\ & \qquad\qquad -(m+1)S(2j+m+1+i)] \\ & \left. + \frac{(m+n-2)!}{m!} [(2j+m+2)S(m+i) - (2j+m+n-1)S(m+1+i)] \right\} \end{aligned} \tag{4.17}$$

We observe that the eigenvalue does not depend on the indices q_1, q_2. Thus the eigenvalue $\overset{(n)}{E}{}^{jm}$ is degenerate, with multiplicity $(2j+1)^2$.

5. REMARKS

The result of the quantization depends on the choice of representation of $SU(2,2)$. The question arises how to interpret this choice. According to Berezin's interpretation,[6] the label n of the representation depends on the parameter h, which plays the role of the Planck constant. By taking the limit $h \to 0$, the correspondence principle is obtained. From this point of view, the relation between this parameter and the Planck constant in (2.12) is not clear.

The Hamiltonian of the $SU(2,2)$-harmonic oscillator may also be interpreted as the generalization of Born's quantum metric operator, which plays a crucial role in the reciprocity theory. This fact may encourage us to interpret the spectrum of the quantum Hamiltonian in the spirit of this theory.

Acknowledgments

The author is grateful to Prof. A. Jadczyk for inspiration, suggestions and helpful discussions, and he would like to thank Prof. R. Coquereaux for critical comments.

References

1. R. Coquereaux, A. Jadczyk, Conformal theories, curved phase spaces, relativistic wavelets and the geometry of complex domains, *Rev. Math. Phys.* 2:1-44 (1990)
2. A. Odzijewicz, A conformal holomorphic field theory, *Commun. Math. Phys.* 107:561-575 (1986)
3. A. Odzijewicz, On reproducing kernels and quantization of states, *Commun. Math. Phys.* 114:577-597 (1988)
4. A. Jadczyk, Born's reciprocity in the conformal domain, *in*: "Spinors, Twistors, Clifford Algebras and Quantum Deformations", Z. Oziewicz et al., eds., pp. 129-140, Kluwer, Dordrecht (1993)
5. Y.S. Kim, M.E. Noz, "Phase Space Picture of Quantum Mechanics", World Scientific, Singapore (1991)
6. F.A. Berezin, Quantization, *Math. USSR Izvestja* 8, No 5 (1974); Quantization in complex symmetric spaces, *Math. USSR Izvestja* 9, No. 2 (1975)
7. J.R. Klauder, B.S. Skagerstam, "Coherent States", World Scientific, Singapore (1985)
8. M.I. Monastyrsky, A.M. Perelomov, Coherent states and bounded homogeneous domains, *Rep. Math. Phys.* 6:1-14 (1974)
9. W. Rühl, Distributions on Minkowski space and their connection with analytic representations of the conformal group, *Commun. Math. Phys.* 27:53-86 (1972)
10. W. Mulak, Quantum SU(1,1)-harmonic oscillator, *Rep. Math. Phys.* 33:155-161 (1993).

GEOMETRO-STOCHASTIC QUANTIZATION AND QUANTUM GEOMETRY

Eduard Prugovečki

Department of Mathematics
University of Toronto
Toronto, Canada M5S 1A1

Abstract

The most basic features of the geometro-stochastic method of quantization are outlined in the nonrelativistic and special relativistic regime. Their adaptation to the general relativistic regime leads to the replacement of the classical frame bundles, which underlie the formulation of parallel transport in classical general relativity, with quantum frame bundles. This gives rise to quantum geometries for quantum field theory in curved spacetime, in which quantum frames take over the role played by complete sets of observables in conventional quantum theory. The ensuing quantum-geometric mode of propagation in general relativistic quantum bundles is implemented by path integration methods based on parallel transport along broken paths consisting of arcs of geodesics of the Levi-Civita connection, and can be further extrapolated to a geometric formulation of quantum gravity.

1. INTRODUCTION

The method of geometro-stochastic quantization was developed[1–4] originally in order to deal with still unresolved problems[5,6] in relativistic quantum particle and field localization. It eventually led,[7,8] however, to a geometric framework for quantum general relativity, capable of resolving foundational problems in quantum field theory (QFT) in curved spacetime and in quantum gravity.[9,10] In its use of phase space representations of the Galilei and the Poincaré group,[11] the geometro-stochastic method of quantization shares[12] some features with the well-known geometric method of Kostant[13] and Souriau;[14] whereas, in its use of coherent states, it displays common features (see Ref. 15, Sec. 16.2) with the method of Berezin.[16] However, in the ultimate analysis,[9,10] it transcends both these methods, since it proves applicable to the general relativistic regime. Indeed, at the physical level, geometro-stochastic quantization leads to a concept of quantum frame capable of taking over in the general relativistic regime the role played in conventional quantum theory by complete sets of observables. In QFT in curved spacetime, this mediates the formulation of quantum geometries that are Hilbert or pseudo-Hilbert fibre bundles associated with Poincaré frame bundles over curved spacetime. In turn, this enables a purely geometric formulation[17] of propagation of quantum fields in such bundles by means of path integrals that conform to the strong equivalence principle of general relativity. This kind of propagation is based on quantum connections whose connection coefficients are provided by the Levi-Civita connection form. Moreover, locally it is governed by energy density operators which arise from well-defined stress-energy tensor operators. The relation of this quantum-geometric framework for QFT to the conventional framework is arrived at by establishing the existence of action integrals on the basis of purely geometric considerations. Moreover, a consistent physical interpretation of this framework is obtained by proving the existence of Poincaré gauge invariant and locally conserved probability currents. In Sec. 2 we briefly outline those basic features of the geometro-stochastic method of quantization in the nonrelativistic regime which establish its physical consistency with orthodox quantum mechanics. We then describe in Sec. 3 the special relativistic counterparts of these features. In Sec. 4 we extrapolate this framework into a geometric framework for quantum field theory in curved spacetime, and in Sec. 5 we describe the basic features of the quantum-geometric mode of propagation within such a framework. We then indicate in the last section how these ideas can be applied to quantum gravity.

2. GEOMETRO-STOCHASTIC QUANTIZATION IN THE NONRELATIVISTIC REGIME

As is well-known, the canonical quantization of a classical Hamiltonian system carried out in Cartesian coordinates can be at odds with that in other general canonical coordinate. The fundamental physical idea underlying the geometro-stochastic method of quantization is that *it is not the choice of coordinates, but the choice of reference frames, that is of primary significance to the quantization procedure.* This suggests basing quantization on a group-theoretical foundation, in which the *operational* procedures required in the construction of inertial frames receives a natural interpretation, directly related to the group of physical operations describing changes of such frames. For the sake of simplicity, we shall concentrate on the spinless case. In the nonrelativistic regime the Galilei group plays the role of fundamental kinematical group, and the elements $(\mathbf{q}, \mathbf{p})$ of the familiar nonrelativistic phase space Γ are represented by the Cartesian coordinate triples $\mathbf{q}, \mathbf{p} \in \mathbf{R}^3$ in relation to a classical inertial frame of reference $\boldsymbol{u}$. The geometro-stochastic quantization scheme first introduced in Appendix C of Ref. 3 requires that representations of the canonical commutation relations with respect to $\boldsymbol{u}$ are to be derived from the irreducible subrepresentation of the unitary ray representation of the Galilei group in $L^2(\Gamma)$. If we denote by $(b, \mathbf{a}, \mathbf{v}, R)$ the generic element of that group, where b indicates a time translation, $\mathbf{a}$ a space translation, $\mathbf{v}$ a velocity boost, and R a rotation, then such a representation is obtained by assigning to it the unitary operator

$$(U(b, \mathbf{a}, \mathbf{v}, R)\psi)(\mathbf{q}, \mathbf{p}, t) = \\ e^{i\left(-\frac{1}{2}m\mathbf{v}^2(t-b)+m\mathbf{v}\cdot(\mathbf{q}-\mathbf{a})\right)}\,\psi(R^{-1}[\mathbf{q}-\mathbf{a}-\mathbf{v}(t-b)], R^{-1}[\mathbf{p}-m\mathbf{v}], t-b), \quad (2.1)$$

acting on wave functions $\psi(\mathbf{q}, \mathbf{p}, t)$ which are, for each fixed value of $\mathbf{p} \in \mathbf{R}^3$, solutions of the free Schrödinger equation in $\mathbf{q}$ and t for the mass m, expressed in Planck natural units.

The harmonic analysis[11] of the above representation reveals an infinity of irreducible subrepresentations, all of which are unitarily equivalent to the well-known irreducible ray representations of the Galilei group in configuration or momentum space. Each such irreducible phase space subrepresentation can be uniquely characterized by means of a unique resolution generator $\xi \in L^2(\Gamma)$ that gives rise, in conjunction with the elements $U(b, \mathbf{a}, \mathbf{v}, R)$ of that representation, to the generalized coherent states

$$\xi_{\mathbf{q},\mathbf{p}} = U(0, \mathbf{q}, \mathbf{p}/m, I)\xi \in L^2(\Gamma)\,, \qquad (\mathbf{q}, \mathbf{p}) \in \Gamma, \tag{2.2}$$

which, in turn, supply the orthogonal projector

$$\mathbf{P}_\xi = \int_{\mathbf{R}^6} |\xi_{\mathbf{q},\mathbf{p}}\rangle \; d\mathbf{q}d\mathbf{p} \; \langle \xi_{\mathbf{q},\mathbf{p}}|, \tag{2.3}$$

of $L^2(\Gamma)$ onto the subspace which carries that subrepresentation. The family of generalized coherent states in (2.2) is therefore uniquely assigned to the global classical frame of reference $\boldsymbol{u}$, and can be envisaged[4,9] as constituting a quantum frame, operationally obtained by subjecting identical duplicates of a quantum test body to the basic kinematical procedures of spacetime translation, spatial rotation and velocity boost, in order to obtain an array of kinematically correlated microdetectors that can be then used for the spatio-temporal localization of quantum systems. The transition amplitudes

$$\psi(\mathbf{q}, \mathbf{p}) = \langle U(0, \mathbf{q}, \mathbf{p}/m, I)\xi \mid \psi\rangle\,, \qquad \psi \in \mathbf{P}_\xi L^2(\Gamma), \tag{2.4}$$

between the normalized quantum state vector ψ of a quantum system and the constituents of such a quantum frame are directly related[9] to the purely geometric concept of Fubini-Study distance between their representative rays in the projective space of unit Hilbert rays corresponding to elements in that subspace. The square of the absolute value in (2.4) can be interpreted as the probability density for detection in relation to these quantum frame elements. Indeed, there is[2,3,11] a unitary map which assigns to each time-dependent wave function $\psi(\mathbf{q}, \mathbf{p}, t)$ in (2.1) a unique solution $\psi(\mathbf{x}, t)$ of the free Schrödinger equation representing a wave function in configuration space. It can be then easily proved[4] that the probability density derivable from this phase-space wave function is related to the probability density derivable from the corresponding configuration space wave function as follows:

$$\rho(\mathbf{q}, t) = \int_{\mathbf{R}^3} |\psi(\mathbf{q}, \mathbf{p}, t)|^2 d\mathbf{p} = (2\pi)^3 \int_{\mathbf{R}^3} |\psi(\mathbf{x}, t)|^2 \, |\xi(\mathbf{x} - \mathbf{q})|^2 \, d\mathbf{x} \tag{2.5}$$

Hence, in the sharp-point limit

$$(2\pi)^3 \, |\xi(\mathbf{x} - \mathbf{q})|^2 \quad \longrightarrow \quad \delta^3(\mathbf{x} - \mathbf{q}) \tag{2.6}$$

physically corresponding to the limiting choice of *perfectly* pointlike quantum test bodies, the probability density in (2.5) converges to the probability density for infinitely precise position measurement outcomes originally postulated by Born,

$$\rho(\mathbf{q}, t) \quad \longrightarrow \quad |\psi(\mathbf{q}, t)|^2 \tag{2.7}$$

at all those points at which this configuration space wave function is continuous.

Of special importance is the case of resolution generators which in the configuration space representation of the Galilei group are supplied by the following ground-state wave function,

$$\xi^{(\ell)}(\mathbf{x}) = (8\pi^3\ell^2)^{-3/4} \exp(-\mathbf{x}^2/4\ell^2)\,, \qquad \ell > 0 \tag{2.8}$$

of a nonrelativistic harmonic oscillator. Indeed, the special relativistic counterparts of such a resolution generator is the ground states of a corresponding relativistic harmonic oscillator, which can be interpreted[4] as the ground state of the quantum metric operator introduced by Born,[5] in which ℓ plays the role of fundamental length. The ensuing probability current,

$$\mathbf{j}_\ell(\mathbf{q},t) = \int_{\mathbf{R}^3} \frac{\mathbf{p}}{m} \left|\psi_\ell(\mathbf{q},\mathbf{p},t)\right|^2 d\mathbf{p}, \tag{2.9}$$

constructed[1] in complete analogy with its counterpart in classical statistical mechanics, is Galilei-covariant and conserved. Moreover, in the case of a configuration space wave functions $\psi(\mathbf{x},t)$ with continuous first partial derivatives in the $\mathbf{x}$-variable, this new quantum probability current converges in the sharp-point limit to the conventional probability current in configuration space,

$$\mathbf{j}_\ell(\mathbf{q},t) \underset{\ell\to+0}{\longrightarrow} (2im)^{-1}\psi^*(\mathbf{q},t)\overleftrightarrow{\nabla}\,\psi(\mathbf{q},t), \tag{2.10}$$

thus further substantiating the earlier described physical interpretation of the transition amplitude in (2.4).

On account of (2.3), the following family of transition probability amplitudes for free quantum propagation,

$$K_\ell(\mathbf{q}'',\mathbf{p}'',t'';\mathbf{q}',\mathbf{p}',t') = \left\langle U(t'',\mathbf{q}'',\mathbf{p}''/m\,,I)\xi^{(\ell)}\middle|U(t',\mathbf{q}',\mathbf{p}'/m\,,I)\xi^{(\ell)}\right\rangle, \tag{2.11}$$

displays the basic properties of a free propagator:

$$K_\ell(\mathbf{q}'',\mathbf{p}'',t'';\mathbf{q}',\mathbf{p}',t') = \int_{\mathbf{R}^6} K_\ell(\mathbf{q}'',\mathbf{p}'',t'';\mathbf{q},\mathbf{p},t)K_\ell(\mathbf{q},\mathbf{p},t;\mathbf{q}',\mathbf{p}',t')\,d\mathbf{q}d\mathbf{p}. \tag{2.12}$$

This phase-space free propagator can be used[4,18] in the formulation of path integrals for quantum propagators in the presence of interactions. Upon suitable renormalization, these phase space free propagators converge in the sharp-point limit to the well-known free Feynman propagator[19] for nonrelativistic quantum point particles:

$$(\pi/2\ell^2)^{3/2}K_\ell(\mathbf{q}'',\mathbf{p}'',t'';\mathbf{q}',\mathbf{p}',t') \underset{\ell\to+0}{\longrightarrow} K(\mathbf{q}'',t'';\mathbf{q},t). \tag{2.13}$$

However, the above required renormalization constant obviously diverges in this sharp-point limit. Closer scrutiny reveals[4,18] that the basic formulae,

$$K_\ell(\mathbf{q}'',\mathbf{p}'',t'';\mathbf{q}',\mathbf{p}',t') = \lim_{\varepsilon\to+0}\int K_\ell(\mathbf{q}_N,\mathbf{p}_N,t_N;\mathbf{q}_{N-1},\mathbf{p}_{N-1},t_{N-1})$$
$$\times\prod_{n=1}^{N-1} K_\ell(\mathbf{q}_n,\mathbf{p}_n,t_n;\mathbf{q}_{n-1},\mathbf{p}_{n-1},t_{n-1})\,d\mathbf{q}_n d\mathbf{p}_n\,, \qquad \varepsilon=(t''-t)/N, \tag{2.14}$$

on which path integration in general relies, are mathematically well-defined in case of the phase space propagators in (2.11). On the other hand, in the case of the Feynman propagators, which rely on pointlike localization in configuration space, the same type of formulae,[19]

$$K(\mathbf{x}'',t'';\mathbf{x},t) = \lim_{\varepsilon\to+0}\int K(\mathbf{x}(t_N);\mathbf{x}(t_{N-1}))\prod_{n=1}^{N-1}K(\mathbf{x}(t_n);\mathbf{x}(t_{n-1}))\,d\mathbf{x}(t_n), \tag{2.15}$$

are merely formal, since the above Lebesgue integrals exist if and only if their integrand are integrable in the absolute sense – which is not the case in (2.15). Hence, the mathematically rigorous treatment of Feynman path integrals require analytic continuations to imaginary time, leading to the well-known Feynman-Kac formula; whereas those based on (2.14) are mathematically well-defined in real time. This constitutes the basis of the usefulness of their special relativistic counterparts, described in the next section, and presents decided advantages in the general relativistic regime, which will be our main concern in the remaining three sections.

3. GEOMETRO-STOCHASTIC QUANTIZATION IN THE SPECIAL RELATIVISTIC REGIME

In the special relativistic regime the role played in geometro-stochastic quantization by the Galilei group is taken over by the Poincaré group. Similarly, that of nonrelativistic phase space is taken over by the conventional relativistic phase space, whose elements are labelled by the 4-vectors q and $p = mv$, with p restricted to the forward and backward mass hyperboloids in the case of particles and antiparticles, respectively. Hence, in the case of particles of rest mass m and zero spin, the Hilbert space $L^2(\Gamma)$ is replaced with the Hilbert space $L^2(\Sigma)$ of wave functions $\varphi(q, v)$ which are solutions in q of the free Klein-Gordon equation for rest mass m at each fixed value of v on the forward 4-velocity hyperboloid $\boldsymbol{V}^+$, and for which the inner product is given by the following integral over a hypersurface $\Sigma = \sigma \times \boldsymbol{V}^+$ in relativistic phase space

$$\langle\varphi|\varphi'\rangle = \int_\Sigma \varphi^*(q,v)\varphi'(q,v)\, d\Sigma(q,v)\,, \qquad d\Sigma(q,v) = 2v_\mu\delta(v^2-1)d\sigma^\mu(q)d^4v, \quad (3.1)$$

where σ denotes a maximal spacelike hypersurface in Minkowski space. The above integration is carried out with respect to the unique, modulo a multiplicative constant, Lorentz covariant measure over Σ, so that the measure element $m^3 d\Sigma$ reduces to $d\mathbf{q}d\mathbf{p}$ along any hypersurface Σ for which σ corresponds to a constant value of q^0 in the chosen Lorentz frame of reference $\boldsymbol{u}$.

The nonrelativistic representations in (2.1) is now replaced by a representation which assigns to the generic element (a, Λ) of the Poincaré group, represented by a spacetime translation a and Lorentz transformation Λ, the following operator,

$$(U(a,\Lambda)\varphi)(q,v) = \varphi(\Lambda^{-1}(q-a), \Lambda^{-1}v). \quad (3.2)$$

Just as in the nonrelativistic case, the harmonic analysis[11] of this representation reveals a host of irreducible subrepresentations that are unitarily equivalent to well-known momentum-space representations. And, again, each such subrepresentation is uniquely characterized by a resolution generator η providing the orthogonal projector of $L^2(\Sigma)$ onto the subspace which carries it, since it gives rise to the generalized coherent states

$$\eta_\zeta = U(q, \Lambda_v)\eta \in L^2(\Sigma)\,, \qquad \zeta = (q, v), \quad (3.3)$$

where Λ_v denotes the Lorentz boost to the 4-velocity v, and that projector can be expressed in the form

$$\mathbf{P}_\eta = \int_\Sigma |\eta_\zeta\rangle\, d\Sigma(\zeta)\, \langle\eta_\zeta|\,. \quad (3.4)$$

Hence, by an extrapolation of the interpretation of (2.4) in the nonrelativistic case,

$$\varphi(q,v) = \langle U(q,\Lambda_v)\eta \mid \varphi\rangle\,, \qquad \varphi \in \mathbf{P}_\eta L^2(\Sigma), \quad (3.5)$$

can be interpreted as a probability amplitude of detection of a quantum particle in a state described by the normalized vector φ in relation to the constituents of a quantum a special relativistic quantum frame associated to the Lorentz frame of reference $\boldsymbol{u}$.

We note that the above probability amplitude depends on the mean 4-velocity v of each micro-detector in the quantum frame relative to the corresponding classical frame. This is in complete accord with on of Born's key observations on an underlying reciprocity between spacetime and 4-momentum relationships in nature, succinctly presented in the following quotation: "Ordinary relativity is based on the invariance of the 4-dimensional distance, or its square $R = x_k x^k$. Can one really define the distance of two particles in sub-atomic dimensions independently of their velocity? This seems to me not evident at all." (Ref. 5, p. 208)

Born's idea of a quantum metric operator $D^2 = Q^2 + P^2$, which incorporates a fundamental length ℓ that limits spatio-temporal localizability, leads (cf. Ref. 4, Sec. 4.5) to the *fundamental* quantum frames whose constituents are described by the generalized coherent states in (3.3) represented by the wave functions

$$\boldsymbol{\Phi}^u_{\ell,m;\zeta}(q,v) = \tilde{Z}^{-2}_{\ell,m} \int_{k^0>0} \exp\{[i(q-q') - \ell(v'+v)]\cdot k\}\ \delta(k^2 - m^2)\, d^4k. \tag{3.6}$$

The normalization constant in (3.6) can be expressed in terms of the modified Bessel function K_2,

$$\tilde{Z}_{\ell,m} = 8\pi^4 \mathrm{K}_2(2\ell m)/\ell m^2 = (4\pi^4/\ell^3 m^4) + O(\ell^{-1}) \underset{\ell\to+0}{\longrightarrow} +\infty, \tag{3.7}$$

and it diverges in the sharp-point limit $\ell \to 0$. However, that value cannot be arbitrarily adjusted, since it is fixed by the fact that the operator in (3.4) is an orthogonal projection operator, so that

$$\mathbf{P}^u_{\ell,m} = \int \left|\boldsymbol{\Phi}^u_{\ell,m;\zeta}\right\rangle d\Sigma(\zeta) \left\langle \boldsymbol{\Phi}^u_{\ell,m;\zeta}\right| = \left(\mathbf{P}^u_{\ell,m}\right)^* = \left(\mathbf{P}^u_{\ell,m}\right)^2. \tag{3.8}$$

As a consequence, no counterparts of (2.7) and (2.10) exist. This is in agreement with the fact that, as pointed out by Wigner in the context of conventional relativistic quantum theory, "every attempt to provide a precise definition of a position coordinate [for a quantum point particle] stands in direct contradiction with special relativity" (Ref. 6, p. 313).

Despite this fact, a consistent special relativistic quantum theory of spacetime localization is feasible if a *fixed* value $\ell > 0$ is adopted as a fundamental length, since then the following current, constructed in analogy to the one in (2.9),

$$j^\mu_\ell(q) = 2\int_{v^0>0} v^\mu \left|\varphi_\ell(q,v)\right|^2 \delta(v^2-1)\, d^4v\,, \qquad \varphi_\ell(q,v) = \left\langle \Phi^u_{\ell,m;q,v} \,\middle|\, \varphi\right\rangle, \tag{3.9}$$

is Poincaré-covariant as well as conserved, and it has a positive-definite timelike component, so that it provides a *bona fide* relativistic *probability* current. This does not mean that the usual treatment of the Klein-Gordon equation has to be altogether abandoned in the present framework, since the following counterpart,

$$J^\mu_\ell(q) = i\hat{Z}_{\ell,m} \int_{v^0>0} \varphi^*_\ell(q,v)\overleftrightarrow{\partial}^\mu \varphi_\ell(q,v)\, \delta(v^2-1)\, d^4v\,, \qquad \partial_\mu = \partial/\partial q^\mu, \tag{3.10}$$

of the familiar Klein-Gordon current also exists, and it is Poincaré-covariant as well as conserved. However, as it is the case in the conventional treatment of the Klein-Gordon equation, this latter current is *not* a probability current, since its timelike component is *not* positive definite even for wave functions corresponding to positive energies. Hence, as in the conventional approach, its use is restricted to the formulation of quantum field theoretical couplings of Klein-Gordon fields to other fields in the quantum-geometric field theory discussed in subsequent sections.

The normalization constant in (3.10) is uniquely determined by the following alternative form,

$$\langle \varphi_\ell | \varphi'_\ell \rangle = i\hat{Z}_{\ell,m} \int_{v^0>0} \varphi^*_\ell(q,v)\overleftrightarrow{\partial}_\mu \varphi'_\ell(q,v)\, d\sigma^\mu(q)\, \delta(v^2-1)\, d^4v, \tag{3.11}$$

of the restriction of the inner product in (3.1) to subspace of $L^2(\Sigma)$ onto which the projector in (3.8) projects. Its value is finite for all $\ell > 0$, but it diverges in the sharp-point limit:

$$\hat{Z}_{\ell,m} = \mathrm{K}_2(2\ell m)/m\mathrm{K}_1(2\ell m) = (1/\ell m^2) + O(1) \underset{\ell\to+0}{\longrightarrow} +\infty. \tag{3.12}$$

As shown in Refs. 9,10, the divergence in the sharp-point limit $\ell \to 0$ of the two normalization constants in (3.7) and (3.12) provides an explanation of the root causes of the divergences that plague the conventional formulation of QFT. In essence, this explanation is that, as maintained already in the 1930s by Bohr, Born, Dirac, Heisenberg, and other founders of quantum mechanics, there exists a fundamental length in nature. Hence, the appearance of divergences in conventional QFT is simply a manifestation of that fact.

In the present context, as a consequence of (3.8), we have

$$K_{\ell,m}(q'',v'';q',v') = \left\langle \boldsymbol{\Phi}^u_{q'',v''} \middle| \boldsymbol{\Phi}^u_{q',v'} \right\rangle = \int K_{\ell,m}(q'',v'';q,v)K_{\ell,m}(q,v;q',v')\, d\Sigma(q,v). \tag{3.13}$$

Hence, as was the case in (2.11) and (2.12), the above transition probability amplitudes display the basic properties of a propagator, and can be used[4,18] in the formulation of path integrals. On the other hand, in the sharp-point limit $\ell \to 0$ we can carry out their renormalization, so as to obtain,

$$2(2\pi)^{-3}\tilde{Z}^2_{\ell,m}K_{\ell,m}(q'',p'';q',p') \underset{\ell\to+0}{\longrightarrow} K_F(q''-q') \;, \qquad q''^0 > q'^0, \tag{3.14}$$

i.e., so that the forward time-ordered phase space propagator specified by (3.13) converges in that limit to the corresponding forward part of the time-ordered Feynman propagator for Klein-Gordon particles, given by

$$K_F(x''-x') = 2(2\pi)^{-3}\int_{k^0>0} \exp[ik\cdot(x'-x'')]\;\delta(k^2-m^2)\,d^4k \;, \qquad x''^0 > x'^0. \tag{3.15}$$

This basic feature contributes to the *formal* term-by-term convergence of formal perturbative expansions of transition amplitudes for collision processes in quantum-geometric field theory to their conventional counterparts for models involving Klein-Gordon quantum fields. Such term-by-term convergence can be achieved after other factors that diverge in the sharp-point limit have also been discarded namely those divergent factors that appear in the formal sharp-point limit due to the independence of free Feynman propagators on the mean 4-velocity variable v.[9,10,17]

Totally analogous comparisons between the quantum-geometric and the conventional approach to QFT can be drawn[9,10] in the case of the Dirac equation, quantum electromagnetic fields, etc. However, in the presence of gravity, the retention of a fixed value $\ell > 0$ for the fundamental length becomes mandatory, since basic measurement-theoretical considerations[20,21] show that Planck's length imposes a lower bound on ℓ.

4. QUANTUM GEOMETRY IN CURVED SPACETIME

The basic physical idea in extending the geometro-stochastic method of quantization to the semiclassical general relativistic regime, in which a Lorentzian manifold $(\mathbf{M}, \boldsymbol{g}^{\mathrm{L}})$ is assumed to be *a priori* given, is based on the observation that, due to the strong equivalence principle, in classical general relativity (CGR) the role played by the Minkowski space $M^4 \cong (\mathbf{R}^4, \boldsymbol{\eta})$ is taken over by the fibres $T_x\mathbf{M}$ of the tangent bundle $T\mathbf{M}$, whose typical fibre is $\mathbf{R}^4$; moreover, the role of the global Lorentz frames $\mathcal{L} = \{\boldsymbol{e}_i(x_0)|i = 0, 1, 2, 3\}$ with their origins at points $x_0 \in M^4$, is taken over by the local Lorentz frames within the Lorentz frame bundle $L\mathbf{M}(\boldsymbol{g}^{\mathrm{L}})$, which has the Lorentz group SO(3,1) as its structure group. Hence, all the tensor fields that appear in CGR represent sections of various tensor bundles associated with $L\mathbf{M}(\boldsymbol{g}^{\mathrm{L}})$. The resulting Lorentz gauge invariance is underlined by the fact that the Levi-Civita connection employed in CGR is compatible with the metric $\boldsymbol{g}^{\mathrm{L}}$.

Mathematically, this Lorentz gauge invariance can be always extended into Poincaré gauge invariance, based on the Poincaré frame bundle $P\mathbf{M}(\boldsymbol{g}^{\mathrm{L}})$, that has the Poincaré group ISO(3,1) as its structure group. The Levi-Civita connection on $L\mathbf{M}(\boldsymbol{g}^{\mathrm{L}})$ is then extended into a connection on $P\mathbf{M}(\boldsymbol{g}^{\mathrm{L}})$, so that the operator forms ∇ for covariant differentiation acquire four additional terms containing into the connection 1-forms[22]

$$\tilde{\boldsymbol{\theta}}^i = \boldsymbol{\theta}^i + (\nabla \boldsymbol{a})^i, \qquad \boldsymbol{s} : x \mapsto (\boldsymbol{a}, \boldsymbol{e}_i), \qquad \boldsymbol{a} = a^i \boldsymbol{e}_i, \tag{4.1}$$

for any section $\boldsymbol{s}$ of $P\mathbf{M}(\boldsymbol{g}^{\mathrm{L}})$ – where the canonical (i.e., soldering) forms $\boldsymbol{\theta}^i$ are provided by the coframes $\{\boldsymbol{\theta}^i|i = 0, 1, 2, 3\}$ dual to the Lorentz frames $\{\boldsymbol{e}_i|i = 0, 1, 2, 3\}$ in $\boldsymbol{s}$, and each Poincaré frame $\{(\boldsymbol{a}, \boldsymbol{e}_i)|i = 0, 1, 2, 3\}$ is obtained by translating the corresponding local Lorentz frame by the 4-vector $\boldsymbol{a}$ within the tangent space within which that frame lies. However, such an extension is not necessary for CGR itself. On the other hand, it is essential from the quantum point of view, on account of the basic fact that the kinematics and dynamics incorporated into such equations of motion as those of Klein-Gordon, Dirac, etc., which govern quantum propagation, gives rise to generators of infinitesimal spacetime translations for unitary representations of the Poincaré group.

Hence, in order to describe quantum-geometrically the propagation of massive spin-0 quantum fields, we construct the Fock-Klein-Gordon bundle

$$\mathcal{E}(\mathbf{M}, \boldsymbol{g}^{\mathrm{L}}) = P\mathbf{M}(\boldsymbol{g}^{\mathrm{L}}) \times_{\mathbf{G}} \mathcal{F}, \qquad \mathbf{G} = \mathrm{ISO}(3,1), \tag{4.2}$$

associated with the principal bundle $P\mathbf{M}(\boldsymbol{g}^{\mathrm{L}})$. The typical fibre of this quantum bundle is the Fock space

$$\mathcal{F} = \oplus_{n=0}^{\infty} \mathcal{F}_n, \qquad \mathcal{F}_n = \mathbf{F} \underset{S}{\otimes} \cdots \underset{S}{\otimes} \mathbf{F}, \qquad \mathbf{F} = \mathbf{P}^{u_0}_{\ell,m} L^2(\Sigma), \tag{4.3}$$

obtained from symmetrized tensor products of the Hilbert space determined for the standard frame $\boldsymbol{u}_0$ in $(\mathbf{R}^4, \boldsymbol{\eta})$ by (3.5) and (3.8). Hence, the representation provided by (3.2) gives rise within this Fock space to the representation

$$\boldsymbol{U}(a, \Lambda) = \bigoplus_{n=0}^{\infty} U(a, \Lambda)^{\otimes n}, \qquad (a, \Lambda) \in \mathrm{ISO}(3,1), \tag{4.4}$$

which can be used in the standard manner in the construction of the bundle $\mathbf{G}$-product according in (4.2), so that the Poincaré group becomes the structure group of $\mathcal{E}(\mathbf{M}, \boldsymbol{g}^{\mathrm{L}})$. We observe that the fibre $\mathcal{F}_x$ over $x \in \mathbf{M}$ contains the vacuum subfibre $\mathcal{F}_{0;x}$ spanned by a *local* Fock vacuum state vector $\boldsymbol{\Psi}_{0;x}$, which is left invariant by the counterpart of (4.4) in $\mathcal{F}_x$.

The operational description of measurements of spatio-temporal separations of events in a neighborhood of some base point $x \in \mathbf{M}$ where the relative curvature effects are small, so that classically they can be described exclusively in terms of Riemann normal coordinates for which $dx_k dx^k = \eta_{ik} dx^i dx^k$ at $x \in \mathbf{M}$, can be now supplemented,[10] in accordance with Born's[5] epistemic ideas, by fundamental uncertainties described by the local quantum fluctuation amplitudes

$$\Delta_x^{(+)}(\boldsymbol{\zeta}';\boldsymbol{\zeta}) = -i\ \boldsymbol{\Phi}_{\ell,m;\zeta}^{u(x)}(\zeta'),\ \boldsymbol{\zeta} = (\boldsymbol{a}+q^i\boldsymbol{e}_i, v^i\boldsymbol{e}_i) \in T_x\mathbf{M}\times \boldsymbol{V}_x^+\ , \quad \zeta = q - i\ell v \in \mathbf{C}^4. \tag{4.5}$$

According to the interpretation of (3.5), these fluctuations result from the transition probability amplitudes of the geometrically local quantum frames associated by the construction in (4.2) to each Poincaré frame $\boldsymbol{u}(x) = (\boldsymbol{a}(x), \boldsymbol{e}_i(x))$, that can be used to define those Riemann normal coordinates by means of the exponential map at x.

On account of (3.8) and (3.11), we have

$$2\int \Delta_x^{(+)}(\boldsymbol{\zeta}';\boldsymbol{\zeta})\,\Delta_x^{(+)}(\boldsymbol{\zeta};\boldsymbol{\zeta}'')\,\delta(v^2-1)\,v_k\,d\sigma^k(q)d^4v \;=\; -i\Delta_x^{(+)}(\boldsymbol{\zeta}';\boldsymbol{\zeta}''), \tag{4.6}$$

so that the 2-point functions in (4.5) can play also the role of local propagators. Hence, with their help we can define annihilation operators by the following action upon n-exciton state vectors $\boldsymbol{\Psi}_{n;x} \in \mathcal{F}_{n;x}$:

$$\left(\varphi^{(-)}(x;\boldsymbol{\zeta})\boldsymbol{\Psi}_{n;x}\right)_{n-1}(\boldsymbol{\zeta}_1,\ldots,\boldsymbol{\zeta}_{n-1}) = in^{1/2}\int \Delta_x^{(+)}(\boldsymbol{\zeta};\boldsymbol{\zeta}_n)\boldsymbol{\Psi}_{n;x}(\boldsymbol{\zeta}_1,\ldots,\boldsymbol{\zeta}_n)\,d\Sigma(\boldsymbol{\zeta}_n). \tag{4.7}$$

In conventional QFT the counterparts of these operators, and of their adjoints that represent creation operators, have to be smeared with test functions, since in that context such operators can be proven to exist only in the sense of operator-valued distributions (Ref.23, Sec. 10.4). On the other hand, the non-singularity of the 2-point functions in (4.5), which is a consequence of the presence of the fundamental length ℓ, makes the annihilation operators in (4.7), as well as of their adjoints $\varphi^{(+)}(x;\boldsymbol{\zeta})$ representing creation operators, mathematically well-defined without any need for "smearing" with test functions. Hence, the Klein-Gordon quantum frame field

$$\varphi(x;\boldsymbol{\zeta}) = \varphi^{(+)}(x;\boldsymbol{\zeta}) + \varphi^{(-)}(x;\boldsymbol{\zeta})\ , \qquad \boldsymbol{\zeta}\in T_x\mathbf{M}\times\boldsymbol{V}_x^+\ , \qquad x\in\mathbf{M}, \tag{4.8}$$

is well-defined as it stands. As a consequence it can be proved[8−10] that when the Poincaré group acts upon the elements $\boldsymbol{s}(x)$ of a section $\boldsymbol{s}$ of $P\mathbf{M}(\boldsymbol{g}^{\mathrm{L}})$, the infinitesimal generators of the counterparts

$$\boldsymbol{U}_{s(x)}(a,\Lambda) = \bigoplus_{n=0}^{\infty} U_{s(x)}(a,\Lambda)^{\otimes n}\ , \qquad\qquad \boldsymbol{s}(x)\in P\mathbf{M}(\boldsymbol{g}^{\mathrm{L}}), \tag{4.9}$$

of the representation in (4.4) are given by the Bochner integrals

$$\boldsymbol{P}_{j;s(x)} = \int :T_{jk}[\varphi(x;\boldsymbol{\zeta})]:\ d\sigma^k(q)\,\delta(v^2-1)\,d^4v\ , \qquad \boldsymbol{Q}^j_{s(x)} = q^j - (i/m)\partial/\partial v_j, \tag{4.10}$$

$$\boldsymbol{M}^{ij}_{s(x)} = \int :\boldsymbol{Q}^i_{s(x)}T^{jk}[\varphi(x;\boldsymbol{\zeta})] - \boldsymbol{Q}^j_{s(x)}T^{ik}[\varphi(x;\boldsymbol{\zeta})]:\ d\sigma_k(q)\,\delta(v^2-1)\,d^4v, \tag{4.11}$$

whose integrands are *bona fide* operator-valued functions, given by the normally-ordered values of the following *bona fide* operator-valued stress-energy tensor:

$$T_{jk}[\varphi] = \hat{Z}_{\ell,m}\left(\varphi_{,j}\varphi_{,k} + \tfrac{1}{2}\eta_{jk}(m^2\varphi^2 - \eta^{il}\varphi_{,i}\varphi_{,l})\right)\ , \qquad \varphi_{,j} = \partial\varphi/\partial q^j. \tag{4.12}$$

However, as can be seen from (3.12), divergencies manifest themselves in the sharp-point limit $\ell\to+0$.

In each Fock fibre $\mathcal{F}_x$ we can introduce the Glauber-type coherent states

$$\boldsymbol{\Phi}_{\mathbf{f}} = \exp\left[-\tfrac{1}{2}\langle \mathbf{f} \mid \mathbf{f}\rangle + \varphi^{(+)}(\mathbf{f})\right] \boldsymbol{\Psi}_{0;x}\,, \quad \varphi^{(+)}(\mathbf{f}) = \int \varphi^{(+)}(x;\boldsymbol{\zeta})\,\mathbf{f}(\boldsymbol{\zeta})\,d\Sigma(\boldsymbol{\zeta})\,, \quad \mathbf{f} \in \mathcal{F}_{1;x}, \tag{4.13}$$

defined by the strongly convergent power series for the above exponential. These coherent states constitute a family of second-quantized frames, in the sense that they give rise to the following continuous resolution of the identity operator $\mathbf{1}_x$ in each fibre $\mathcal{F}_x$,

$$\int_{\mathbf{F}_x} |\boldsymbol{\Phi}_{\mathbf{f}}\rangle\; d\mathbf{f}\, d\mathbf{f}^*\, \langle \boldsymbol{\Phi}_{\mathbf{f}}| \;=\; \mathbf{1}_x\,, \qquad \mathbf{F}_x = \mathcal{F}_{1;x}, \tag{4.14}$$

where the above functional integral is defined by the method of Berezin.[24] These continuous resolutions of the identity are essential for the derivation of action integrals for quantum-geometric propagation (cf. Sec. 5), together with the fact that the elements of these second-quantized frames are *bona fide* eigenvectors of the annihilation operators in (4.7):

$$\varphi^{(-)}(x;\boldsymbol{\zeta})\boldsymbol{\Phi}_{\mathbf{f}} = \mathbf{f}(\boldsymbol{\zeta})\,\boldsymbol{\Phi}_{\mathbf{f}}\;, \qquad \mathbf{f} \in \mathbf{F}_x. \tag{4.15}$$

This quantum-geometric propagation emerges from the fact that the quantum bundle $\mathcal{E}(\mathbf{M},\boldsymbol{g}^{\mathrm{L}})$ is associated to $P\mathbf{M}(\boldsymbol{g}^{\mathrm{L}})$, and consequently each connection on $P\mathbf{M}(\boldsymbol{g}^{\mathrm{L}})$ gives rise to parallel transport in $\mathcal{E}(\mathbf{M},\boldsymbol{g}^{\mathrm{L}})$. In particular, this is the case with the earlier discussed extension to $P\mathbf{M}(\boldsymbol{g}^{\mathrm{L}})$ of the Levi-Civita connection on $L\mathbf{M}(\boldsymbol{g}^{\mathrm{L}})$. Therefore, for any choice of Poincaré gauge given by a section $\boldsymbol{s}$ of $P\mathbf{M}(\boldsymbol{g}^{\mathrm{L}})$, the corresponding parallel transport determines the quantum connection

$$\boldsymbol{\nabla} = \boldsymbol{d} + i\tilde{\boldsymbol{\theta}}^i \boldsymbol{P}_{i;s} + \tfrac{i}{2}\tilde{\omega}_{jk}\boldsymbol{M}_s^{jk}\;, \qquad \boldsymbol{d} = \boldsymbol{\theta}^i \boldsymbol{\partial}_{e_i}, \tag{4.16}$$

whose connection 1-forms are those derived from that extension, and whose infinitesimal generators for spacetime translations and Lorentz transformations are the in (4.10) and (4.11). Thus, Poincaré gauge covariance is embedded into the parallel transport resulting from this above quantum connection.

5. QUANTUM-GEOMETRIC PROPAGATION

In conventional QFT[16] in a stationary but nonstatic classical spacetime represented by a Lorentzian manifold $(\mathbf{M},\boldsymbol{g}^{\mathrm{L}})$, free-fall quantum field propagation gives rise to *ex nihilo* particle production. This violates local energy-momentum conservation as well as the strong equivalence principle of general relativity. Indeed, according to that principle all observers in free fall are *inertial*, and therefore the situation should for them locally appear the same as for inertial observers in Minkowski space, where such *ex nihilo* particle production is not in evidence.

This difficulty can be overcome by adapting to fields defined on quantum bundles the original form of the path-integration method advanced by Feynman. In that original form[25,26] the path-integral depiction of quantum propagation was formulated in spacetime, rather than in the nowadays more popular momentum representation. Its two central idea were that such propagation takes place over broken polygonal paths in nonrelativistic or relativistic flat spacetimes, and that all the observed probability transition amplitudes result from the superposition of *spacetime* propagators over such paths.

The quantum-geometric adaptation[7–10] of these ideas to a curved spacetime represented by a Lorentzian manifold $(\mathbf{M},\boldsymbol{g}^{\mathrm{L}})$ is based on replacing the straight lines of such broken polygonal paths with the arcs of geodesics of the Lorentzian metric $\boldsymbol{g}^{\mathrm{L}}$, and on deriving spacetime propagators by extending the strong equivalence principle from the classical to the quantum regime. The latter feature entails that free-fall quantum-geometric propagation is governed by the Levi-Civita connection determined by $\boldsymbol{g}^{\mathrm{L}}$, and that *ex nihilo* particle production is thereby avoided. Moreover, a central role in such quantum-geometric evolution is played by the *local* proper energy density and 3-momentum of the nongravitational sources associated, as is the case in classical canonical gravity, with a geometrodynamic evolution of spacetime that is mathematically equivalent to a foliation of Lorentzian manifold $(\mathbf{M}, \boldsymbol{g}^{\mathrm{L}})$ into a family of reference hypersurfaces Σ_t with unit future-pointing normals $\boldsymbol{n}(x) \in T_x\mathbf{M}$.

The presence of the fundamental length ℓ makes the definition of such geometrically *local* quantities not only mathematically consistent, but also in keeping with the uncertainty principle, since the envisaged spacetime localization is not infinitely precise, but exhibits stochastic fluctuations described by the quantum fluctuations amplitudes in (4.5). Hence, let us introduce the quantum-geometric field

$$\boldsymbol{\varphi}(x, \boldsymbol{v}) := \varphi(x; (\mathbf{0}, \boldsymbol{v})) \ , \qquad \boldsymbol{v} = v^i \boldsymbol{e}_i(x) \in \boldsymbol{V}_x^+, \tag{5.1}$$

whose values are actually independent of the chosen Poincaré frame $(\boldsymbol{a}(x), \boldsymbol{e}_i(x)) \in P\mathbf{M}(\boldsymbol{g}^{\mathrm{L}})$, and let us consider for any section $\boldsymbol{s} = \{\boldsymbol{a}(x), \boldsymbol{e}_i(x) | x \in \mathbf{M}\}$ the local energy-momentum density operators

$$\hat{\boldsymbol{P}}_{j;s}(x) = n^k(x) \int_{v^o>0} : T_{jk}[\boldsymbol{\varphi}(x; \boldsymbol{v})] : \ \delta(v^2 - 1)\, d^4 v, \tag{5.2}$$

where $n^k(x), k = 0, 1, 2, 3$, are the components of the normal $\boldsymbol{n}(x)$ with respect to the local Lorentz frame $\{\boldsymbol{e}_i(x)\}, x \in \Sigma_t$, and the stress-energy operator is the one that appears in (4.12). If the Poincaré gauge represented by $\boldsymbol{s}$ is adapted to the family of reference hypersurface Σ_t, in the sense that the timelike components of its vierbeins are orthogonal to those hypersurfaces, then the proper energy operator $\boldsymbol{\rho}$ and the 3-momentum operators $\boldsymbol{j}_a$ of the quantum-geometric field $\boldsymbol{\varphi}$ can be defined as in classical canonical gravity:

$$\boldsymbol{\rho}(x) = \hat{\boldsymbol{P}}_{0;s}(x) \ , \qquad \boldsymbol{j}_a(x) = \hat{\boldsymbol{P}}_{a;s}(x) \ , \qquad a = 1, 2, 3. \tag{5.3}$$

We note that, as opposed to the quantum frame field in (4.8), the quantum-geometric field in (5.2) does not depend on the Poincaré gauge variables $q^j, j = 0, 1, 2, 3$. The reason is that, although these variables are required for the definition of infinitesimal spacetime translations at each base location $x \in \mathbf{M}$, they cannot play a direct physical role, as it becomes evident as soon as the transition to the special relativistic regime is performed. Indeed, in case that $(\mathbf{M}, \boldsymbol{g}^{\mathrm{L}})$ is the Minkowski space $\mathbf{M}^4 \cong (\mathbf{R}^4, \boldsymbol{\eta})$, a global Lorentz frames $\mathcal{L} = \{\boldsymbol{e}_i(x_0) | i = 0, 1, 2, 3\}$ with its origin at $x_0 \in \mathbf{M}^4$ corresponds to the following cross-section of the Poincaré frame bundle PM^4,

$$\boldsymbol{s}_0(\mathcal{L}) = \left\{ (\boldsymbol{a}(x), \boldsymbol{e}_i(x)) \,\middle|\, \boldsymbol{a}(x) = -x^i \boldsymbol{e}_i(x) \in T_x M^4, \ \ x = x^i \boldsymbol{e}_i(x_0) \in M^4 \right\}, \tag{5.4}$$

since each tangent space $T_x M^4$ can be identified with M^4 itself. Hence, each special-relativistic state vector $\varphi \in \mathbf{F}$ can be then identified with a cross-section $\{\boldsymbol{\Psi}_{1;x} \in \mathcal{F}_x | x \in \mathbf{M}\}$ of the quantum bundle $\mathcal{E}(M^4)$ whose local state vectors $\boldsymbol{\Psi}_{1;x}$ have the coordinate wave functions that satisfy

$$\Psi_{1;x}(-i\ell v) \ = \ \varphi(-a(x) - i\ell v) \ , \quad -a(x) = (x^0, ..., x^3) \in \mathbf{R}^4, \tag{5.5}$$

in relation to the Poincaré frames in $\boldsymbol{s}_0(\mathcal{L})$, since by analytic continuation the values of $\Psi_{1;x}(q - ilv)$ are uniquely determined at all $q \in \mathbf{R}^4$. In view of the Poincaré gauge invariance of the quantum geometry framework, this identification is invariant under changes of global Lorentz frames $\mathcal{L}$, since all such changes induce corresponding changes in the global Poincaré gauges of the type (5.4). Hence, the special relativistic framework of Sec. 3 is recovered as a special case of the present Poincaré gauge invariant framework, and the gauge variables q^j are seen to be physically redundant.

This suggests the construction of the Fock space

$$\hat{\mathcal{F}} = \oplus_{n=0}^{\infty} \hat{\mathcal{F}}_n \ , \qquad \hat{\mathcal{F}}_n = \hat{\mathbf{F}} \underset{S}{\otimes} \cdots \underset{S}{\otimes} \hat{\mathbf{F}} \ , \qquad \hat{\mathbf{F}} = \left\{ \hat{f} \mid f \in \mathbf{F} \right\}, \tag{5.6}$$

whose state vectors consist of n-exciton field modes $\hat{\Psi}_n \in \hat{\mathcal{F}}_n$ that are related to the corresponding $\Psi_n \in \mathcal{F}_n$ by the equalities

$$\hat{\Psi}_n(v_1, ..., v_n) = \Psi_n(-a(x) - i\ell v_1, ..., -a(x) - i\ell v_n) \ , \qquad \Psi_n \in \mathcal{F}_n. \tag{5.7}$$

By analytic continuation, these equalities establish isomorphisms between $\hat{\mathcal{F}}_n$ and $\mathcal{F}_n$ for $n = 1, 2, \ldots$. In turn, these isomorphisms become isometries if the restrictions of (3.2) to Lorentz transformations are employed to introduce in $\hat{\mathbf{F}}$ Perelomov-type generalized coherent states as in (3.3). We then have that

$$\langle \hat{\varphi}_1 | \hat{\varphi}_2 \rangle = Z_{\ell,m} \int_{v^o > 0} \hat{\varphi}_1^*(v) \hat{\varphi}_2(v) \, \delta(v^2 - 1) \, d^4 v = \langle \varphi_1 | \varphi_2 \rangle , \quad \hat{\varphi}_1, \hat{\varphi}_2 \in \hat{\mathbf{F}} \ , \quad \varphi_1, \varphi_2 \in \mathbf{F}, \tag{5.8}$$

where the value of the renormalization constant $Z_{\ell,m}$ can be determined[10,17] by the method presented in Ref. 15.

We can now define a bundle of localized quantum field excitation modes of φ,

$$\hat{\mathcal{E}}(\mathbf{M}, \boldsymbol{g}^{\mathrm{L}}) = L\mathbf{M}(\boldsymbol{g}^{\mathrm{L}}) \times_{\mathbf{G}} \hat{\mathcal{F}} \ , \qquad \mathbf{G} = \mathrm{SO}(3,1), \tag{5.9}$$

which has the Lorentz group as its structure group. On account of (5.7) its Lorentz gauge covariance can be extended into the Poincaré gauge covariance of the Fock-Klein-Gordon bundle in (4.2), so that the parallel transport in the latter can be transferred to the former. We can use this fact to define the quantum-geometric evolution between two such hypersurfaces $\Sigma_{t'}$ and $\Sigma_{t''}$ of a geometrodynamic spacetime evolution given by a foliation of the Lorentzian manifold $(\mathbf{M}, \boldsymbol{g}^{\mathrm{L}})$ into synchronous reference hypersurfaces Σ_t labeled by a globally defined parameter t.

Let us therefore consider for some integer N the family of reference hypersurfaces Σ_{t_n}, $n = 1, ..., N$, which are such that $\Sigma_{t'} = \Sigma_{t_0}$ and $\Sigma_{t''} = \Sigma_{t_N}$, and that $t_n - t_{n1} = (t'' - t')/N = \varepsilon$. Let us denote by $\mathbf{S}_n$ the segment in the base manifold $(\mathbf{M}, \boldsymbol{g}^{\mathrm{L}})$ between the hypersurface $\Sigma_{t_{n-1}}$ and Σ_{t_n}, to which we shall refer as its inflow and outflow hypersurface, respectively. For a choice of cross-section $\boldsymbol{s} = \{\boldsymbol{a}(x), \boldsymbol{e}_i(x) | x \in \mathbf{M}\}$ of the Poincaré frame bundle $P\mathbf{M}$ we then introduce above each one of the inflow-outflow hypersurfaces corresponding to $n = 1, ..., N-1$, the coherent field-mode sections

$$\hat{\boldsymbol{\Phi}}^{s}_{\varphi_n(x_n)} \in \hat{\mathcal{F}}_{x_n} \ , \qquad \varphi_n(x_n) \in \mathbf{F}_{x_n} \ , \qquad \varphi_n \in \mathbf{F} \ , \qquad x_n \in \Sigma_{t_n}, \tag{5.10}$$

determined[10] from single-boson states φ_n in the typical fibre $\mathbf{F}$ in a manner which naturally extrapolates the construction in (5.5) from flat to curved spacetimes. Let us then consider all the broken paths between $\Sigma_{t'}$ and $\Sigma_{t''}$ which consist of geodesic arcs $\gamma(x_{n-1}, x_n)$ connecting points x_{n-1} on inflow hypersurfaces $\Sigma_{t_{n-1}}$ with points x_n on outflow hypersurfaces Σ_{t_n} of the base-segments $\mathbf{S}_n$, $n = 1, ..., N$.

As mentioned earlier, quantum-geometric propagation proceeds by parallel transport which abides by the strong equivalence principle and the geodesic postulate. Consequently, if $\tau_\gamma(x_n, x_{n-1})$ denotes the operator for parallel transport along $\gamma(x_{n-1}, x_n)$ determined by the quantum connection in (4.16), the quantum-geometric free-fall propagator is given by

$$K(\varphi(x''); \varphi(x)) = \qquad (5.11)$$
$$\lim_{\varepsilon \to +0} \prod_{n=1}^{N} \int_{\hat{\mathcal{E}}_1(\Sigma_{t_n})} \mathcal{D}\varphi_n \left\langle \hat{\boldsymbol{\Phi}}^S_{\varphi_n(x_n)} \right| e^{-i\varepsilon \int d\sigma(x_n) \boldsymbol{\rho}(x_n)} \tau_\gamma(x_n, x_{n-1}) \hat{\boldsymbol{\Phi}}^S_{\varphi_{n-1}(x_{n-1})} \Big\rangle,$$

where $\boldsymbol{\rho}(x_n)$ is given by (5.1)-(5.3), and $d\sigma(x_n)$ is the Riemannian measure element determined by the 3-metric induced by the Lorentzian metric $\boldsymbol{g}^{\mathrm{L}}$ on the spacelike reference hypersurface Σ_{t_n}. The above functional integration is to be carried out over all the coherent field modes above Σ_{t_n} with respect to the functional "measure"

$$\mathcal{D}\varphi_n = \prod\nolimits_{x_n \in \Sigma_{t_n}} \mathcal{D}[\varphi_n(x_n)], \qquad (5.12)$$

in the sense of Riemannian rather than Lebesgue integration,[10,17] so that no transition to imaginary time and the "Euclidean regime" is required – as in the case with the Feynman-Kac formula. Rather, the smoothness of coherent sections secures the existence of such integrals over physical families of stochastic paths in *real* spacetime.

The key formula (4.15), together with other properties of coherent states, enables the recasting of (5.11) in the form of the path integral

$$K(\varphi(x''); \varphi(x)) = \int \mathrm{D}\varphi \, \exp[iS(\varphi)] \;, \qquad \mathrm{D}\varphi = \prod\nolimits_{t''>t>t'} \prod\nolimits_{x \in \Sigma_t} \mathcal{D}\varphi(x), \qquad (5.13)$$

based on an action that can be expressed terms of a Lagrangian density:

$$S(\varphi) = \int_{t'}^{t''} dt \int_{\Sigma_\tau} d\sigma(x) \int \delta(v^2 - 1) \, d^4v \, \mathcal{L}_0(\varphi_t(x, v), \varphi_{t-0}(x, v)), \qquad (5.14)$$

$$\mathcal{L}_0 = \tfrac{i}{2} Z_{\ell,m} \left[\overline{\varphi}_t(x, v) \, \dot{\varphi}_t(x, v) - \dot{\overline{\varphi}}_t(x, v) \, \varphi_{t-0}(x, v) \right] - T_{00} [\overline{\varphi}_t(x, v) + \varphi_{t-0}(x, v)]. \quad (5.15)$$

Thus, in quantum-geometric field theory Lagrangians and action integrals are *derived* from geometric physical principles, rather than being postulated. Nevertheless. it can be shown[10,17] that field interactions can be incorporated without difficulty in the quantum-geometric framework, and that in the special relativistic regime agreement with conventional QFT can be achieved by taking the sharp-point limit $\ell \to 0$ in a *formal* perturbation expansion.

6. GEOMETRIC QUANTUM GRAVITY

The extrapolation of quantum-geometric propagation to quantum gravity, where matter fields are in *mutual* interaction with a quantum gravitational field $\boldsymbol{g}^{\mathrm{Q}}$, has to take into account the fact that already in classical gravity such mutual interactions between "matter" fields (including those describing nongravitational radiation) and gravitational fields display diffeomorphism invariance.[27] Thus, if $Riem^{\mathrm{L}}\mathbf{M}$ denotes the family of all Lorentzian metrics in $\mathbf{M}$, and if $Diff\mathbf{M}$ is the diffeomorphism group on $\mathbf{M}$, then a CGR model in $\mathbf{M}$ is actually represented by an equivalence class g^{L} of Lorentzian metrics in $\mathbf{M}$, which constitutes a gauge orbit $(\mathbf{M}, g^{\mathrm{L}})$ within the principal bundle $Diff\mathbf{M} \to Riem^{\mathrm{L}}\mathbf{M} \to Riem^{\mathrm{L}}\mathbf{M}/Diff\mathbf{M}$.

The extrapolation of this fact to quantum gravity can be achieved by viewing such a CGR gauge orbit as providing a *metrization* of $\mathbf{M}$, which is mathematically represented by the reduction of the general linear frame bundle $GL\mathbf{M}$ to the family of all Lorentz frame bundles $L\mathbf{M}(\boldsymbol{g}^{\mathrm{L}})$ with $\boldsymbol{g}^{\mathrm{L}} \in g^{\mathrm{L}}$. Physically, such a metrization is tantamount to an operational verification aimed at establishing which *physical* representatives of linear frames $\{\boldsymbol{e}_0(x), \ldots, \boldsymbol{e}_3(x)\}$ in $GL\mathbf{M}$ are actually local Lorentz frames in a *physically existing* metric. Indeed, given the fact that the *existence* of such a metric is a feature of the physical reality around us, rather than a matter of mere choice left to the discretion of the experimenter or of the observer of a given physical occurrence, from the point of view of CGR such a procedure amounts to verifying which ones of the linear *macro*-frames, consisting of "rigid rods" and "standard clocks", are Lorentzian in the operational sense originally stipulated by Einstein. However, if a relabeling of all frames, test bodies and fields is carried out, this *physical* verification cannot distinguish between all the various choices of *mathematical* labels $\{\boldsymbol{e}_0(x), \ldots, \boldsymbol{e}_3(x)\}$, if such relablings are related by means of diffeomorphisms that give rise to various equivalent Lorentzian metric representatives. Moreover, even in those cases where the mathematical labeling of frames, test bodies and fields is kept fixed, by the strong equivalence principle physically equivalent descriptions of natural phenomena are still obtained for physically distinct choices of inertial Lorentz frames, if those frames in free fall are related by Lorentz gauge transformations.

The adaptation of these basic facts to quantum *micro*-frames leads,[9,10] in broad outline, to the following formulation of *quantum* geometrodynamic evolution: the presence of a specific state of the quantum gravitational field $\boldsymbol{g}^{\mathrm{Q}}$ within a base segment $\mathbf{S}_n$ dictates the quantum-geometric evolution of all the quantum "matter" fields from the inflow hypersurface $\Sigma_{t_{n-1}}$ to the outflow hypersurface Σ_{t_n} of that base-segment; the values of the local quantum energy-momentum density operators of all the quantum "matter" fields in the resulting states of those fields along Σ_{t_n} then creates a gauge orbit $(\mathbf{S}_{n+1}, g^{\mathrm{M}})$ by metrizing the quantum frame bundles over the subsequent base segment $\mathbf{S}_{n+1}$; this determines the formation of local states of the quantum gravitational field $\boldsymbol{g}^{\mathrm{Q}}$ within that base-segment (whose mean values provide the Lorentzian metrics in g^{M}), and singles out within the quantum gravitational fibres above its points x the multi-graviton states of quantum gravitational radiation. This process then repeats itself from segment to segment, with the limit to "infinitely-thin" segments being eventually taken.

Of course, the presence of constraints and quantum gravitational self-interaction has to be taken into account before this sketchy picture of quantum-geometric gravitational evolution can be completed into a legitimate extrapolation to the quantum regime of Einstein's formulation of classical gravity. This involves the extension of the base-segments $\mathbf{S}_n$ into supermanifolds $\boldsymbol{S}_n$ over which a quantum gravitational superbundle can be defined, and on whose sections a quantum gravitational gauge supergroup can act. This quantum gravitational supergroup incorporates diffeomorphism invariance as well as Poincaré gauge invariance.

The resulting framework shares features with both covariant and canonical quantum gravity, but it is also fundametally distinct in many respects. Thus, in geometric quantum gravity equivalence classes g^{M} of mean metrics are generated by the quantum geometrodynamic evolution of gravitational fields in mutual interaction with "matter" fields, so that no "background" metric is prescribed – as is the case in covariant quantum gravity. Moreover, the presence of *quantum* (super)-frames also resolves the "issue of time", that is much-debated in contemporary canonical gravity.[27] This solution emerges from de Broglie's fundamental idea[28] that, on account of its rest mass m, each massive elementary quantum object represents a natural clock with period $T = 2\pi/m$ in Planck natural units. The replacement of global with local frames insures that such a proper time is kept *locally*, in accordance with the fundamental features of Einstein's *general* relativistic conceptualization of an ultimately valid description of nature. On the other hand, the use of bundles of quantum frames or superframes which incorporate the fundamental length ℓ into their very structure removes the need for de Broglie's "pilot waves". Thus, the replacement of classical geometries, upon which de Broglie founded his ideas, with quantum geometries, obviates the need for extraneous assumptions about *quantum* reality in the same manner in which the introduction of the geometry of Minkowski space obviated the need for regarding Lorentz contractions of (classically conceptualized) electrons as real phenomena.

The mathematical techniques required for the implementation of these ideas on geometric quantum gravity cannot be outlined within the restricted space of the present review article. Consequently, the reader interested in them is directed to Chapter 8 in Ref. 10.

References

1. E. Prugovečki, *Ann. Phys. (N.Y.)* 110:102 (1978)
2. E. Prugovečki, *J. Math. Phys.* 19:2260 (1978)
3. E. Prugovečki, *Phys. Rev. D* 18:3655 (1978)
4. E. Prugovečki, "Stochastic Quantum Mechanics and Quantum Spacetime", Reidel, Dordrecht (1984); reprinted with corrections (1986)
5. M. Born, *Rev. Mod. Phys.* 21:463 (1949)
6. E. P. Wigner, *in*: "Quantum Theory and Measurement", J. A. Wheeler and W. H. Zurek (eds.), Princeton University Press, Princeton (1983), pp. 260-314
7. E. Prugovečki, *Class. Quantum Grav.* 4:1659 (1987)
8. E. Prugovečki, *Nuovo Cimento A* 97:597, 837 (1987)
9. E. Prugovečki, "Quantum Geometry", Kluwer, Dordrecht (1992)
10. E. Prugovečki, "Principles of Quantum General Relativity", World Scientific, Singapore (1995)
11. S. T. Ali and E. Prugovečki, *Acta Appl. Math.* 6:1, 19, 47 (1986)
12. S. T. Ali and G. G. Emch, *J. Math. Phys.* 27:2936 (1986)
13. B. Kostant, "Quantization and Unitary Representations", *Springer Lecture Notes in Mathematics*, vol. 170, New York (1970)
14. J.-M. Souriau, "Structure des systmes dynamiques", Dunod, Paris (1970)
15. A. M. Perelomov, "Generalized Coherent States and Their Applications", Springer, Berlin (1986)
16. F. A. Berezin, *Commun. Math. Phys.* 40:153 (1975)
17. E. Prugovečki, *Class. Quantum Grav.* 11:1981 (1994)
18. E. Prugovečki, *Nuovo Cimento A* 61,:85 (1981)
19. L. S. Schulman, "Techniques and Applications of Path Integration", Wiley, New York (1981)

20. B. S. DeWitt, *in*: "Gravitation: An Introduction to Current Research", L. Witten (ed.), Wiley, New York (1962)

21. H.-H. von Borzeszkowski and H.-J. Treder, "The Meaning of Quantum Gravity", Kluwer, Dordrecht (1988)

22. W. Drechsler, *Fortschr. Phys.* 23:449 (1984)

23. N. N. Bogolubov, A. A. Logunov and I. T. Todorov, "Introduction to Axiomatic Quantum Field Theory", Benjamin, Reading, Mass. (1975)

24. F. A. Berezin, "The Method of Second Quantization", Academic Press, New York (1966)

25. R. P. Feynman, *Rev. Mod. Phys.* 20:367 (1948)

26. R. P. Feynman, *Phys. Rev.* 76:749, 769 (1949)

27. A. Ashtekar and J. Stachel (eds.), "Conceptual Problems in Quantum Gravity", Birkhäuser, Boston (1991)

28. L. de Broglie, *in*: "Perspectives in Quantum Theory", W. Yourgrau and A. van der Merwe (eds.), Dover, New York (1979).

PREQUANTIZATION

D. J. Simms

School of Mathematics, Trinity College
Dublin 2, Ireland

Abstract

We give an exposition of the concept of prequantization, which enables standard constructions for cotangent bundles to be extended to symplectic manifolds whose symplectic forms represents an integral de Rham cohomology class.

1. INTRODUCTION

The concept of geometric quantization as developed by Kostant[1] and Souriau[2] is now thirty years old. Since it is based on the symplectic structure of phase space, rather than the structure of the cotangent bundle of configuration space, it has gained in relevance over the years, as symplectic manifolds have arisen as moduli spaces.

The starting point of geometric quantization is the concept of prequantization which is the most fundamental point of contact between the classical mechanical notions of Hamiltonian vector fields and classical Poisson brackets on one side, and linear operators and operator commutators on the other.

For a cotangent bundle the associated symplectic form $\omega = \sum dp_i \wedge dq_i$ is exact with $\omega = d\alpha$ where α is the canonical 1-form, and the action S is the integral of the 1-form $\alpha - H dt$.

Prequantization gives a method of considering Dirac amplitude $\exp iS$ for the wider class of symplectic manifolds for which ω is not necessarily exact, but does represent an integral de Rham cohomology class.

More recently a particular impetus to these ideas has been given by Witten's treatment of the Jones polynomial. This gives rise to the moduli space of flat G-bundles over a a compact surface of fixed genus, for G a compact simple Lie group. The moduli space in this case has been shown by Atiyah and Bott to be a symplectic manifold whose symplectic form represents an integral de Rham cohomology class. For a recent relevant article and bibliography see J.-L. Brylinski and D. McLaughlin.[3]

2. THE LINE BUNDLE

Let M be a symplectic manifold with symplectic form ω. Thus ω is a closed 2-form, $d\omega = 0$, and nondegenerate.

Quantization, Coherent States, and Complex Structures
Edited by J.-P. Antoine *et al.*, Plenum Press, New York, 1995

By prequantization we mean the selection of a complex line bundle L over M with connection ∇ and invariant hermitian structure $(.|.)$ having ω as curvature form:

$$\pi : L \rightarrow M$$

with projection π.

We denote by $\mathcal{S}_L$ the space of smooth sections of the line bundle L and by $\mathcal{V}_L$ the space of smooth vector fields on M. The connection ∇ is a $C^\infty(M)$ -map:

$$\mathcal{V}_L \rightarrow End(\mathcal{S}_L); \qquad \xi \mapsto \nabla_\xi$$

which assigns to each smooth vector field ξ on M a linear operator ∇_ξ on the sections of L with the Leibniz properties:

1. $$\nabla_\xi(\phi s) = (\xi\phi)s + \phi\nabla_\xi s$$

2. $$\xi(s|t) = (\nabla_\xi s|t) + (s|\nabla_\xi t)$$

and the curvature property:

3. $$[\nabla_\xi, \nabla_\eta] - \nabla_{[\xi,\eta]} = 2\pi i\omega(\xi,\eta).$$

Here s and t are sections of L, $(s|t)$ is their hermitian inner product, and ξ and η are vector fields on M. $\nabla_\xi s$ is called the covariant derivative of the section s along the vector field ξ.

To express sections of L locally, on an open set U of M, as complex valued functions on U, we choose as 'reference frame' (or 'gauge') a nowhere vanishing section s over U. Thus s is a smooth function on U whose value $s(x)$ belongs to $\pi^{-1}(x)$ at each point $x \in U$. Then ∇ is determined on U by a 'connection 1-form' α_s given by

$$\nabla_\xi s = 2\pi i < \alpha_s, \xi > s.$$

For any $\phi \in C^\infty(U)$ we have:

$$\nabla_\xi(\phi s) = (\xi\phi)s + \phi\nabla_\xi s = (\xi\phi)s + \phi 2\pi i < \alpha_s, \xi > s = ([\xi + 2\pi i < \alpha_s, \xi >]\phi)s.$$

Thus ∇_ξ is represented locally by the operator:

$$\xi + 2\pi i < \alpha_s, \xi >$$

with respect to the reference frame s. This is the familiar $\partial_\mu + A_\mu$ as in minimal coupling.

By the curvature condition we have:

$$2\pi i\omega(\xi,\eta)s = \nabla_\xi(\nabla_\eta s) - \nabla_\eta(\nabla_\xi s) - \nabla_{[\xi,\eta]}s$$

and therefore:

$$2\pi i\omega(\xi,\eta)s = \nabla_\xi(2\pi i < \alpha_s, \eta > s) - \nabla_\eta(2\pi i < \alpha_s, \xi > s) - 2\pi i < \alpha_s, [\xi,\eta] > s.$$

Therefore:

$$\omega(\xi,\eta)s = [\xi < \alpha_s, \eta > -\eta < \alpha_s, \xi > - < \alpha_s, [\xi,\eta] >]s = d\alpha_s(\xi,\eta)s.$$

Therefore:

$$d\alpha_s = \omega.$$

This indicates that the connection ∇, with local representative the 1-form α_s, takes the role of the canonical 1-form $\alpha = pdq$ carried by the cotangent bundle.

We consider now the dependence of α_s on the reference frame s. If r is a second reference frame then $r = gs$ where g is a nowhere vanishing complex valued function. From

$$2\pi i < \alpha_r, \xi > r = \nabla_\xi r$$

it follows that:

$$2\pi i < \alpha_r, \xi > gs = \nabla_\xi(gs) = (\xi g)s + g\nabla_\xi s = [< dg, \xi > + g2\pi i < \alpha_s, \xi >]s.$$

Therefore

$$\alpha_r = \frac{1}{2\pi i}\frac{dg}{g} + \alpha_s.$$

Now $\dfrac{dg}{g}$ is closed since, locally, it is $d\log g$ so that

$$d\alpha_r = d\alpha_s$$

on their common domain, as expected, and equal to ω locally..

Now the connection 1-forms $\{\alpha_s\ ;\ s \text{ a reference frame}\}$ derive from a single 1-form α on the total space of the principal bundle L^* associated to L. This 1-form α is the canonical equivalent of the 1-form pdq carried by the cotangent bundle.

To see how α arises, we note first that L has fibre $\mathbf{C}$ and group $\mathbf{C}^* = GL(1, \mathbf{C}) =$ non-zero complex numbers. The principal bundle L^* has fibre $\mathbf{C}^* \subset \mathbf{C}$ so $\mathbf{L}^* \subset \mathbf{L}$.

A reference frame s on an open set U represents $\pi^{-1}U$ as a product

$$\pi^{-1}U \to U \times \mathbf{C}$$

by $zs(x) \mapsto (x, z)$.

If r is another reference frame, domain V, $r = gs$, then:

$$(U \cap V) \times \mathbf{C} \to \pi^{-1}(U \cap V) \to (U \cap V) \times \mathbf{C}$$

by:

$$(x, z) \mapsto zr(x) = zgs(x) \mapsto (x, zg)$$

where the first map is given by the reference frame r and the second map is given by s.

Therefore the 1-form $\alpha_s + \dfrac{1}{2\pi i}\dfrac{dz}{z}$ pulls back under the composition of these two maps to the 1-form:

$$\alpha_s + \frac{1}{2\pi i}\frac{d(zg)}{zg} = \alpha_s + \frac{1}{2\pi i}\frac{dg}{g} + \frac{1}{2\pi i}\frac{dz}{z} = \alpha_r + \frac{1}{2\pi i}\frac{dz}{z}.$$

Thus there is a unique 1-form α on L^* which corresponds to $\alpha_s + \dfrac{1}{2\pi i}\dfrac{dz}{z}$ in the reference frame s, for each s.

We note that:

1. $s^*\alpha = \alpha_s$

2. $s^* d\alpha = d\alpha_s = \omega$

3. $d\alpha = \pi^* \omega$

We note further that for L and ∇ to exist we require ω to represent an *integral* de Rham cohomology class. This is a *quantization* condition which must be satisfied by the symplectic form. To see this we choose reference frames $\{s_i\}$ whose domains $\{V_i\}$ form a contractible open cover of M. Then $s_i = g_{ij} s_j$ (say) and:

$$\alpha_i = \alpha_j + \frac{1}{2\pi i} d \log g_{ij}.$$

Then:

$$n_{ijk} = \frac{1}{2\pi i} [\log g_{ij} + \log g_{jk} - \log g_{ik}]$$

is a Čech 2-cocycle and is integral since:

$$\exp(2\pi i n_{ijk}) = 1.$$

We note that the objects α_i, g_{ij}, n_{ijk} which appear in the above calculation are representatives of sheaf cohomology classes which are used in the passage from the closed 2-form ω to the corresponding real cohomology class which is associated to ω by de Rham's theorem.

The short exact sequence of sheaves:

$$0 \to Z^1 \to \Omega^1 \to Z^2 \to 0$$

where Z^r is the sheaf of germs of closed r-forms on M, and Ω^r is the sheaf of germs of r-forms on M, gives rise to a long exact sequence:

$$0 \to H^0(Z^1) \to H^0(\Omega^1) \to H^0(Z^2) \to H^1(Z^1) \to 0$$

The map $H^0(Z^2) \to H^1(Z^1)$ corresponds to $\omega \mapsto \alpha_i - \alpha_j$. The short exact sequence of sheaves:

$$0 \to \mathbf{R} \to \Omega^0 \to Z^1 \to 0$$

gives rise to a long exact sequence:

$$.... \to H^1(\mathbf{R}) \to H^1(\Omega^0) \to H^1(Z^1) \to H^2(\mathbf{R}) \to ...$$

The map $H^1(Z^1) \to H^2(\mathbf{R})$ corresponds to $\alpha_i - \alpha_j \mapsto n_{ijk}$. The short exact sequence of sheaves:

$$0 \to \mathbf{Z} \to \mathbf{C} \to \mathbf{C}^* \to 0$$

gives rise to a long exact sequence:

$$.... \to H^1(\mathbf{Z}) \to H^1(\mathbf{C}) \to H^1(\mathbf{C}^*) \to H^2(\mathbf{Z}) \to ...$$

The map $H^1(\mathbf{C}^*) \to H^2(\mathbf{Z})$, which is an isomorphism which maps the equivalence class of a line bundle to its first Chern class, corresponds to $g_{ij} \mapsto n_{ijk}$.

3. THE QUANTUM VECTOR FIELD

Each smooth function ϕ on the symplectic manifold M defines a Hamiltonian vector field ξ_ϕ on M by:

$$\xi_\phi \lrcorner \omega + d\phi = 0.$$

The Hamiltonian vector fields ξ_ϕ give $C^\infty(M)$ the structure of a Lie algebra by the Poisson bracket given by:

$$[\phi, \psi] = \omega(\xi_\phi, \xi_\psi).$$

Here and in the following we use the notation[4] $\xi \lrcorner \beta$ to denote the contraction of a vector field ξ with a differential form β. ξ_ϕ represents the classical evolution, and lifts to a *quantum vector field* η_ϕ on L^*.

η_ϕ is defined as the unique vector field which lifts ξ_ϕ:

$$\pi_* \eta_\phi = \xi_\phi$$

and satisfies:

$$< \alpha, \eta_\phi >= \pi^* \phi.$$

ξ_ϕ preserves ω since the Lie derivative of ω along ξ_ϕ is:

$$\mathcal{L}_{\xi_\phi} \omega = d[\xi_\phi \lrcorner \omega] + \xi \lrcorner d\omega = d[-d\phi] = 0.$$

η_ϕ is $\mathbf{C}^*$-invariant and preserves α:

$$\mathcal{L}_{\xi_\phi} \alpha = d[\eta_\phi \lrcorner \alpha] + \eta_\phi \lrcorner d\alpha = d\pi^*\phi + \eta_\phi \lrcorner \pi^*\omega = \pi^*(d\phi + \xi_\phi \lrcorner \omega) = 0.$$

In fact $C^\infty(M) \to \mathcal{E}(L, \alpha)$; $\phi \mapsto \eta_\phi$ is a Lie algebra isomorphism of the Poisson bracket algebra onto the Lie algebra of vector fields on L^* which are $\mathbf{C}^*$-invariant and preserve α.

To see this we note that $\phi \mapsto \xi_\phi$ is a Lie algebra homomorphism from $C^\infty(M)$ to the Hamiltonian vector fields on M, since:

$$[\xi_\phi, \xi_\psi] \lrcorner \omega = \mathcal{L}_{\xi_\phi}(\xi_\psi \lrcorner \omega) - \xi_\psi \lrcorner \mathcal{L}_{\xi_\phi} \omega$$

which, since $\mathcal{L}_{\xi_\phi} \omega = 0$, equals:

$$\xi_\phi \lrcorner d(\xi_\psi \lrcorner \omega) + d(\xi_\phi \lrcorner \xi_\psi \lrcorner \omega) = \xi_\phi \lrcorner d(-d\psi) + d(\omega(\xi_\psi, \xi_\phi))$$

and this, since $d(-d\psi) = 0$, equals:

$$-d[\phi, \psi].$$

Therefore

$$[\xi_\phi, \xi_\psi] = \xi_{[\phi,\psi]}$$

Also

$$\pi_*[\eta_\phi, \eta_\psi] = [\pi_*\eta_\phi, \pi_*\eta_\psi] = [\xi_\phi, \xi_\psi] = \xi_{[\phi,\psi]}$$

and

$$\pi^*[\phi, \psi] = \pi^*[\omega(\xi_\phi, \xi_\psi)] = \pi^*\omega(\eta_\phi, \eta_\psi) = d\alpha(\eta_\phi, \eta_\psi)$$

Therefore

$$\pi^*[\phi, \psi] = \eta_\phi < \alpha, \eta_\psi > -\eta_\psi < \alpha, \eta_\phi > - < \alpha, [\eta_\phi, \eta_\psi] >$$

which equals:

$$\pi^*[\phi,\psi] - \pi^*[\psi,\phi] - < \alpha, [\eta_\phi, \eta_\psi] >$$

Therefore

$$< \alpha, [\eta_\phi, \eta_\psi] >= \pi^*[\phi,\psi]$$

So

$$[\eta_\phi, \eta_\psi] = \eta_{[\phi,\psi]}$$

Now η_ϕ is a $\mathbf{C}^*$-invariant vector field on the principal bundle L^*, so it acts on the space $\mathcal{S}_L$ of sections of the associated bundle L. Thus we have a Lie algebra representation:

$$\phi \rightarrow \eta_\phi$$

of the Poisson bracket Lie algebra $C^\infty(M)$ on the space of sections of the associated bundle. This representation is called *prequantization.*

If ϕ is the classical Hamiltonian then η_ϕ is the differentiated form of the Dirac amplitude. To see this we take a reference frame s_i. A section s of L is represented by a function ψ_i:

$$s = \psi_i s_i$$

In time t the flow of ξ_ϕ carries m to m_t (say) in M. The flow of η_ϕ carries the section s to a section s_t:

$$s_t = \psi_{it} s_i$$

and:

$$\psi_{it}(m) = \exp[i \int_0^t (\xi_\phi \lrcorner \alpha_i - \phi) dt] \psi_i(m_t)$$

We thus see that in the 'gauge' s_i the action of the quantum vector field η_ϕ appears, when integrated, as a combination of the Hamiltonian flow ξ_ϕ on the base space M, together with a multiplying factor equal to $\exp iS$ where S is the action as measured in the 'gauge' s_i.

References

1. B. Kostant, Quantization and unitary representations, *in*: "Lectures in Modern Analysis III", C. T. Taam, ed., Lecture Notes in Mathematics, Volume 170, Springer, Berlin (1970)

2. J.-M. Souriau, "Structure des Systèmes Dynamiques", Dunod, Paris, (1970)

3. J.-L. Brylinski and D. McLaughlin, Unitary representations of the Teichmüller group, *in*: "Lie Theory and Geometry: In Honor of Bertram Kostant", J.-L.Brylinski, R. Brylinski, V. Guillemin, V. Kac ed., Progress in Mathematics, Volume 123, Birkhäuser, Boston-Basel-Berlin (1994)

4. N.M.J. Woodhouse, "Geometric Quantization", second edition, Oxford University Press, Oxford (1992).

CLASSICAL YANG-MILLS AND DIRAC FIELDS IN THE MINKOWSKI SPACE AND IN A BAG

Jędrzej Śniatycki

Department of Mathematics and Statistics
University of Calgary
Calgary, Alberta, Canada

Abstract

Extended and reduced phase spaces for minimally interacting Yang-Mills and Dirac fields in the Minkowski space-time and in a bag are discussed.

1. INTRODUCTION

This lecture is based on joint research with Günter Schwarz devoted to the study of the canonical structure of minimally interacting Yang-Mills and Dirac fields,[1-5] and also on some earlier works.[6,7] We are going to discuss the general structure of the theory. At present we have results for Yang-Mills and Dirac fields in the Minkowski space-time, and in spatially bounded domains under the M.I.T. bag boundary conditions.

We start with the study of the existence and uniqueness theorems for the evolution equations in order to determine the extended phase space P of the system. The constraint equation is preserved by the evolution, hence the constraint set it defines is stable under the evolution.

For a given extended phase space P, the group of gauge symmetries $GS(P)$ consists of gauge transformations preserving its structure. Hence, the topology of the extended phase space determines the topology of the group of gauge symmetries.

Extended phase spaces under considerations are weakly symplectic, and the action of the group $GS(P)$ of gauge symmetries of P is Hamiltonian. The corresponding momentum map J has values in the dual $gs(P)^*$ of the Lie algebra $gs(P)$ of $GS(P)$. The constraint equation defines an ideal $gs(P)_0$ in $gs(P)$ such that the constraint set is the zero level of the restriction J_0 of the momentum map J to $gs(P)_0$. J_0 is the momentum map for the Hamiltonian action in P of the normal subgroup $GS(P)_0$ of $GS(P)$ generated by $gs(P)_0$. The reduced phase space $\check{P}$ is defined as the space of $GS(P)_0$-orbits in the constraint set ${J_0}^{-1}(0)$.

Let P_M denote the extended phase space in the Minkowski space-time theory. We find that the action of $GS(P_M)_0$ is free and proper, the constraint set ${J_0}^{-1}(0)$ is a submanifold of P_M, and the reduced phase space $\check{P}_M$ is a weakly symplectic quotient manifold of ${J_0}^{-1}(0)$ with an exact symplectic form.

Quantization, Coherent States, and Complex Structures
Edited by J.-P. Antoine *et al.*, Plenum Press, New York, 1995

For the bag boundary conditions the extended phase space is denoted by P_B. The action of $GS(P_B)_0$ is not free and the the constraint set has singularities. However, the action of $GS(P_B)_0$ is proper, and the reduced phase space is partitioned by weakly symplectic manifolds with exact symplectic forms. Each part corresponds to a definite mode of symmetry breaking with vector bosons vanishing identically.

Colour charges in the reduced phase space are labelled by the elements of the colour algebra

$$colour(P) = gs(P)/gs(P)_0.$$

For the bag boundary conditions the colour algebra is trivial, $colour(P_B) = \{0\}$, and the theory does not allow for an invariant definition of colour charges.

In the Minkowski space-time theory the colour algebra is isomorphic to the structure algebra,

$$colour(P_M) \simeq g.$$

For colour labels in the centre of the colour algebra, there exist well defined colour charge densities depending locally on the fields. If the colour label is not in the centre of the colour algebra, every invariant definition of colour charge densities has to depend non-locally on the fields. The corresponding quantum operators could not be observable since they would violate the locality axiom of quantum field theory.

2. THE NOTION OF AN EXTENDED PHASE SPACE

In gauge theory the phase space consists of the gauge equivalence classes of solutions of the fields equations. Since this definition may lead to singularities, it is convenient to proceed in stages. Separating the field equations into the evolution equations and constraint equations one is lead to the notion of an extended phase space consisting of the Cauchy data, on a given Cauchy surface, which admit solutions of the evolution equations. It is determined by the existence and uniqueness theorem of the theory. Since gauge invariance implies uniqueness of solutions up to a gauge transformation, uniqueness results require an imposition of a gauge condition. Thus, an extended phase space depends on the choice of the Cauchy surface, the choice of a gauge condition, the choice of splitting of the field equations into the evolution equations and the constraint equations, and the choice of the function space in which we can prove the existence and uniqueness theorems for the evolution equations.

3. FIELD EQUATIONS

We choose the usual $(3+1)$ splitting of Minkowski space $\mathbf{R}^4 = \mathbf{R} \times \mathbf{R}^3$, which leads to the splitting of the Yang-Mills field A_μ into the scalar potential $\Phi = A_0$, the vector potential $A = A_i dx^i$,

$$A_\mu = (\Phi, A),$$

and the representation of the field strength $F_{\mu\nu}$ in terms of the "electric" field E and the "magnetic" field B with components

$$E_j = F_{0j} = \partial_0 A_j - \partial_j \Phi + [\Phi, A_j],$$

$$B_j = \tfrac{1}{2}\varepsilon_j{}^{kl} F_{kl} = (\operatorname{curl} A)_j + [A\times, A]_j,$$

where we use the Euclidean metric in $\mathbf{R}^3$ to identify vector fields and forms, and $\times$ denotes the cross product.

Let M be a contractible domain in $\mathbf{R}^3$ with smooth boundary ∂M describing the spatial extent of the system under consideration. For the M.I.T. bag model M is bounded, and it describes the interior of the bag. For the Minkowski space theory $M = \mathbf{R}^3$ and ∂M describes the sphere of directions at spatial infinity. The dynamical variables of the theory are A, E, and the spinor field Ψ on M. A and E are time dependent vector fields on M with values in the structure algebra g of the theory, while Ψ is a time dependent spinor field with values in the space V of the fundamental representation of the structure group G. The scalar potential Φ is a time dependent function from M to g. The field equations split into the evolution equations

$$\partial_t A = E + \operatorname{grad}\Phi - [\Phi, A],$$

$$\partial_t E = \operatorname{curl} B - [A\times, B] - [\Phi, E] + J,$$
$$\partial_t \Psi = -\gamma^0(\gamma^j\partial_j + im + \gamma^0\Phi + \gamma^j A_j)\Psi,$$

and the constraint equation

$$\operatorname{div} E + [A; E] = J^0.$$

The source terms can be described in terms of an orthonormal basis $\{T_a\}$ in the Lie algebra g of the structure group G,

$$J^0 = J^{0a}T_a = \Psi^\dagger T^a \Psi T_a,$$

$$J^k = J^{ka}T_a = \Psi^\dagger(\gamma^0\gamma^k) \otimes T^a\Psi T_a.$$

where the latin indices are lifted in terms of an ad-invariant metric on g, existence of which is assured by the assumption the G is compact. It should be noted that the evolution component of the field equations given here could be modified by adding to it terms which vanish when the constraint equation is satisfied.

4. EXISTENCE AND UNIQUENESS RESULTS

The problem of existence and uniqueness of solutions of Yang-Mills equations in Minkowski space-time have been studied by several authors.[8-13] Using the temporal gauge condition

$$\Phi = 0,$$

we have extended the results of Eardley and Moncrief[10] to the phase space

$$P_M = \{(A, E, \Psi) | A \in H^2(\mathbf{R}^3),\ E \in H^1(\mathbf{R}^3),\ \Psi \in H^2(\mathbf{R}^3)\},$$

where $H^k(\mathbf{R}^3)$ denotes the Sobolev space of fields on $\mathbf{R}^3$, which are square integrable together with their derivatives up to the order k.

For a spatially bounded domain M, we use the gauge condition

$$E^L = -\operatorname{grad}\Phi,$$

where E^L is the longitudinal part of E in the Helmholz-Hodge decomposition.[14] It enables us to obtain existence and uniqueness results for the evolution equations in the extended phase space P_B consisting of $(A, E, \Psi) \in H^2(M) \times H^1(M) \times H^2(M)$ which satisfy the boundary conditions

$$nE = 0,\ \ tB = 0,\ \ in_j\gamma^j\Psi|_{\partial M} = \Psi|_{\partial M},$$

$$nA = 0, \;\; in_k\gamma^k\{\gamma^0(\gamma^j\partial_j + im)\Psi\}|_{\partial M} = \{\gamma^0(\gamma^j\partial_j + im)\Psi\}|_{\partial M},$$

where nA and nE are the normal components of A and E, respectively, tB is the tangential component of B, and $\Psi|_{\partial M}$ denotes the restriction of Ψ to ∂M. The first three conditions coincide with the boundary conditions of the M.I.T. bag model.[15] The condition $nA = 0$ gives a partial gauge fixing, while the last condition is required in order to ensure that $\Psi \in H^2(M)$ for all time.

Both proofs are based on the general theory of non-linear semigroups.[16]

5. SYMPLECTIC STRUCTURE

The extended phase spaces P_M and P_B are endowed with weakly symplectic forms given by

$$\omega = d\theta,$$

where θ is a 1-form such that

$$\langle\theta(A, E, \Psi)|(\delta A, \delta E, \delta\Psi)\rangle = \int_M (E \cdot \delta A + \Psi^\dagger \delta\Psi) d_3 x.$$

6. GAUGE SYMMETRIES

The group $GS(P)$ of gauge symmetries of the theory with an extended phase space P are defined as gauge transformations which preserve **P**. In our case they are given by

$$A \to \phi A \phi^{-1} + \phi \,\mathrm{grad}\, \phi^{-1}, \;\; E \to \phi E \phi^{-1}, \;\; \Psi \to \phi\Psi \;,$$

where ϕ is a map from M into the structure group G of the theory presented as a matrix group. The Lie algebra $gs(P)$ of $GS(P)$ consists of maps $\xi : M \to g$. Their action in P is given by

$$A \to A - D_A\xi, \;\; E \to E - [E, \xi], \;\; \Psi \to \Psi + \xi\Psi \;,$$

where

$$D_A\xi = \mathrm{grad}\, \xi + [A, \xi]$$

is the covariant differential of ξ with respect to the connection defined by A. It gives rise to a vector field

$$\xi_P(A, E, \Psi) = -D_A\xi\frac{\delta}{\delta A} - [E, \xi]\frac{\delta}{\delta E} + \xi\Psi\frac{\delta}{\delta\Psi}$$

on P.

The topology of the group $GS(P)$ is determined by the requirement that the action of $GS(P)$ in P should be continuous. In particular, the action of the algebra $gs(P)$ should be continuous. For the bag boundary conditions $gs(P_B)$ is a closed subspace of the Sobolev space $H^3(M)$,

$$gs(P_B) = \{\xi : M \to g \,|\, \xi \in H^3(M), n \cdot \mathrm{grad}\, \xi = 0\} \;.$$

In the Minkowski space theory

$$gs(P_M) = \{\xi : \mathbf{R}^3 \to g \,|\, \mathrm{grad}\, \xi \in H^2(M)\} \;,$$

so that it is an intersection of the Beppo Levi spaces $BL_1(L^2(\mathbf{R}^3))$, $BL_2(L^2(\mathbf{R}^3))$ and $BL_3(L^2(\mathbf{R}^3))$, where $BL_m(L^2(\mathbf{R}^3))$ is the space of functions with square integrable derivatives of order m.[17] This space is bigger than the Sobolev space of infinitesimal gauge symmetries considered in Refs. 8 and 18.

The topology of the Lie algebras $gs(P_B)$ and $gs(P_M)$ determines the topology of the connected groups $GS(P_B)$ and $GS(P_M)$, respectively.[8] With this topology the actions of $GS(P_B)$ and $GS(P_M)$ are proper and admit slices.[4,5,19]

In the following we shall write P for P_B and P_M, if the result are the same for both spaces. The action of $GS(P)$ in P preserves the 1-form θ. Hence, it is Hamiltonian. For every $\xi \in gs(P)$, the corresponding momentum J_ξ given by

$$J_\xi(A, E, \Psi) = \langle \theta | \xi_P(A, E, \Psi) \rangle = \int_M (E \cdot D_A \xi + \Psi^\dagger \xi \Psi) d_3 x \ .$$

Integrating by parts we obtain

$$J_\xi(A, E, \Psi) = \int_{\partial M} nE \cdot \xi dS - \int_M \{(\operatorname{div} E + [A; E])\xi - \Psi^\dagger \xi \Psi\} d_3 x \ ,$$

where dS is the element of surface area. For the Minkowski space theory, $M = \mathbf{R}^3$, and ∂M is the sphere at infinity.

7. CONSTRAINTS

The constraint equations are equivalent to the vanishing the integrand in the volume integral above,

$$(\operatorname{div} E + [A; E])\xi - \Psi^\dagger \xi \Psi = 0 \ \forall \ \xi \in gs(P) \ .$$

We denote by C the constraint set in P,

$$C = \{(A, E, \Psi) \in P \,|\, (\operatorname{div} E + [A; E])\xi - \Psi^\dagger \xi \Psi = 0 \ \forall \ \xi \in gs(P)\}.$$

Then

$$gs(P)_0 = \{\xi \in gs(P) | J_\xi(A, E, \Psi) = 0 \ \forall \ (A, E, \Psi) \in C\}$$

is a closed ideal in $gs(P)$.[2] It follows from the above expression for $J_\xi(A, E, \Psi)$ that $\xi \in gs(P)_0$ if the restriction of ξ to ∂M vanishes. The converse is not true.

For the bag boundary conditions the surface integral vanishes for all ξ in $gs(P_B)$. Hence,

$$gs(P_B)_0 = gs(P_B).$$

In the Minkowski space theory the surface integral has to be considered the limit of the integral over spheres S_r of radius r as $r \to \infty$,

$$\int_{\partial M} nE\xi dS = \lim_{r \to \infty} \int_{S_r} nE\xi dS,$$

and it vanishes for all $(A, E, \Psi) \in C$ if ξ vanishes sufficiently fast at infinity. It follows from the results of Ref.17 that $gs(P_M)$ has a direct sum decomposition

$$gs(P_M) = gs(P_M)_0 \oplus g,$$

where the second term g is interpreted as the space of constant maps from $\mathbf{R}^3$ to g. Thus, $gs(P_M)_0$ is a proper ideal in $gs(P_M)$.

Let $J_0 : P \to gs(P)_0^*$ be such that, for each $(A, E, \Psi) \in P$ and $\xi \in gs(P)_0$,

$$\langle J_0(A, E, \Psi) | \xi \rangle = J_\xi(A, E, \Psi).$$

It follows that the constraint set C is the zero level set of J_0,

$$C = {J_0}^{-1}(0) .$$

This presentation of the constraint set facilitates the study of its structure. It follows from the general theory that the singular component of ${J_0}^{-1}(0)$ consists of points $(A, E, \Psi) \in {J_0}^{-1}(0)$ for which $\xi_P(A, E, \Psi) = 0$ for some non-zero $\xi \in gs(P)_0$.[20] For the Minkowski space theory $\xi \in gs(P_M)_0$ tends to zero as $x \to \infty$. Since $\xi_P(A, E, \Psi) = 0$ only if $D_A \xi = 0$ it implies that $\xi = 0$, and all points of ${J_0}^{-1}(0)$ are regular. Hence, we can conclude that ${J_0}^{-1}(0)$ is a submanifold of P_M. A direct proof of this result, not involving the structure of the symmetry group, was given in Ref.7.

For the bag boundary conditions ${J_0}^{-1}(0)$ has not only a regular component, which is a submanifold of P_B, but also a singular component. Its structure will be discussed later.

8. REDUCTION

We denote by $GS(P)_0$ the connected subgroup of $GS(P)$ with the Lie algebra $gs(P)_0$. Since the constraint set C is the zero level of the momentum map J_0 for the Hamiltonian action of $GS(P)_0$ in P the reduced phase space $\check{P}$ is defined as the space of the $GS(P)_0$ orbits in C,

$$\check{P} = C/GS(P)_0.$$

Since $GS(P)_0$ is a closed subgroup of $GS(P)$, the properness of the action of $GS(P)$ implies that the action of $GS(P)_0$ is also proper. Hence, the quotient topology in $\check{P}$ is Hausdorff.

9. THE REDUCED PHASE SPACE FOR THE MINKOWSKI SPACE THEORY

For the Minkowski space theory the constraint set ${J_0}^{-1}(0)$ a submanifold of P_M. The restriction to ${J_0}^{-1}(0)$ of the symplectic form ω has a null foliation. The reduced phase space $\check{P}_M$ is the space of leaves of this foliation. It is a quotient manifold of ${J_0}^{-1}(0)$, that is the natural projection $\rho : {J_0}^{-1}(0) \to \check{P}_M$ is a submersion. Local charts in $\check{P}_M$ are given by the slices for the action of $GS(P_M)_0$. The restrictions to ${J_0}^{-1}(0)$ of the forms θ and ω project to forms $\check{\theta}$ and $\check{\omega}$ on $\check{P}_M$ making it a weakly symplectic manifold.[5] The evolution equations in $\check{P}_M$ are Hamiltonian.

These results are formally analogous to the results of Mitter and Viallet who considered gauge group action on compact manifolds without boundary with one fixed point.[21]

10. THE REDUCED PHASE SPACE FOR THE BAG BOUNDARY CONDITIONS

The partition of the reduced phase space in Yang-Mills theory was first studied in Ref.22. A more general analysis in terms of the topological structures of the spaces under consideration was given in Ref.20. However, some of the assumptions made in Ref.20 are not satisfied in our case. We have managed to circumvent this difficulty and reproduce the same results.[4]

Since $GS(P_B)_0 = GS(P_B)$, the reduced phase space of P_B is the space $\check{P}_B$ of $GS(P_B)$ orbits in $J^{-1}(0)$,

$$\check{P}_B = J^{-1}(0)/GS(P_B).$$

If all the spaces involved were finite dimensional, properness of the action of $GS(P_B)$ would imply a partition of $\check{P}_B$ by symplectic manifolds.[23] We can obtain a similar result also in our case.

For each $p = (A, E, \Psi) \in P_B$, the isotropy group of p is a compact subgroup GS_p of $GS(P_B)$. Its Lie algebra will be denoted gs_p. For a compact subgroup H of $GS(P_B)$, let P_H be the set of points in P_B with the isotropy group GS_p equal to H,

$$P_H = \{p \in P_B | GS_p = H\}.$$

For every subgroup H of $GS(P_B)$ such that $P_H \neq \emptyset$, the projection of $P_H \cap J^{-1}(0)$ to $\check{P}_B$ by the canonical projection map $\rho : J^{-1}(0) \to \check{P}_B$ is a symplectic manifold

$$\check{P}_H = \rho(P_H \cap J^{-1}(0)),$$

with an exact symplectic form

$$\check{\omega}_H = d\check{\theta}_H.$$

The form $\check{\theta}_H$ on $\check{P}_H$ pulls back to the restriction of θ to $P_H \cap J^{-1}(0)$. If H_1 and H_2 are subgroups of $GS(P_B)$ such that $\check{P}_{H_1} \cap \check{P}_{H_2} \neq \emptyset$, then H_1 is conjugate to H_2, and $(\check{P}_{H_1}, \check{\omega}_{H_1}) = (\check{P}_{H_2}, \check{\omega}_{H_2})$. Thus, the reduced phase space is the union of symplectic manifolds parameterized by the conjugacy classes of compact subgroups H of $GS(P_B)$ such that $P_H \cap J^{-1}(0) \neq \emptyset$.

Each part $\check{P}_H$ in $\check{P}_B$ corresponds to a definite mode of symmetry breaking. In order to describe the symmetry breaking by the Cauchy data $p = (A, E, \Psi)$ in $P_H \cap J^{-1}(0)$, recall that A describes a connection in the trivial principal fibre bundle

$$Q = M \times G,$$

and E and Ψ are sections of bundles associated to Q. Since elements of H are maps from M to G, a choice of a point $x_0 \in M$ gives an isomorphism between H and a subgroup H_0 of G, defined by

$$H_0 = \{\phi(x_0) | \phi \in H\} \ .$$

The set

$$Q_0 = \{(x, g) \in Q | \phi^{-1}(x) g \phi(x) = \phi^{-1}(x_0) g \phi(x_0) \ \forall \ \phi \in H\}$$

is a right principal fibre bundle over M. Its structure group bundle is the centralizer $Z[H_0]$ of H_0 in G,

$$Z[H_0] = \{g \in G | g\phi(x_0) = \phi(x_0) g \ \forall \ \phi \in H\}.$$

Cauchy data (A, E, Ψ) are in P_H if and only if the connection defined by A reduces to a connection in Q_0, and E and Ψ reduce to sections of corresponding bundles associated to Q_0. The reducibility of connection means that, the pullback to Q_0 of the connection form on Q has values in the Lie algebra $z[H_0]$ of $Z[H_0]$.

The centre $C[H_0] = H_0 \cap Z[H_0]$ of H_0 gives rise to further symmetry breaking. As before, we can construct a sub-bundle Q_1 of Q_0 with structure group $C[H_0]$. The connection in Q_0 defined by A decomposes into a connection in Q_1 and the component normal in $z[H_0]$ to the Lie algebra of $C[H_0]$ which corresponds to the vector bosons.

This symmetry breaking is purely intrinsic. It does not require a Higgs field. However, the vector bosons obtained in this way are massless. It is possible that the mass of the vector bosons might appear in quantization as an anomaly.

11. COLOUR CHARGES

Colour charges are the conserved quantities J_ξ corresponding to infinitesimal gauge symmetries $\xi \in gs(P)$. For every $(A, E, \Psi) \in C$,

$$J_\xi(A, E, \Psi) = \int_{\partial M} nE \cdot \xi dS \ ,$$

where dS is the element of surface area. For the bag boundary conditions $nE = 0$ and the colour charges vainsh identically.

In the Minkowski space theory, $M = \mathbf{R}^3$, ∂M is the sphere at infinity, and

$$J_\xi(A, E, \Psi) = \lim_{r \to \infty} \int_{S_r} nE\xi dS.$$

If $\xi - \xi' \in gs(P_M)_0$, and $(A, E, \Psi) \in C$, then $J_\xi(A, E, \Psi) = J_{\xi'}(A, E, \Psi)$. Hence,the colour charges in physical states are labelled by the elements of the quotient

$$colour(P_M) = gs(P_M)/gs(P_M)_0.$$

It is a Lie algebra isomorphic to the structure algebra,

$$colour(P_M) \simeq g.$$

The pull back of J_ξ to C depends only on the class $[\xi] \in colour(P_M)$ of ξ, and it determines a function $J_{[\xi]}$ on C, which induces a function $\check{J}_{[\xi]}$ on the reduced phase space $\check{P}_M$. For every $\pi(A, E, \Psi) \in \check{P}_M$,

$$\check{J}_{[\xi]}(\pi(A, E, \Psi)) = J_\xi(A, E, \Psi)$$

gives the value in the physical state $\pi(A, E, \Psi)$ of the colour charge labelled by $[\xi]$. Note that the class $[\xi] \in colour(P_M)$ of $\xi \in gs(P_M)$ is uniquely determined by the asymptotic behaviour of ξ at infinity.

Since the colour charge $\check{J}_{[\xi]}(\pi(A, E, \Psi))$ can be also expressed as a volume integral

$$\check{J}_{[\xi]}(\pi(A, E, \Psi)) = \int_{\mathbf{R}^3} (E \cdot D_A\xi + \Psi^\dagger \xi \Psi) d_3x,$$

one would like to interpret the integrand

$$\check{j}_{[\xi]} = E \cdot D_A\xi + \Psi^\dagger \xi \Psi$$

as a function on the reduced phase space $\check{P}_M$ giving the physical density of the colour $[\xi]$. However, it depends on the choice of a representative ξ of the class $[\xi]$, even though the integral over $\mathbf{R}^3$ does not. Hence, in order to make $\check{j}_{[\xi]}$ well defined, we have to determine a way of choosing a representative ξ of $[\xi]$ such that the right hand side is $GS(P_M)_0$ invariant. In Ref.2 we have shown that, unless $[\xi]$ is in the centre of *colour*(P_M), the resulting function $\check{\rho}_{[\xi]}$ depends non-locally on (A, E, Ψ). Hence, only colours in the centre of the colour algebra admit local charge densities satisfying local commutation relations.

Suppose that we have $[\xi]$ not in the centre of *colour*(P_M), and we have made a choice of a $GS(P_M)_0$ invariant charge density $\check{j}_{[\xi]}$, which depends non-locally on (A, E, Ψ). Then, the Poisson bracket $\{\check{j}_{[\xi]}(x), \check{j}_{[\xi]}(x')\}$ has in its support points (x, x') such that $x \neq x'$. A quantization of this theory, preserving the classical Poisson brackets up to local Schwinger terms, would yield quantum operators with non-local commutation relations. Considering these operators as observables would violate the locality axiom of quantum field theory.[25] This result suggests that such a theory leads to a confinement of non-abelian charges.

This conclusion, presented at the XIIth Workshop, was criticized by physicists. Their main argument was that, in the Coulomb gauge formulation of electrodynamics, the longitudinal components of the electric field have non-local commutation relations even though the electric field strength is observable. However, the longitudinal component of the electric field is a non-local function of the electric field strength. Hence, it cannot be considered an observable. Observable quantities are the local charge density, given by the divergence of the longitudinal component of the electric field strength, and the total charge in a region, given by its flux through the boundary.

In axiomatic relativistic field theory observables are assumed to form an algebra. That is, products of smeared out field operators are observables, but one does not extend it to arbitrary functions. In particular, non-local functions of field operators cannot be observable, because they would violate the locality axiom of the theory. We have to be careful in distinguishing observables and functions of observables which cannot be observed but can be evaluated.

References

1. J. Śniatycki and G. Schwarz, The existence and uniqueness of solutions of Yang-Mills equations with the bag boundary conditions, *Commun. Math. Phys.* 159:593-304 (1994)

2. J. Śniatycki and G. Schwarz, An invariant argument for confinement, *Rep. Math. Phys.* (to appear)

3. G. Schwarz and J. Śniatycki, Yang-Mills and Dirac fields in a bag, existence and uniqueness results, *Commun. Math. Phys.* (to appear)

4. J. Śniatycki, G. Schwarz and L. Bates, Yang-Mills and Dirac fields in a bag, constraints and reduction, submitted to *Commun. Math. Phys.*

5. G. Schwarz and J. Śniatycki, Yang-Mills and Dirac fields in the Minkowski space-time, in preparation

6. E. Binz, J. Śniatycki and H. Fischer, "Geometry of Classical Fields", North Holland, Amsterdam (1988)

7. J. Śniatycki, Regularity of constraints in the Minkowski space Yang-Mills theory, *Commun. Math. Phys.* 141:593-597 (1991)

8. I. Segal, The Cauchy problem for the Yang-Mills equations, *J. Funct. Anal.* 33:175-194 (1979)

9. J. Ginibre and G. Velo, The Cauchy problem for coupled Yang-Mills and scalar fields in temporal gauge, *Commun. Math. Phys.* 82:1-28 (1981)

10. D.M. Eardley and V. Moncrief, "The global existence of Yang-Mills-Higgs fields in 4-dimensional Minkowski space", *Commun. Math. Phys.* 83:171-191, 193-212 (1982)

11. D. Christodoulou, *C. R. Acad. Sc. Paris, Sér. I,* 293:139 (1981)

12. Y. Choquet-Bruhat and D. Christodoulou, Existence de solutions globales des équations classiques des théories de jauge, *C. R. Acad. Sc. Paris, Sér. I,* 293:195-199 (1981)

13. S. Klainerman and M. Machedon, Finite energy solutions of the Yang-Mills equations in $\mathbf{R}^{3+1}$, preprint, Department of Mathematics, Princeton University

14. R. Abraham, J. Marsden and T. Ratiu, "Manifolds, Tensor Analysis and Applications", Springer Verlag, New York (1988)

15. K. Johnson, The M.I.T. bag model, *Acta Physica Polonica B* 6:865-892 (1975)

16. I. Segal, Non-linear semi-groups, *Ann. of Math.* 78:339-364 (1963)

17. H. Aikawa, On weighted Beppo Levi functions – Integral representations and behavior at infinity, *Analysis*, 9:323-346 (1989)

18. J. Eichhorn, Gauge theory of open manifolds of bounded geometry, preprint, Fachbereich Mathematik, Universität Greifswald

19. W. Kondracki and J. Rogulski, "On the stratification of the orbit space for the action of automorphisms on connections", *Dissertationes Mathematicae*, CCL, Warsaw (1986)

20. J. Arms, J. Marsden and V. Moncrief, Symmetry and bifurcation of momentum maps, *Commun. Math. Phys.* 78:455-478 (1981)

21. P. Mitter and C. Viallet, On the bundle of connections and the gauge orbit manifold in Yang-Mills theory, *Commun. Math. Phys.* 79:457-472 (1981)

22. J. Arms, The structure of the solution set for the Yang-Mills equations, *Math. Proc. Camb. Phil. Soc.* 90:361-372 (1981)

23. R. Sjamaar and E. Lerman, Stratified symplectic spaces and reduction, *Ann. of Math.* 134:375-422 (1991)

24. Y. Kerbrat, H. Kerbrat-Lunc and J. Śniatycki, How to get masses from Kaluza-Klein theory, *J. Geom. Phys.* 6:311-329 (1989)

25. R.F. Streater and A.S. Wightman, "PCT, Spin and Statistics and All That", W.A. Benjamin, New York (1964).

SYMPLECTIC INDUCTION, UNITARY INDUCTION AND BRST THEORY (Summary)

G. M. Tuynman

UFR de Mathématiques, Université de Lille I
F-59655 Villeneuve d'Ascq Cedex, France

The data needed for a unitary representation of a Lie group G are a Hilbert space $\mathcal{H}$ (finite or infinite dimensional) and a (linear) action of G on $\mathcal{H}$. This action should preserve the inner product in $\mathcal{H}$. In the same spirit one can define a symplectic and a hamiltonian representation of G. The data needed are a symplectic manifold (M, ω) and an action of G on M. For a symplectic representation this action should preserve ω; for a hamiltonian representation it should also posses an equivariant moment map (which takes values in the dual Lie algebra of G). In these terms the procedure of geometric quantization can be described as a procedure to transform a hamiltonian representation of G into a unitary representation.

Now let $H \subset G$ be a closed subgroup of G. If we have a unitary representation of H, the well known procedure of unitary induction constructs out of this a unitary representation of G. If on the other hand we have a hamiltonian representation of H, the less well known procedure of symplectic induction constructs out of this a hamiltonian representation of G. It is thus quite natural to ask whether geometric quantization intertwines these two induction procedures.

We show that it is very probable that this is indeed the case. Although the general question is still open, in various special cases this has been proved. Moreover, no counter example is known. The link between this question and BRST theory is as follows. In the context of BRST theory it was found that for non-unimodular groups the standard Dirac procedure for the form of quantum constraints needs to be modified by a term involving the trace of the adjoint representation (which is zero for unimodular groups). Since constraints also appear in the symplectic induction procedure (in the form of a Marsden-Weinstein reduction), this modification also appears there. It then turns out that this modification is crucial to prove that geometric quantization intertwines the two types of induction procedures.

References

1. C. Duval, J. Elhadad and G.M. Tuynman, Symplectic induction and Pukanszky's condition, *J. Diff. Geom.* 36:331 (1992)

2. C. Duval, J. Elhadad and G. M. Tuynman, The BRS method and geometric quantization : some examples, *Commun. Math. Phys.* 126:535 (1990)

3. C. Duval, J. Elhadad, M.J. Gotay, J. Śniatycki and G. M. Tuynman, Quantization and bosonic BRS theory, *Ann. Phys.* 206:1 (1991).

PART II

COHERENT STATES, COMPLEX AND POISSON STRUCTURES

SPIN COHERENT STATES FOR THE POINCARÉ GROUP

S. Twareque Ali[1] and J.-P. Gazeau[2]

[1] Department of Mathematics and Statistics
Concordia University, Montréal, Quebec
Canada H4B 1R6

[2] Lab. de Physique Théorique et Mathématique
Université Paris 7–Denis Diderot, 2, place Jussieu
F - 75251 Paris Cedex 05, France

Abstract

We present here a construction for certain families of coherent states, arising from representations of the Poincaré group, for particles of positive mass and integral or half-integral spin. These coherent states are labelled by points in the classical phase space of the relativistic particle and are associated to affine sections of the Poincaré group, considered as a fibre bundle over the phase space.

1. INTRODUCTION

In the spirit of a general theory, elaborated elsewhere,[1,2] *coherent states* (CS) will be associated in this report to certain *square integrable* representations of groups. The following framework will be adopted: Let G be a locally compact group, $g \mapsto U(g)$ a unitary, irreducible representation (UIR) of G on the (complex, separable) Hilbert space $\mathcal{H}$; let $H \subset G$ be a closed subgroup and suppose that $X = G/H$ carries the *invariant* measure ν. Assume that there exists a (finite) set of vectors η^i, $i = 1, 2, \ldots, n$, in $\mathcal{H}$ and a Borel section $\sigma : X \to G$, such that

$$\sum_{i=1}^{n} \int_X |\eta^i_{\sigma(x)}\rangle\langle\eta^i_{\sigma(x)}| \, d\nu(x) = A, \qquad \eta^i_{\sigma(x)} = U(\sigma(x))\eta^i, \tag{1.1}$$

where A is a bounded positive operator on $\mathcal{H}$, with a densely defined inverse. (The integral in (1.1) is assumed to converge weakly.) We then say that the representation U is *square integrable* mod (H, σ), and call the set of vectors,

$$\mathcal{S} = \{\eta^i_{\sigma(x)} \mid x \in X, \ \ i = 1, 2, \ldots, n\} \subset \mathcal{H}, \tag{1.2}$$

a *family of covariant coherent states* (CS, for short), for the representation U. If A^{-1} is also a bounded operator, we say that the family of CS $\mathcal{S}$ forms a *rank-n frame*, denoted $\mathcal{F}\{\eta^i_{\sigma(x)}, A, n\}$, and if furthermore, A is a multiple of the identity, we say that the frame is *tight*.

The results presented here generalize some previous work[2,3] on $\mathcal{P}^{\uparrow}_{+}(1,1)$ (the Poincaré group in 1-space and 1-time dimensions) as well as some older work[5] on $\mathcal{P}^{\uparrow}_{+}(1,3)$

Quantization, Coherent States, and Complex Structures
Edited by J.-P. Antoine *et al.*, Plenum Press, New York, 1995

(the full Poincaré group). We obtain here families of CS for any UIR of $\mathcal{P}_+^\uparrow(1,3)$, corresponding to mass $m > 0$ and spin $s = 0, \frac{1}{2}, 1, \frac{3}{2}, \ldots$. Just as in the $\mathcal{P}_+^\uparrow(1,1)$ situation, each family of CS is associated to a particular affine section, and one always obtains a frame in this manner. In this note we only report on the main results. A fuller discussion, along with proofs of theorems, will be published elsewhere.[4]

The notation adopted is as follows: The full Poincaré group $\mathcal{P}_+^\uparrow(1,3)$ is the 2-fold covering group,

$$\mathcal{P}_+^\uparrow(1,3) = T^4 \oslash SL(2,\mathbb{C}), \tag{1.3}$$

where $T^4 \simeq \mathbb{R}_{1,3}$ is the group of space-time translations. Elements of $\mathcal{P}_+^\uparrow(1,3)$ will be denoted by

$$(a, A), \qquad a = (a_0, \mathbf{a}) \in \mathbb{R}_{1,3}, \qquad A \in SL(2,\mathbb{C}). \tag{1.4}$$

The multiplication law is

$$(a, A)(a', A') = (a + \Lambda a', AA') \tag{1.5}$$

where $\Lambda \in \mathcal{L}_+^\uparrow(1,3)$ (the proper, orthochronous Lorentz group) is the Lorentz transformation corresponding to A. The metric tensor is $g_{00} = 1 = -g_{11} = -g_{22} = -g_{33}$. Also, denote by

$$\mathcal{V}_m^+ = \{k = (k_0, \mathbf{k}) \in \mathbb{R}_{1,3} \mid k^2 = k_0^2 - \mathbf{k}^2 = m^2\}, \tag{1.6}$$

the *forward mass hyperboloid* (we take $c = \hbar = 1$).

In the Wigner realization, the unitary, irreducible representation U_W^s of $\mathcal{P}_+^\uparrow(1,3)$ for a particle of mass $m > 0$ and spin $s = 0, \frac{1}{2}, 1, \frac{3}{2}, 2, \ldots$, is carried by the Hilbert space

$$\mathcal{H}_W^s = \mathbb{C}^{2s+1} \otimes L^2(\mathcal{V}_m^+, \frac{d\mathbf{k}}{k_0}), \tag{1.7}$$

and is defined by

$$(U_W^s(a, A)\phi)(k) = e^{ik\cdot a}\, \mathcal{D}^s(h(k)^{-1} A h(\Lambda^{-1}k))\phi(\Lambda^{-1}k), \qquad k \cdot a = k_0 a_0 - \mathbf{k}\cdot\mathbf{a}, \tag{1.8}$$

where $\mathcal{D}^s$ is the $(2s+1)$-dimensional irreducible spinor representtation of $SU(2)$ (carried by $\mathbb{C}^{2s+1}$) and

$$k \to h(k) = \frac{m\mathbf{I}_2 + \sigma\cdot\overline{k}}{\sqrt{2m(k_0 + m)}}, \qquad (\overline{k} = (k_0, -\mathbf{k})), \tag{1.9}$$

is the image in $SL(2,\mathbb{C})$ of the Lorentz boost Λ_k, which brings the four vector $(m, \mathbf{0})$ to the 4-vector k in $\mathcal{V}_m^+$ (σ is a 4-vector of Pauli matrices).

The matrix form of the Lorentz boost is

$$\Lambda_k = \frac{1}{m}\begin{pmatrix} k_0 & \mathbf{k}^\dagger \\ \mathbf{k} & mV_k \end{pmatrix} = \Lambda_k{}^\dagger, \tag{1.10}$$

where V_k is the 3×3 symmetric matrix

$$V_k = \mathbf{I}_3 + \frac{\mathbf{k}\otimes\mathbf{k}^\dagger}{m(k_0 + m)} = V_k{}^\dagger. \tag{1.11}$$

2. COMPUTATION OF COHERENT STATES

As in the case of the $\mathcal{P}_+^\uparrow(1,1)$ group, CS for the $\mathcal{P}_+^\uparrow(1,3)$ group will be associated to the phase space Γ of a classical relativistic particle:

$$\Gamma = \mathcal{P}_+^\uparrow(1,3)/T \times SU(2), \tag{2.1}$$

T denoting the subgroup of time translations. For $A \in SL(2,\mathbb{C})$, let

$$A = h(k)R(k), \quad R(k) \in SU(2), \tag{2.2}$$

be its Cartan decomposition. An arbitrary element $(a, A) \in \mathcal{P}_+^\uparrow(1,3)$ has the left coset decomposition,

$$(a, A) = \left((0, \mathbf{a} - \frac{a_0\mathbf{k}}{k_0}),\ h(k)\right)\left((\frac{ma_0}{k_0}, \mathbf{0}),\ R(k)\right), \tag{2.3}$$

according to (2.1). Thus, the elements in Γ have the global coordinatization, $(\mathbf{q}, \mathbf{p}) \in \mathbb{R}^6$:

$$\left.\begin{aligned} \mathbf{q} &= \mathbf{a} - \frac{a_0\mathbf{k}}{k_0} \\ \mathbf{p} &= \mathbf{k} \end{aligned}\right\}. \tag{2.4}$$

In terms of these variables, the action of $\mathcal{P}_+^\uparrow(1,3)$ on Γ is given by $(\mathbf{q},\mathbf{p}) \to (\mathbf{q}',\mathbf{p}') = (a, A)(\mathbf{q}, \mathbf{p})$,

$$\left.\begin{aligned} \mathbf{q}' &= \frac{1}{p_0'}\left(p_0'[\mathbf{a} + \underline{\Lambda(0,\mathbf{q})}] - \mathbf{p}'[a_0 + \{\Lambda(0,\mathbf{q})\}_0]\right) \\ \mathbf{p}' &= \underline{\Lambda p}, \quad p = (\sqrt{m^2 + \mathbf{p}^2},\ \mathbf{p}) \end{aligned}\right\}, \tag{2.5}$$

where $\Lambda \in \mathcal{L}_+^\uparrow(1,3)$ is the Lorentz matrix corresponding to A, and $p_0' = (\Lambda p)_0$. The measure $d\mathbf{q}\, d\mathbf{p}$ is invariant under this action.

Next, in terms of these variables let us define the basic section,

$$\left.\begin{aligned} \sigma_0 : \Gamma &\to \mathcal{P}_+^\uparrow(1,3) \\ \sigma_0(\mathbf{q},\mathbf{p}) &= ((0,\mathbf{q}),\ h(p)) \end{aligned}\right\}, \tag{2.6}$$

to be called the *Galilean section*. Any other section $\sigma : \Gamma \to \mathcal{P}_+^\uparrow(1,3)$ is then related to σ_0 in the manner

$$\sigma(\mathbf{q},\mathbf{p}) = \sigma_0(\mathbf{q},\mathbf{p})\ ((f(\mathbf{q},\mathbf{p}),\ \mathbf{0}),\ R(\mathbf{q},\mathbf{p})) \tag{2.7}$$

where $f : \mathbb{R}^6 \to \mathbb{R}$ and $R : \mathbb{R}^6 \to SU(2)$ are smooth functions. As usual, we work with the particular class of *affine sections*, for which

$$f(\mathbf{q},\mathbf{p}) = \varphi(\mathbf{p}) + \mathbf{q}\cdot\boldsymbol{\vartheta}(\mathbf{p}), \tag{2.8}$$

where, $\varphi : \mathbb{R}^3 \to \mathbb{R}$, $\boldsymbol{\vartheta} : \mathbb{R}^3 \to \mathbb{R}^3$ are smooth functions of $\mathbf{p}$ alone. As far as the construction of CS is concerned, φ only introduces an inessential phase, and hence we omit this term. Moreover, we also impose the restriction that $R(\mathbf{q},\mathbf{p}) = R(\mathbf{p})$. Thus writing,

$$\sigma(\mathbf{q},\mathbf{p}) = (\hat{q},\ h(p)R(\mathbf{p})), \qquad \hat{q} = (\hat{q}_0,\ \hat{\mathbf{q}}) \in \mathbb{R}_{1,3} \tag{2.9}$$

we see that

$$\left.\begin{aligned} \hat{q}_0 &= \frac{p_0}{m}\boldsymbol{\vartheta}\cdot\mathbf{q} \\ \hat{\mathbf{q}} &= M(\mathbf{p},\boldsymbol{\vartheta})\mathbf{q} \end{aligned}\right\}, \tag{2.10}$$

where $M(\mathbf{p}, \boldsymbol{\vartheta})$ is the 3×3 real matrix

$$M(\mathbf{p}, \boldsymbol{\vartheta}) = \mathbf{I}_3 + \frac{\mathbf{p} \otimes \boldsymbol{\vartheta}(\mathbf{p})^\dagger}{m}. \tag{2.11}$$

Note that

$$\det\,[M(\mathbf{p}, \boldsymbol{\vartheta})] = 1 + \frac{\mathbf{p} \cdot \boldsymbol{\vartheta}(\mathbf{p})}{m}, \quad \text{and} \quad \|M(\mathbf{p}, \boldsymbol{\vartheta})\| = \max\{1,\ 1,\ 1 + \frac{\mathbf{p} \cdot \boldsymbol{\vartheta}(\mathbf{p})}{m}\}, \tag{2.12}$$

Also, assuming $M(\mathbf{p}, \boldsymbol{\vartheta})$ to be continuous in $\mathbf{p}$ and $\boldsymbol{\vartheta}$, and non-singular for all $(\mathbf{p}, \boldsymbol{\vartheta})$ (i.e., $\mathbf{q} = \mathbf{0} \Leftrightarrow \hat{\mathbf{q}} = \mathbf{0}$ in (2.10)) and since $\det\,[M(\mathbf{0}, \boldsymbol{\vartheta})] = 1$, it follows that

$$\det\,[M(\mathbf{p}, \boldsymbol{\vartheta})] > 0, \quad \forall\,(\mathbf{p}, \boldsymbol{\vartheta}). \tag{2.13}$$

We now take an arbitrary afine section σ, and going back to the Hilbert space $\mathcal{H}^s_W$ in (1.7) choose a set of vectors $\boldsymbol{\eta}^i$, $i = 1, 2, \ldots, 2s+1$, in it to define the formal operator (see (1.1) and (1.8)):

$$A_\sigma = \sum_{i=1}^{2s+1} \int_{\mathbb{R}^6} |\boldsymbol{\eta}^i_{\sigma(\mathbf{q},\mathbf{p})}\rangle\langle\boldsymbol{\eta}^i_{\sigma(\mathbf{q},\mathbf{p})}|\, d\mathbf{q}\, d\mathbf{p}, \qquad \boldsymbol{\eta}^i_{\sigma(\mathbf{q},\mathbf{p})} = U^s_W(\sigma(\mathbf{q}, \mathbf{p}))\boldsymbol{\eta}^i. \tag{2.14}$$

From the general definition, in order for the set of vectors

$$\mathcal{S}_\sigma = \{\boldsymbol{\eta}^i_{\sigma(\mathbf{q},\mathbf{p})} \mid (\mathbf{q}, \mathbf{p}) \in \mathbb{R}^6,\ \ i = 1, 2, \ldots, 2s+1\} \subset \mathcal{H}^s_W, \tag{2.15}$$

to constitute a family of coherent states for the representation U^s_W, the integral in (2.14) must converge weakly, and define A_σ as a bounded operator with inverse. Thus, we need to to determine the convergence of the ordinary integral

$$I_{\phi,\psi} = \sum_{i=1}^{2s+1} \int_{\mathbb{R}^6} \langle\phi|\boldsymbol{\eta}^i_{\sigma(\mathbf{q},\mathbf{p})}\rangle\langle\boldsymbol{\eta}^i_{\sigma(\mathbf{q},\mathbf{p})}|\psi\rangle\, d\mathbf{q}\, d\mathbf{p} \tag{2.16}$$

for arbitrary $\phi, \psi \in \mathcal{H}^s_W$. Detailed computations show that the existence of this integral depends on the positivity of the determinant

$$\det\,[\mathcal{J}(\mathbf{k})] = 1 + \frac{1}{mk_0}\,\boldsymbol{\vartheta}(\mathbf{p}) \cdot [k_0\mathbf{p} - \mathbf{k}p_0]. \tag{2.17}$$

which in turn, imposes restrictions on $\boldsymbol{\vartheta}$ and hence on the 4-vector $\hat{q} = (\hat{q}_0, \hat{\mathbf{q}})$, coming from the affine section σ and defined in (2.10). Indeed, since the matrix $M(\mathbf{p}, \boldsymbol{\vartheta})$ in that equation has an inverse, we easily otain from there,

$$\hat{q}_0 = \boldsymbol{\beta}(\mathbf{p}) \cdot \hat{\mathbf{q}}, \tag{2.18}$$

where $\boldsymbol{\beta}(\mathbf{p})$ is the 3-vector field

$$\boldsymbol{\beta}(\mathbf{p}) = \frac{p_0\boldsymbol{\vartheta}(\mathbf{p})}{m + \mathbf{p} \cdot \boldsymbol{\vartheta}(\mathbf{p})}. \tag{2.19}$$

Solving for $\boldsymbol{\vartheta}(\mathbf{p})$, this gives

$$\boldsymbol{\vartheta}(\mathbf{p}) = \frac{m\boldsymbol{\beta}(\mathbf{p})}{p_0 - \mathbf{p} \cdot \boldsymbol{\beta}(\mathbf{p})}. \tag{2.20}$$

Let us also introduce the *dual* vector fields $\boldsymbol{\beta}^*$ and $\boldsymbol{\vartheta}^*$,

$$\boldsymbol{\beta}^*(\mathbf{p}) = \frac{\mathbf{p} - mV_p\boldsymbol{\beta}(\mathbf{p})}{p_0 - \mathbf{p}\cdot\boldsymbol{\beta}(\mathbf{p})}, \tag{2.21}$$

$$\boldsymbol{\vartheta}^*(\mathbf{p}) = \frac{m\boldsymbol{\beta}^*(\mathbf{p})}{p_0 - \mathbf{p}\cdot\boldsymbol{\beta}^*(\mathbf{p})}, \tag{2.22}$$

where V_p is the matrix defined (1.11). Note that

$$\boldsymbol{\beta}^{**} = \boldsymbol{\beta}, \qquad \boldsymbol{\vartheta}^{**} = \boldsymbol{\vartheta}, \tag{2.23}$$

and

$$\boldsymbol{\vartheta}(\mathbf{p}) = \frac{1}{m}[\mathbf{p} - mV_p\boldsymbol{\beta}^*(\mathbf{p})] = \frac{\mathbf{p} - p_0V_p^{-1}\boldsymbol{\vartheta}^*(\mathbf{p})}{m + \mathbf{p}\cdot\boldsymbol{\vartheta}^*(\mathbf{p})}. \tag{2.24}$$

Furthermore, in terms of $\boldsymbol{\beta}^*$, we can rewrite (2.17) as

$$\det\left[\mathcal{J}(\mathbf{k})\right] = \frac{p_0(\Lambda_k\overline{p})_0}{mk_0}\left[1 + \frac{(\Lambda_k\overline{p})}{(\Lambda_k\overline{p})_0}\cdot\rho(k\to\overline{p})^\dagger\boldsymbol{\beta}^*(\mathbf{p})\right], \tag{2.25}$$

where $\rho(k \to \overline{p})$ is the rotation matrix,

$$\rho(k\to\overline{p}) = \Lambda_p^{-1}\Lambda_k\Lambda_p\Lambda_k^{-1}. \tag{2.26}$$

Thus, the positivity of $\det\left[\mathcal{J}(\mathbf{k})\right]$ would be ensured if the second term within the square brackets in (2.25) does not exceed 1 in magnitude, i.e., if $\|\boldsymbol{\beta}^*(\mathbf{p})\| < 1$, $\forall\mathbf{p}$. Indeed, we have the result

Proposition 2.1. *The following conditions are equivalent*

1. *the 4-vector* $\hat{q} = (\hat{q}_0, \hat{\mathbf{q}})$ *is space-like, i.e.,*

$$|\hat{q}_0|^2 - \|\hat{\mathbf{q}}\|^2 < 0; \tag{2.27}$$

2. *the matrix*

$$S(\mathbf{p},\boldsymbol{\vartheta}) = \mathbf{I}_3 + \left[\boldsymbol{\vartheta}\otimes(\frac{\mathbf{p}}{m} - \frac{\boldsymbol{\vartheta}}{2})^\dagger + (\frac{\mathbf{p}}{m} - \frac{\boldsymbol{\vartheta}}{2})\otimes\boldsymbol{\vartheta}^\dagger\right] \tag{2.28}$$

is positive definite for all $p \in \mathcal{V}_m^+$;

3. *for all unit vectors* $\hat{e} \in \mathbb{R}^3$ *and all* $\mathbf{p} \in \mathbb{R}^3$,

$$|\hat{e}\cdot\left(\boldsymbol{\vartheta}(\mathbf{p}) - \frac{\mathbf{p}}{m}\right)| < \frac{1}{m}[(\hat{e}\cdot\mathbf{p})^2 + m^2]^{\frac{1}{2}}; \tag{2.29}$$

4. *for all* $p \in \mathcal{V}_m^+$,

$$p_0\|\boldsymbol{\vartheta}(\mathbf{p})\| < m + \mathbf{p}\cdot\boldsymbol{\vartheta}(\mathbf{p}) = |m + \mathbf{p}\cdot\boldsymbol{\vartheta}(\mathbf{p})|; \tag{2.30}$$

5. *for all* $\mathbf{p} \in \mathbb{R}^3$, *the 3-vector field* $\boldsymbol{\beta}$ *obeys*

$$\|\boldsymbol{\beta}(\mathbf{p})\| < 1; \tag{2.31}$$

6. *for all* $\mathbf{p} \in \mathbb{R}^3$, *the 3-vector field* $\boldsymbol{\beta}^*$ *obeys*

$$\|\boldsymbol{\beta}^*(\mathbf{p})\| < 1. \tag{2.32}$$

Note that (2.29) shows, in particular, that

$$\hat{q} \text{ space-like } \Rightarrow \|\boldsymbol{\vartheta}(\mathbf{p}) - \frac{\mathbf{p}}{m}\| < \frac{p_0}{m}, \tag{2.33}$$

for all $p \in \mathcal{V}_m^+$. As a consequence of this proposition we have the next result:

Proposition 2.2. *The determinant* det $[\mathcal{J}(\mathbf{k})]$ *is strictly positive for all* $\mathbf{k}$, $\mathbf{p} \in \mathbf{R}^3$ *if and only if the 4-vector* $\hat{q} = (\hat{q}_0, \hat{\mathbf{q}})$ *is space-like, i.e., if and only if any one of the equivalent conditions in Proposition 2.1 is satisfied.*

Assuming that the above conditions are fulfilled by the affine sections σ that we choose, in order to construct the CS (2.15), we may return to the evaluation of the integral (2.16) and obtain estimates on $|I_{\phi,\psi}|$, therby arriving at conditions under which (2.14) would define a bounded invertible operator A_σ. This would ensure that (2.15) does indeed define a family of CS for the representation U_W^s of $\mathcal{P}_+^\uparrow(1,3)$. It will be necessary to put additional restrictions on the vectors $\boldsymbol{\eta}^i$, defining the CS, in order to obtain our final result in Proposition 2.3 below:

1. Assumption of *rotational invariance*:

$$\mathcal{D}^s(R)(\sum_{i=1}^{2s+1} |\boldsymbol{\eta}^i\rangle\langle\boldsymbol{\eta}^i|)\mathcal{D}^s(R)^\dagger = \sum_{i=1}^{2s+1} |\boldsymbol{\eta}^i\rangle\langle\boldsymbol{\eta}^i|, \tag{2.34}$$

   for all $R \in SU(2)$, which implies that we may take

$$\boldsymbol{\eta}^i = \hat{\mathbf{e}}_i \otimes \eta, \qquad i = 1, 2, \ldots, 2s+1, \tag{2.35}$$

   where $\hat{\mathbf{e}}_i$ are the unit vectors (δ_{ij}), $j = 1, 2, \ldots, 2s+1$, in $\mathbb{C}^{2s+1}$ and $\eta \in L^2(\mathcal{V}_m^+, d\mathbf{k}/k_0)$. We assume, furthermore, that the function $|\eta(k)|^2$ itself is also rotationally invariant, i.e., $|\eta(\rho k)|^2 = |\eta(k)|^2$, $\forall \rho \in SO(3)$.

2. Assumption of *admissibility*:

$$\int_{\mathcal{V}_m^+} |\eta(k)|^2 d\mathbf{k} < \infty \tag{2.36}$$

Also, let us define

$$\langle P_0 \pm \|\mathbf{P}\|\rangle_\eta = \int_{\mathcal{V}_m^+} (p_0 \pm \|\mathbf{p}\|)\, |\eta(p)|^2 \, \frac{d\mathbf{p}}{p_0}. \tag{2.37}$$

We may now state the main result of this paper:

Proposition 2.3. *Let* $\boldsymbol{\eta}^i$, $i = 1, 2, \ldots, 2s+1$, *satisfy the condition of rotational invariance. Then, for each* $\boldsymbol{\beta}$ *satisfying the conditions of Proposition 2.1 the set of vectors* $\mathcal{S}_\sigma$ *in (2.15) is a family of spin-s coherent states which defines a rank* $(2s+1)$-*frame,* $\mathcal{F}\{\boldsymbol{\eta}^i_{\sigma(\mathbf{q},\mathbf{p})}, A_\sigma, 2s+1\}$ *if and only if* η *satisfies the condition of admissibility. Furthermore, one has the estimates*

$$\frac{(2\pi)^3}{m}\langle P_0 - \|\mathbf{P}\|\rangle_\eta \leq \|A_\sigma\| \leq \frac{(2\pi)^3}{m}\langle P_0 + \|\mathbf{P}\|\rangle_\eta. \tag{2.38}$$

Some special cases of the sections σ are of particular interest:

(1) The Galilean section σ_0 :
For this section

$$\left.\begin{array}{ll} \boldsymbol{\beta}(\mathbf{p}) = \boldsymbol{\beta}_0(\mathbf{p}) = \mathbf{0}, & \boldsymbol{\vartheta}(\mathbf{p}) = \boldsymbol{\vartheta}_0(\mathbf{p}) = \mathbf{0} \\ \boldsymbol{\beta}_0^*(\mathbf{p}) = \dfrac{\mathbf{p}}{p_0}, & \boldsymbol{\vartheta}_0^*(\mathbf{p}) = \dfrac{\mathbf{p}}{m} \end{array}\right\}. \tag{2.39}$$

Here, $\|\boldsymbol{\beta}_0(\mathbf{p})\| < 1$, $\|\boldsymbol{\beta}_0^*(\mathbf{p})\| < 1$, $\forall\mathbf{p}$. The frame obtained, using this section is tight:

$$A_\sigma = A_0 = \frac{(2\pi)^3}{m}\langle P_0\rangle_\eta I. \tag{2.40}$$

(2) The Lorentz section σ_ℓ :
This time

$$\boldsymbol{\beta}(\mathbf{p}) = \boldsymbol{\beta}_\ell(\mathbf{p}) = \boldsymbol{\beta}_0^*(\mathbf{p}), \qquad \boldsymbol{\vartheta}(\mathbf{p}) = \boldsymbol{\vartheta}_\ell(\mathbf{p}) = \boldsymbol{\vartheta}_0^*(\mathbf{p}), \tag{2.41}$$

in other words, the Galilean and Lorentz sections are duals to each other. Furthermore, for this section one again gets a tight frame:

$$A_\sigma = A_\ell = (2\pi)^3 m\langle P_0^{-1}\rangle_\eta I. \tag{2.42}$$

(3) The symmetric section σ_s :
This section is self-dual, being given by

$$\boldsymbol{\beta}(\mathbf{p}) = \boldsymbol{\beta}_s(\mathbf{p}) = \boldsymbol{\beta}_s^*(\mathbf{p}) = \frac{\mathbf{p}}{m+p_0}, \qquad \boldsymbol{\vartheta}(\mathbf{p}) = \boldsymbol{\vartheta}_s(\mathbf{p}) = \boldsymbol{\vartheta}_s^*(\mathbf{p}) = \frac{\mathbf{p}}{m+p_0}. \tag{2.43}$$

Again, $\|\boldsymbol{\beta}_s(\mathbf{p})\| < 1$, $\forall\mathbf{p}$. For this section the frame is never tight, and in fact, the spectrum of the operator $A_\sigma = A_s$ is the entire interval $[(2\pi)^3\|\eta\|^2,\ (2\pi)^3\langle P_0\rangle_\eta/m]$.

References

1. S. T. Ali, J-P. Antoine and J-P. Gazeau, Square integrability of group representations on homogeneous spaces. I. Reproducing triples and frames, *Ann. Inst. H. Poincaré* 55:829-855 (1991)
2. S. T. Ali, J-P. Antoine and J-P. Gazeau, Square integrability of group representations on homogeneous spaces. II. Coherent and quasi-coherent states. The case of the Poincaré group, *Ann. Inst. H. Poincaré* 55:857-890 (1991)
3. S. T. Ali, J-P. Antoine and J-P. Gazeau, Relativistic quantum frames, *Ann. Phys.(NY)* 222:38-88 (1993)
4. S. T. Ali, J-P. Gazeau, and M. R. Karim, Frames, the β-duality in Minkowski space, and spin coherent states, preprint, Concordia University (1995)
5. E. Prugovečki, Dirac dynamics on stochastic phase spaces for spin-$\frac{1}{2}$ particles, *Rep. Math. Phys.* 17:401-417 (1980).

COHERENT STATES AND GLOBAL DIFFERENTIAL GEOMETRY

Stefan Berceanu [1,2]

[1] Equipe de Physique Mathématique et Géométrie
Institut de Mathématique, CNRS - Université Paris 7-Denis Diderot
Case 7012, Tour 45-55, 5e étage
2, place Jussieu, F- 75251 Paris Cedex 05, France
E-mail: Berceanu@mathp7.jusssieu.fr
[2] Permanent address: Institute of Atomic Physics
Institute of Physics and Nuclear Engineering
Department of Theoretical Physics
P. O. Box MG-6, Bucharest-Magurele, Romania
E-mail: Berceanu@Roifa.Bitnet

Abstract

The relationship between coherent states and geodesics is emphasized. It is found that $CL_0 = \Sigma_0$, where CL_0 is the cut locus of 0 and Σ_0 is the locus of coherent vectors othogonal to $|0>$. The result is proved for manifolds on which the exponential from the Lie algebra to the Lie group equals the geodesic exponential. The conjugate loci on hermitian symmetric spaces are analyzed also in the context of the coherent state approach. The results are illustrated on the complex Grassmann manifold.

1. INTRODUCTION

Coherent states[1] are an excelent interplay of classical and quantum mechanics.[2] The local construction of Perelomov's homogeneous coherent states[3] has already been globalized, including for Kählerian non-homogeneous manifolds.[4] On the other hand, the geometric quantization program[5] yields - at least in principle - a tool towards the quantization program of Dirac on differentiable manifolds. In fact, the coherent state approach offers a straightforward recipe for geometric quantization.[6] However, there are interesting problems in these two classical fields that have not been attacked yet. One of them is *the relationship between coherent states and geodesics* (see Remark 3 in Ref. 7). The aim of this talk is to explore further this relationship. It is found that for some manifolds there is an intimate connection between *the cut locus* CL_0[8] of a point on the manifold $\widetilde{\mathbf{M}}$ corresponding to a fixed coherent vector, say $|0>$, and *the polar divisor* Σ_0, i.e. the locus of coherent vectors orthogonal to $|0>$. Despite the fact that the equality $CL_0 = \Sigma_0$ is proved only for manifolds for which the exponential exp from the Lie algebra to the Lie group equals the (geodesic) exponential[9] Exp (the well known case of Riemannian symmetric spaces[10-12] is included), this remark is attractive even from a pure mathemathical point of view, due to the lack of methods to characterize the cut locus as an object of global differential geometry.[13]

Some of the results presented here have been already briefly announced as part of a tentative to find a geometrical characterization of Perelomov's construction of the coherent state manifold as a Kählerian embedding into a projective space.[14]

2. THE COHERENT STATE MANIFOLD

Let us consider a quantum system with symmetry, i.e. a triplet $(\mathbf{K}, G, \pi)$, where π is a unitary irreducible representation of the Lie group G on the Hilbert space $\mathbf{K}$. Let us consider the orbit $\widetilde{\mathbf{M}} = \{\tilde{\pi}(g)|\widetilde{\psi_0} > \ | g \in G\}$, where $\tilde{\pi}$ is the projective representation of G induced by π, $|\psi_0 > \in \mathbf{K}$ is fixed and $\xi : \mathbf{K} \rightarrow \mathbf{PK}$ is the projection $\xi(|\psi>) = |\widetilde{\psi}>$. Then we have the diffeomorphism $G/K \approx \widetilde{\mathbf{M}}$, where K is the stationary group of the state $|\widetilde{\psi_0}>$. The quantum mechanical framework may be realized as the elementary G-space[15] $(\mathbf{PK}, \omega_{FS}, \rho')$, where ω_{FS} is the Fubini-Study (Kähler) fundamental two-form on a projective space, here $\mathbf{PK}$, and ρ' is the isomorphism of the Lie algebra $\mathbf{g}$ of G into the algebra of smooth functions on $\mathbf{PK}$.

The keystone of the coherent state approach is to find a Hilbert space $\mathbf{L}$ and a Kählerian embedding $i : \widetilde{\mathbf{M}} \hookrightarrow \mathbf{PL}$.[14,16] Then $\widetilde{\mathbf{M}}$ is called the *coherent state manifold* and $(\widetilde{\mathbf{M}}, \omega, \rho)$ is a hamiltonian G-space, with $\omega = \omega_{FS|\widetilde{\mathbf{M}}} = i^*\omega_{FS}$, $\rho = \rho'_{|\widetilde{\mathbf{M}}}$.

If G is a compact connected simply connected Lie group and $|\widetilde{\psi_0}> \equiv |j>$, a (anti-)dominant weight vectors, then i is indeed a Kählerian embedding.[15] So, choosing the representation $\pi = \pi_j$ and the Hilbert space $\mathbf{K_j}$ of holomorphic sections with base $\widetilde{\mathbf{M}}$, it turns out that $\mathbf{L} = \mathbf{K}_j^*$, and the Borel-Weil-Bott theorem solves the requantization problem.[6]

Now we briefly discuss the embedding i for a compact complex manifold $\widetilde{\mathbf{M}}$. Then *the condition for the existence of the embedding i is equivalent with the requirement for the manifold to be Hodge,*[17] *which is the same condition as prequantization in geometric quantization.*[18] For example, for hermitian symmetric spaces, the condition $\omega \in H^2(\widetilde{\mathbf{M}}, \mathbf{Z})$ is sufficient, in virtue of a theorem of Harish-Chandra,[7,5,19] which in the compact case is just the Borel-Weil-Bott theorem. Let now $\xi_0 : \mathbf{M}' \rightarrow \widetilde{\mathbf{M}}$ be a holomorphic line bundle. Another way to express the condition to have the embedding $i : \widetilde{\mathbf{M}} \hookrightarrow \mathbf{PL}$ is that the line bundle $\mathbf{M}'$ be positive, or – equivalently – be ample.[20] The last condition means that there exists an integer m_0 such that for $m \geq m_0$, $M \equiv M'^m = i^*[1]$. We use the notation $[r] = H^r, r \in \mathbf{Z}$, where H is the hyperplan bundle over $\mathbf{PL}$. Here ξ_0 is the positive line bundle appearing in the Kodaira embedding theorem, and the embedding i, in the notation of Griffith and Harris,[21] is $i \equiv i_{\mathbf{M}} : x \rightarrow i_{\mathbf{M}}(x) = [s_1(x), ..., s_N(x)]$. The line bundle $\mathbf{M}$ is furnished by the coherent state approach and is called *coherent vector manifold.*[22] As a consequence of the Kodaira embedding theorem, the Kodaira vanishing[20] theorem implies that in the sum giving the generalized Euler-Poincaré characteristic,[17] only the zero term is present, and the dimension of the representation π_j is given by the Riemann-Roch-Hirzebruch theorem.[14,17,23]

We restrict ourself to coherent state manifolds of flag type, i.e. $\widetilde{\mathbf{M}} \approx G/K \approx G^c/P$, where G is a compact, connected and simply connected semisimple Lie group, G^c is the complexification of G and P is a parabolic subgroup of G^c.[7] Then $\mathbf{M}'$ is the holomorphic line bundle $\mathbf{M}' = \xi_0^{-1}(\widetilde{\mathbf{M}}) \to G_c/P$ associated by the holomorphic caracter $\chi = \chi_j$ of P to the principal bundle $P \to G^c \to G^c/P$.

In fact, if $(\mathbf{M}', \omega, J)$ is a compact Kähler manifold, then $(\mathbf{M}', \nabla, h)$ is a quantization bundle over $\widetilde{\mathbf{M}}$,[5] where J is the complex structure and h is the hermitian form on the tautological line bundle [-1] over **PL**. On [-1], h is given by $h : z \to |z|^2$. If $\varphi_i : V_i \times \mathbf{C} \to \xi_0^{-1}(V_i)$ is a local trivialization of the holomorphic line bundle $\mathbf{M}' \to \widetilde{\mathbf{M}}$, then a global section is given by

$$|s_i(m)>= (g_i(Z_i), f_{s_i}(Z_i)) = (g_i(Z_i), < s_i|Z_i >),$$

where $m = g_i(Z_i) \in V_i$ are matrix elements determined by the local coordinates Z_i. Then the scalar product on the line bundle $\mathbf{M}' \to \widetilde{\mathbf{M}}$ is given by[6,22]

$$< s_i|s_i' >=< f_{s_i}, f_{s_i'} >= \int_{\widetilde{\mathbf{M}}} h_X(s_i(X), s_i'(X)) \frac{\omega^n(X)}{n!} \ .$$

This scalar product is also a hermitian scalar product of sections with base $\widetilde{\mathbf{M}}$ in the $D_{\widetilde{\mathbf{M}}}$- module of differentiable operators on $\widetilde{\mathbf{M}}$.[7]

There is an open covering of $\widetilde{\mathbf{M}}$ by $V_s = \pi_j(s) V_0, s \in \Sigma$, where Σ is a finite set from the quotient $W(G)/W(K)$ of Weyl groups.[22] The coherent state vectors corresponding to the points of the neighborhood $V_0 \subset \widetilde{\mathbf{M}}$ around $Z = 0$ are

$$|Z, j >= \exp \sum_{\varphi \in \Delta_n^+} (Z_\varphi F_\varphi^+)|j >, \ |\underline{Z} >=< Z|Z >^{-1/2} |Z >\in \mathbf{M} \ , \tag{2.1}$$

where Δ_n^+ are the positive noncompact roots,[7] $Z \in \mathbf{C}^d$ are local coordinates and d is the dimension of the manifold $\widetilde{\mathbf{M}}$.

We introduce also the notation

$$|B, j >\equiv |\underline{Z, j} >= \exp \sum_{\varphi \in \Delta_n^+} (B_\varphi F_\varphi^+ - \bar{B}_\varphi F_\varphi^-)|j > . \tag{2.2}$$

Note that

$$F_\varphi^+|j >\neq 0, \ F_\varphi^-|j >= 0, \ \ \varphi \in \Delta_n^+ \ . \tag{2.3}$$

3. THE CUT LOCUS, CONJUGATE LOCUS AND COHERENT STATES

1. Let V be a compact Riemannian manifold. Let C_p denote the set of vectors $X \in V_p$ (the tangent space at $p \in V$) for which $\mathrm{Exp}_p X$ is singular. A point q in V (resp. V_p) is *conjugate* to p if it is in $\mathrm{Exp} C_p$ (resp. C_p).[9]

The point $p \in V$ is in the *cut locus* CL_0 of $0 \in V$ if it is the point nearest to $0 \in V$ on the geodesic joining 0 with p, beyond which the geodesics ceases to minimize its arc length.[8]

The relative position of CL_0 and C_0 is given by Thm. 7.1, p. 97, in Ref. 8. For simply connected symmetric spaces, the cut locus is identified with the first conjugate point.[24]

Most considerations in this Section concern only manifolds with the property

$$(A) \qquad \mathrm{Exp}|_{\lambda(e)} = \lambda \circ \exp|_{\mathbf{m}} ,$$

where e is the unit element in G/K, λ is the canonical projection $\lambda : G \to G/K$, and $\mathbf{g} = \mathbf{k} \oplus \mathbf{m}$ is the orthogonal decomposition with respect to the B-form.

In fact, (A) expresses the fact that *the geodesics in* $\widetilde{\mathbf{M}}$ *are images of one-parameter subgroups of* $\widetilde{\mathbf{M}} \approx G/K$. The symmetric spaces have property (A) (cf. Thm. 3.3, p. 208, in Ref. 9).

We shall also be concerned with manifolds $\widetilde{\mathbf{M}}$ verifying the following condition:

(B) On the Lie algebra $\mathbf{g}$ of G there exists an $Ad(G)$-invariant, symmetric, non-degenerate bilinear form B such that the restriction of B to the Lie algebra $\mathbf{k}$ of K is non-degenerate as well

We point out that *if the homogeneous space* $\widetilde{\mathbf{M}} \approx G/K$ *verifies* (B), *then it also verifies* (A) (cf. Corollary 2.5, Thm. 3.5 and Corollary 3.6, Chapter X, in Ref. 8). Thimm,[25] Kowalski[26] and Montgomery[27] exhibit examples of homogeneous spaces verifying (B).

2. In Ref. 7 we made the following remark, which is in fact E. Cartan's theorem (see e.g. Thm. 3.3, p. 208, in Ref. 9) on geodesics of symmetric spaces expressed in the coherent state setting:

Remark 1. – *The vector* $|tB, j\rangle = \exp \pi^{*\prime}(tB)|j\rangle \in \mathbf{M}, B \in \mathbf{m}$, *describes trajectories in* $\mathbf{M}$ *corresponding to the image in the manifold of coherent states* $\widetilde{\mathbf{M}} \hookrightarrow \mathbf{PL}$ *of geodesics through the identity coset element on the symmetric space* $X \approx G/K$. *The dependence* $Z(t) = Z(tB)$ *appearing when one passes from Eq.(2.2) to Eq.(2.1) describes a geodesic.*

We shall reformulate Remark 1 in a way which is very useful, even for practical calculations.

Remark 2. – *For an n-dimensional manifold* $X \approx G/K$ *which has hermitian symmetric structure, the parameters* B_φ *in formula* (2.2) *of normalized coherent states are normal coordinates in the normal neighborhood* $V_0 \approx \mathbf{C}^n$ *around the point* $Z_\varphi = 0$ *on the manifold* X.

Proof. Let $\mathbf{m}^\pm$ the $\pm i$ eigenspaces of J and $M^\pm$ the subgroups of G^c corresponding to $\mathbf{m}^\pm$. Then, by the Harish-Chandra embedding theorem,[7,19] $b : \mathbf{m}^+ \to X_c = G^c/P$, $b(X) = \exp(X)P$ is a complex analytic diffeomorphism of $\mathbf{m}^+$ onto a dense subset of X_c (that contains X_n) and the Remark follows because the requirement (A) is fulfilled for the symmetric spaces. So we have also proved the following

Theorem 1. – *Let* $\widetilde{\mathbf{M}}$ *be a coherent state manifold with hermitian symmetric space structure, parametrized in* V_0 *around* $Z = 0$ *as in Eqs.* (2.1), (2.2). *Then the conjugate locus of the point* 0 *is obtained by setting to zero the Jacobian of the transformation* $Z = Z(B)$.

The situation is very transparent in the case of the complex Grassmann manifold

$X_c = G_n(\mathbf{C}^{m+n})$. There[7] Z and B are $m \times n$ matrices related by the relation

$$Z = B\frac{\text{tg}\sqrt{B^*B}}{\sqrt{B^*B}}, \tag{3.1}$$

where B^* denotes the hermitian conjugate of the matrix B.

3. First, let us introduce a notation for the *polar divisor* of $|0> \in \mathbf{M}$:

$$\Sigma_0 = \{|\psi> \,||\psi> \in \mathbf{M}, <0|\psi> = 0\}. \tag{3.2}$$

This denomination is inspired from that used by Wu[28] in the case of a Grassmann manifold. We shall now prove the following

Theorem 2. – *Let* $\widetilde{\mathbf{M}}$ *be a homogeneous manifold* $\widetilde{\mathbf{M}} \approx G/K$. *Suppose that there exists a unitary irreducible representation* π_j *of* G *such that, in a neigborhoud* V_0 *around* $Z = 0$, *the coherent states are parametrized as in Eq.* (2.1). *Then the manifold* $\widetilde{\mathbf{M}}$ *can be represented as the disjoint union*

$$\widetilde{\mathbf{M}} = V_0 \cup \Sigma_0. \tag{3.3}$$

Moreover, if the condition (B) *is satisfied, then*

$$\Sigma_0 = CL_0. \tag{3.4}$$

Proof. We can take $|\psi> = |\psi(Z)> \in \mathbf{M}$ such that the parameters Z are in $\mathbf{C}^n$ as in formula (2.1). Now, the second relation (2.3) implies that $<0|\psi> = 1$ for $|\psi> \in \xi_0^{-1}(V_0)$. It follows that the equation

$$\cos\theta = 0, \tag{3.5}$$

where

$$\cos\theta = \frac{|<0|\psi>|}{||0||^{1/2}||\psi||^{1/2}} = ||\psi||^{-1/2}, \tag{3.6}$$

does not have solutions for $|\psi> \in \xi_0^{-1}(V_0)$, and the representation (3.3) follows.

To prove relation (3.4) if (B) is true, use is made of Thm. 7.4 and the subsequent remark on p.100 of Ref. 8, which essentially says that any Riemannian manifold $\widetilde{\mathbf{M}}$ is the disjoint union of the cut locus (closed cell) and the largest open cell of $\widetilde{\mathbf{M}}$ on which normal coordinates can be defined. But $Z \in \mathbf{C}^n$ for points of V_0 corresponding to the largest normal coordinates $B \in \mathbf{m}$, because (B) implies (A).

In addition, we shall prove a Corollary of Theorem 2. Let us introduce the Cayley distance[29] in the projective space. Let $(\cdot,\cdot)$ be the scalar product in $\mathbf{K}$. If $\xi : \mathbf{K}\setminus\{0\} \rightarrow \mathbf{PK}$, $\xi : \omega \rightarrow [\omega]$, then the Cayley distance is

$$d_c([\omega'],[\omega]) = \arccos\frac{|(\omega',\omega)|}{||\omega'||||\omega||}. \tag{3.7}$$

Before proving the Corollary, we shall present

Remark 3 (Geometrical significance of transition amplitudes for coherent states). – *Let* $|\underline{Z}> \in \mathbf{M}, Z \in V_0$ *as in* (2.1) *and* $i : \widetilde{\mathbf{M}} \hookrightarrow \mathbf{PL}$ *the embedding of the* *coherent state manifold into the projective space. Then the angle* $\theta = \theta(Z, Z')$ *defined by*

$$\theta \equiv \arccos | < \underline{Z'}|\underline{Z} > | \tag{3.8}$$

equals the geodesic distance between $i(Z)$ *and* $i(Z')$,

$$\theta = d_c(i(Z'), i(Z)). \tag{3.9}$$

More generally, the (Cauchy) formula is true

$$< \underline{Z'}|\underline{Z} >= (i(Z'), i(Z)). \tag{3.10}$$

Proof. The relation (3.10) is an immediate consequence of the fact that the complex analytic line bundle $\mathbf{M}$ over $\widetilde{\mathbf{M}}$ is *projectively induced*,[17] i.e. $\mathbf{M} = i^*[1]$.[23,14]

Combining Remark 3 and Theorem 2, we get the following

Corollary. – *Suppose that* $\widetilde{\mathbf{M}}$ *is an homogeneous manifold satisfying (B) and admitting the embedding* $i : \widetilde{\mathbf{M}} \hookrightarrow \mathbf{PL}$. *Let* $0, Z \in \widetilde{\mathbf{M}}$. *Then* $Z \in CL_0$ *iff the Cayley distance between the images* $i(0), i(Z) \in \mathbf{PL}$ *is* $\pi/2$:

$$d_c(i(0), i(Z)) = \pi/2. \tag{3.11}$$

4. THE COMPLEX GRASSMANN MANIFOLD REVISITED

The results of Section 3 will be illustrated on the example of the complex Grassmann manifold. The calculation of the cut locus on $G_n(\mathbf{C}^{n+m})$ was announced by Wong[10] and now more proofs[12,13] are available. Also Wong[11] has announced the conjugate locus on the Grassmann manifold, but, as far as I know, the proof has not been published. The results of Wong on conjugate locus on Grassmann manifold were contested by Sakai,[12] who showed that the conjugate locus is larger than the part found out by Wong. In a subsequent paper I shall present my own proofs of the results of Wong on conjugate locus using Theorem 1,[30] and also another proof of the results of Sakai in the tangent space.

1. Let $\mathbf{O}$ be the n-plane passing through the origin of $\mathbf{C}^N (N = n+m)$ corresponding to $Z = 0$ in $V_0 \subset G_n(\mathbf{C}^N)$ in the representation (2.1). Then $Z \in V_0 \approx \mathbf{C}^{nm}$ iff there are n vectors $\vec{z}_1, \ldots, \vec{z}_n \in \mathbf{C}^N$ such that $Z = \vec{z}_1 \wedge \ldots \vec{z}_n \neq 0$. Fixing the canonical basis $\vec{e}_1, \ldots, \vec{e}_N$ for $\mathbf{C}^N$, then

$$\vec{z}_i = \vec{e}_i + \sum_{\alpha=n+1}^{N} Z_{i\alpha}\vec{e}_\alpha, \; i = 1, \ldots, n. \tag{4.1}$$

If the weight j is taken as[7]

$$j = (1, \ldots, 1, 0, \ldots, 0) = (1^m, 0^n), \tag{4.2}$$

then we have the *equality of the scalar product* $< \cdot|\cdot >$ *of coherent vectors from* $\mathbf{M}$ *and of the hermitian scalar product* $((\cdot, \cdot))$[31] *in the holomorphic line bundle* det*

$$< Z'|Z >= ((\hat{Z}'^t, \hat{Z}^t)) = \det((\vec{z'}_i, \vec{z}_j))_{i,j=1,\ldots,n} = \det(\mathbf{1}_n + ZZ'^*) \; , \tag{4.3}$$

where t denotes the transpose. We have also used the notation

$$\hat{Z} = (\mathbf{1}_n, Z), \tag{4.4}$$

where Z is an $n \times m$ matrix and $\mathbf{1}_n$ is the unity $n \times n$ matrix.

So, *the parameters Z in formula* (2.1) *for the Grasmann manifold of coherent states, are the Pontrjagin*[32] *coordinates Z^t in formula* (4.1).

Let us also introduce the Plücker coordinates $Z^{i_1 \ldots i_n}$, i.e.

$$Z = \sum_{1 \le i_1 < \ldots < i_n \le N} Z^{i_1 \ldots i_n} \vec{e}_{i_1} \wedge \ldots \wedge \vec{e}_{i_n} \, . \tag{4.5}$$

Let $i : G_n(\mathbf{K}) \hookrightarrow \mathbf{PL}$ be the Plücker embedding, where $\mathbf{K} = \mathbf{C}^N$, $\mathbf{L} = \mathbf{C}^{*N(m)}$, $N(m) = \binom{N}{n} - 1$. Using the notation of Section 2 for $G_n(\mathbf{C}^N)$, then $\mathbf{M}' = \mathbf{M} = i^*[1]$, i.e. the line bundle det* is not only ample, but very ample.[20]

The (Binet-)Cauchy formula[33] mentioned in Eq.(3.10) reads explicitely

$$\det((\vec{z'}_i, \vec{z}_j))_{i,j=1,\ldots,n} = \sum_{1 \le i_1 < \ldots < i_n \le N} z^{i_1 \ldots i_n} \overline{z'^{i_1 \ldots i_n}} \, . \tag{4.6}$$

2. Consider the following sequences of integers

$$\omega = \{0 \le \omega(1) \le \ldots \le \omega(n) \le m\}; \; \sigma(i) = \omega(i) + i, \; i = 1, \ldots, n. \tag{4.7}$$

The Schubert varieties are defined as[32]

$$Z(\omega) = \left\{ X \in G_n(\mathbf{C}^{n+m}) | \dim(X \cap \mathbf{C}^{\sigma(i)}) \ge i \right\}. \tag{4.8}$$

The "jumps" sequence is introduced as

$$I_\omega = \{i_0 < i_1 < \ldots < i_{l-1} < i_l = n\}, \tag{4.9}$$

$$\omega(i_h) < \omega(i_{h+1}), \omega(i) = \omega(i_{h-1}), i_{h-1} < i \le i_h, \; h = 1, \ldots, l. \tag{4.10}$$

Let us consider the subset of generic elements of $Z(\omega)$[32]

$$Z'(\omega) = \left\{ X \subset G_n(\mathbf{C}^{n+m}) | \dim(X \cap \mathbf{C}^{\sigma(i_h)}) = i_h, i_h \in I_\omega \right\}. \tag{4.11}$$

Introduce also the notation

$$V_l^p = \left\{ Z \subset G_n(\mathbf{C}^{n+m}) | \dim(Z \cap \mathbf{C}^p) \ge l \right\}, \tag{4.12}$$

$$W_l^p = V_l^p - V_{l+1}^p = \left\{ Z \subset G_n(\mathbf{C}^{n+m}) | \dim(Z \cap \mathbf{C}^p) = l \right\}, \tag{4.13}$$

$$\omega_l^p = (\underbrace{p-l, \ldots, p-l}_{l}, \underbrace{m, \ldots, m}_{n-l}). \tag{4.14}$$

Then[11,30,34]

$$V_l^p = Z(\omega_l^p); \;\; W_l^p = Z'(\omega_l^p) \, . \tag{4.15}$$

Let $\mathbf{O}^\perp$ denote the orthogonal complement of the n-plane $\mathbf{O}$ in $\mathbf{C}^N$.

Remark 4 (Wong[10]). – *The cut locus of the point* $\mathbf{O}$ *is given by*

$$\begin{aligned} CL_0 &= \Sigma_0 = V_1^m = Z(\omega_1^m) = Z(m-1, m, \ldots, m) \\ &= \left\{X \subset G_n(\mathbf{C}^{n+m}) | \dim(X \cap \mathbf{O}^{\perp}) \geq 1\right\}. \end{aligned} \tag{4.16}$$

Proof. An immediate proof can be obtained using the results of Wu[28] concerning to the polar divisor Σ_0 on the Grassmann manifold and the theorems in Husemoller[35] characterizing the canonical (universal, det) bundle on $G_n(\mathbf{C}^N)$.

The results of Wong[11] concerning to the conjugate locus are proved[30] using the dependence $Z = Z(B)$ from Eq.(3.1) and Jordan's stationary angles[36] between two n-planes. The stationary angles between two n-planes appear in the expression[37]

$$\cos\theta = \prod_{i=1}^{n} \cos\theta_i\ , \tag{4.17}$$

where θ is the distance on the geodesics in the Plücker embedding, given by the formula

$$\cos\theta = \frac{|\det(\mathbf{1} + ZZ'^{*})|}{|\det(\mathbf{1} + ZZ^{*})|^{1/2} |\det(\mathbf{1} + Z'Z'^{*})|^{1/2}}\ . \tag{4.18}$$

In the proof[30] we use the following

Remark 5. – *The squares* $\cos^2\theta_i$ *of the stationary angles between the* n*-planes* Z_1, Z_2 *with* $((Z_1, Z_2)) \neq 0$ *are given by the eigenvalues of a matrix* W *which, for* $Z_1, Z_2 \in V_0$ *is*

$$W = (\mathbf{1} + Z_1 Z_1^*)^{-1}(\mathbf{1} + Z_1 Z_2^*)(\mathbf{1} + Z_2 Z_2^*)^{-1}(\mathbf{1} + Z_2 Z_1^*). \tag{4.19}$$

This remark is proved using the technique of Rosenfel'd.[38] It can be proved[30] that the eigenvalues of W appear also in the distance on the complex Grassmann manifold.

Note also that if the expression (3.1) of the dependence $Z = Z(B)$ is introduced in the formula of the distance between the points $Z = 0$ and $Z \in V_0$ on the Grassmann manifold, then

$$d^2 = \sum |B_{ij}|^2. \tag{4.20}$$

The author is indebted to the Organizing Committee for inviting him to the Workshop. The constant supervision of Professor L. Boutet de Monvel is kindly acknowledged. Discussions during the Workshop, especially with Professors D. Simms and A. M. Perelomov are acknowledged. The author also expresses his thanks to Professors K. Teleman, S. Kobayashi, M. Berger and Th. Hangan for suggestions.

References

1. J. R. Klauder, B. S. Skagerstam (Eds.), "Coherent States", Word Scientific, Singapore (1985)
2. S. Berceanu, From quantum mechanics to classical mechanics and back, via coherent states, *in* "Quantization and Infinite-Dimensional Systems (Proc. Białowieza 1994)", p. 155, J-P. Antoine et al. (eds.), Plenum Press, New York (1994).
3. A. M. Perelomov, Coherent states for arbitrary Lie groups, *Commun. Math. Phys.* 26:222 (1972)
4. J. R. Rawnsley, Coherent states and Kähler manifolds, *Quart. J. Math. Oxford* 28:403 (1977)

5. B. Kostant, Quantization and unitary representations, *in*: "Lecture Notes in Mathematics", Vol 170, p. 87; C. T. Taam (ed.), Springer-Verlag, Berlin (1970)

6. V. Ceausescu, A. Gheorghe, Classical limit and quantization of Hamiltonian systems, *in*: "Symmetries and Semiclassical Features of Nuclear Dynamics", Lecture Notes in Physics, Vol 279, p. 69; Springer-Verlag, Berlin (1987)

7. S. Berceanu, L. Boutet de Monvel, Linear dynamical systems, coherent state manifolds, flows and matrix Riccati equation, *J. Math. Phys.* 34:2353 (1993)

8. S. Kobayashi, K. Nomizu, "Foundations of Differential Geometry", Vol II, Interscience, New York (1969)

9. S. Helgason, "Differential Geometry, Lie groups and Symmetric Spaces", Academic Press, New York (1978)

10. Y. C. Wong, Differential Geometry of Grassmann manifolds, *Proc. Nat. Acad. Sci. U.S.A.* 57:589 (1967)

11. Y. C. Wong, Conjugate loci in Grassmann manifold, *Bull. Am. Math. Soc.* 74:240 (1968)

12. T. Sakai, On cut loci on compact symmetric spaces, *Hokkaido Math. J.* 6:136 (1977)

13. S. Kobayashi, On conjugate and cut loci, *in* "Global Differential Geometry", M.A.A. Studies in Mathematics, 27, S. S. Chern (ed.), p. 140 (1989)

14. S. Berceanu, The coherent states: old geometrical methods in new quantum clothes, preprint Bucharest, Institute of Atomic Physics, FT-398-1994

15. V. Guillemin, S. Sternberg, "Symplectic Technics in Physics", Cambridge University, Cambridge (1984)

16. A. Odzijewicz, Coherent states and geometric quantization, *Commun. Math. Phys.* 150:385 (1992)

17. F. Hirzebruch, "Topological Methods in Algebraic Geometry", Springer-Verlag, Berlin (1966).

18. H. Woodhouse, "Geometric Quantization", Oxford University, Oxford (1980)

19. A. W. Knapp, "Representation Theory of Semisimple Lie Groups", Princeton University Press, Princeton (1986)

20. B. Shiffman, A. J. Sommese, "Vanishing Theorems on Complex Manifolds", Progress in Mathematics, Vol. 56, Birkhäuser, Boston (1985)

21. P. Griffith, J. Harris, "Principles of Algebraic Geometry", Wiley, New-York (1978)

22. S. Berceanu, A. Gheorghe, On the construction of perfect Morse functions on compact manifolds of coherent states, *J. Math. Phys.* 28:2899 (1987)

23. M. Cahen, S. Gutt, J. Rawnsley, Quantization on Kähler manifolds, II, *Trans. Math. Soc.* 337:73 (1993)

24. R. Crittenden, Minimum and conjugate points in symmetric spaces, *Canad. J. Math.* 14:320 (1962)

25. A. Thimm, Integrable geodesic flows on homogeneous spaces, *Ergod. Theory Dyn. Syst.* 1:495 (1981)

26. O. Kowalski, "Generalized Symmetric Spaces", Lecture Notes in Mathematics 805, Springer-Verlag, Berlin (1980)

27. R. Montgomery, Isoholonomic problems and some applications, *Comm. Math. Phys.* 128:565 (1990)

28. H. H. Wu, "The Equidistribution Theory of Holomorphic Curves", Annals of Maths. Studies 164, Princeton Univ. Press, Princeton (1970)

29. A. Cayley, A sixth memoir upon quantics, *Phil. Trans. Royal. Soc. London.* 149:61 (1859).

30. S. Berceanu, On the geometry of complex Grassmann manifold, its noncompact dual and coherent states (in preparation)

31. S. S. Chern, "Complex Manifolds without Potential Theory", Van Nostrand, Princeton (1967)

32. L. C. Pontrjagin, Charakteristiceskie tzikly differentziruemyh mnogobrazia, *Mat. sb.* 21:233 (1947)

33. F. P. Gantmacher, "Teoria Matritz", Nauka, Moskwa (1966)

34. Y. C. Wong, A class of Schubert varieties, *J. Diff. Geom.* 4:37 (1970)

35. D. Husemoller, "Fibre Bundles", Mc Graw-Hill, New York (1966)

36. C. Jordan, Sur la géométrie à n dimensions, *Bull. Soc. Math. France* t. lll:103 (1875)

37. B. Rosenfel'd, Vnnutrenyaya geometriya mnojestva m-mernyh ploskastei n-mernova ellipticeskovo prostranstva, *Izv. Akad. Nauk. SSSR, ser. mat.* 5:353 (1941)

38. B. Rosenfel'd, "Mnogomernye Prostranstva", Nauka, Moskwa (1966); "Neevklidovy Prostranstva", Nauka, Moskwa (1969).

NATURAL TRANSFORMATIONS OF LAGRANGIANS INTO p-FORMS ON THE TANGENT BUNDLE

Jacek Dębecki

Institute of Mathematics, Jagiellonian University
ul. Reymonta 4, 30-059 Kraków, Poland

Abstract

This paper presents without proofs some theorems giving a complete characterization of natural transformations of finite order of Lagrangians into p-forms on the tangent bundle over n-dimensional manifolds for $n \geq p+1$ (except for the case $p=0$, $n=1$).

1. INTRODUCTION

In this paper we will investigate natural transformations of Lagrangians into p-forms on the tangent bundle.

The most important examples of these natural transformations are well known in theoretical mechanics. Namely, let M be a differentiable manifold and let $L : TM \longrightarrow \mathbf{R}$ be a smooth function on the tangent bundle. The function L is called Lagrangian. If we denote by C_M the Liouville vector field on TM and by J_M the canonical tangent structure on TM, then we can define the following forms on TM: the 0-form $E_M(L) = C_M(L) - L$ called energy, the Poincaré-Cartan 1-form $\alpha_M(L) = dL \circ J_M$ and the Poincaré-Cartan 2-form $\omega_M(L) = d\big(\alpha_M(L)\big)$. It is important that $E_M(L)$, $\alpha_M(L)$, $\omega_M(L)$ can be obtained in the same way for an arbitrary manifold M and an arbitrary Lagrangian L and that they are defined without use of a local system of coordinates on M. Therefore E, α, ω may be regarded as invariants of Lagrangian systems. We will call invariants of this kind the natural transformations of Lagrangians into p-forms on the tangent bundle (Definition 1, page 2).

In Ref.2 a complete characterization of natural transformations of finite order of Lagrangians into p-forms on the tangent bundle over n-dimensional manifolds was given for $p = 0$, 1 and $n \geq p+2$. In Ref.1 a similar characterization was also announced for $p=2$ and $n \geq p+2$. In this paper we give without proofs a complete classification of these natural transformations for all p and $n \geq p+1$ (except for only one case, namely $p=0$ and $n=1$).

The main result of this paper splits into two parts in a natural way.

The first part is the classification of natural transformations in the case $n \geq p+2$. In this case we assert that every natural transformation is a combination of some standard operations such as the Liouville field, exterior differentiation, compositing of a 1-form on the tangent bundle with the canonical tangent structure, form multiplication and addition (Theorem 1, page 5).

The second part, which deals with the case $n = p+1$, is more complicated. In this case there are new natural transformations. The classification is less precise than in the previous case. Namely, we find a basis of the module of all natural transformations for $n \geq p+2$, but we find generators of this module for $n = p+1$, because in this case the module is in general not free.

The first step of the proofs of our theorems consist in showing a relation between natural transformations and equivariant maps. We can establish this relation only for $n \geq p+1$. Hence this paper provides no characterization of natural transformations for $n \leq p$. The relation between natural transformations and equivariant maps is also used to construct the new natural transformations for $n = p+1$. Consequently this relation is briefly described in this paper, although the proofs of theorems are omited.

The author wishes to express his thanks to Professor Jacek Gancarzewicz for suggesting the problem, many stimulating conversations and his active intrest in writing this paper. The author is also indebted to Manuel de León and Włodzimierz Mikulski who are the coauthors of Ref.2, which contains a lot of ideas used in this paper.

This work was supported by a grand of KBN No.: 2 P 301 030 04.

2. BASIC DEFINITIONS

We assume throughout the article that all manifolds and maps are always of class C^∞. Let TM denote the tangent bundle to a differentiable manifold M and let Tf denote the derivative of a smooth map f. An arbitrary smooth function $L : TM \longrightarrow \mathbf{R}$ is called a Lagrangian.

Let $F_p(TM)$ denote the set of all p-forms on TM. Of course $F_0(TM)$ is the set of all Lagrangians. An arbitrary p-form on the tangent bundle will be always regarded as a smooth function $(T \times \ldots \times T)TM \longrightarrow \mathbf{R}$, which is skew-symmetric p-linear on each fibre.

For every open subset $U \subset \mathbf{R}^n$ the set TU will be identified with $U \times \mathbf{R}^n$ and the set $(T \times \ldots \times T)TU$ will be identified with $U \times \mathbf{R}^n \times \mathbf{R}^n \times \mathbf{R}^n \times \ldots \times \mathbf{R}^n \times \mathbf{R}^n$. The canonical basis in the vector space $\mathbf{R}^n$ will be denoted by $e_1, \ldots, e_n$ and the group of all permutations of the set $\{1, \ldots, p\}$ by S_p.

Fix non-negative integers n and p.

Definition 1 – A family of maps

$$A_m : F_0(TM) \longrightarrow F_p(TM),$$

where M is an arbitrary n-dimensional manifold, is called a *natutal transformation* of Lagrangians into p-forms on the tangent bundle, if the following two conditions hold:

(i) The Naturality Condition: for every embedding $\varphi : M \longrightarrow N$ of two n-dimensional manifolds M, N and for every Lagrangian $L \in F_0(TN)$, we have

$$A_M(L \circ T\varphi) = A_N(L) \circ (T \times \ldots \times T)T\varphi;$$

(ii) The Regularity Condition: For every manifold M, every n-dimensional manifold N, every smooth map

$$M \times TN \ni (\alpha, v) \longrightarrow L_\alpha(V) \in \mathbf{R}$$

and every $Q \in (T \times \ldots \times T)TN$, the map

$$M \ni \alpha \longrightarrow A_N(L_\alpha)(Q) \in \mathbf{R}$$

is also smooth.

It is well known that for every manifold M there are a canonical vector field on TM, called the Liouville field and denoted by C_M, and a canonical tensor field of type (1,1) on TM, called the canonical tangent structure and denoted by J_M. We recall that for each chart $\chi : U \longrightarrow \mathbf{R}^n$ on n-dimensional manifold M the following formulas

$$\begin{aligned} \big((TT\chi)\circ C_M \circ T\chi^{-1}\big)(x,v) &= (x,v,0,v), \\ \big((TT\chi)\circ J_M \circ (TT\chi)^{-1}\big)(x,v,y,w) &= (x,v,0,y) \end{aligned}$$

hold true for all $(x,v) \in T\big(\chi(U)\big)$, $(x,v,y,w) \in TT\big(\chi(U)\big)$.

Let r be a non-negative integer.

Definition 2 – We say that a natural transformation A of Lagrangians into p-forms on the tangent bundle is of order r if, for every n-dimensional manifold M, every vector $v \in TM$ and all Lagrangians $K, L \in F_0(TM)$, the following implication holds:

$$j^r_v K = j^r_v L \Longrightarrow A_M(K)|(T\times\ldots\times T)_v TM = A_M(L)|(T\times\ldots\times T)_v TM.$$

The natural transformation A is called *of finite order* if there exists a non-negative integer r such that A is of order r.

We will restrict our attention to natural transformations of finite order.

Nevertheless, we now give some examples of natural transformation of Lagrangians into Lagrangians which are not of finite order.[2]

Example 1 – Let $f : [0,1] \longrightarrow \mathbf{R}$ be a continuous function. Put

$$A_M(L)(v) = \int_0^1 L\big(f(t)v\big)\,dt$$

for each $v \in TM$.

Example 2 – Set $A_M(L) = d(L \circ 0_M)$, where 0_M is the zero vector field on M.

Example 3 – Let

$$F(x) = \begin{cases} 0, & \text{for } x \le 0; \\ e^{-\frac{1}{x}}, & \text{for } x > 0. \end{cases}$$

Write

$$A_M(L)(v) = \sum_{i=1}^{\infty} f\bigg(L(v) - \sum_{j=1}^{i}\Big(1 + \big(C^j_M(L)(v)\big)^2\Big)\bigg),$$

for each $v \in TM$, where $C^j_M = \underbrace{C_M \circ \ldots \circ C_M}_{j \text{ times}}$.

3. EQUIVARIANT MAPS

We will denote by N_{npr} the set of all natural transformations of order r of Lagrangians into p-forms on the tangent bundle over n-dimensional manifolds.

Let Ω be a set of pairs (α,β) of n-uples of non-negative integers for which the inequality $|\alpha+\beta| \le r$ holds. The set $\mathbf{R}^\Omega$ will be identified with the set of all jets $j^r_{(0,e_n)}L$ of smooth functions $L : T\mathbf{R}^n \longrightarrow \mathbf{R}$.

If $n \geq p+1$, it is possible to define the set E_{npr} as the set consisting of all smooth functions $f : \mathbf{R}^{\Omega} \longrightarrow \mathbf{R}$, which satisfy the following three conditions:

$$f\left(j^r_{(0,e_n)}(L \circ T\varphi)\right) = f(j^r_{(0,e_n)}L) \tag{3.1}$$

for every $L \in F_0(T\mathbf{R}^n)$, every open subset $U \subset \mathbf{R}^n$ such that $0 \in U$ and every embedding $\varphi : U \longrightarrow \mathbf{R}^n$ such that $(T \times \ldots \times T)T\varphi(0, e_n, e_1, \ldots, e_p) = (0, e_n, e_1, \ldots, e_p)$;

$$f\left(j^r_{(0,e_n)}(L \circ T\varphi_t)\right) = t^1 \cdot \ldots \cdot t^p f(j^r_{(0,e_n)}L) \tag{3.2}$$

for every $L \in F_0(T\mathbf{R}^n)$ and every $t \in \mathbf{R}^p$, where $\varphi_t : \mathbf{R}^n \longrightarrow \mathbf{R}^n$ is given by $\varphi_t(x) = (t^1x^1, \ldots, t^px^p, x^{p+1}, \ldots, x^n)$;

$$f\left(j^r_{(0,e_n)}(L \circ T\varphi_\sigma)\right) = \operatorname{sgn}\sigma f(j^r_{(0,e_n)}L) \tag{3.3}$$

for every $L \in F_0(T\mathbf{R}^n)$ and every $\sigma \in S_p$, where $\varphi_\sigma : \mathbf{R}^n \longrightarrow \mathbf{R}^n$ is given by $\varphi_\sigma(x) = (x^{\sigma^{-1}(1)}, \ldots, x^{\sigma^{-1}(p)}, x^{p+1}, \ldots, x^n)$.

The elements of the set E_{npr} will be called equivariant maps.

It is easy to see that N_{n0r} and E_{n0r} are rings, N_{npr} is a module over N_{n0r} and E_{npr} is a module over E_{n0r}.

Let $A \in N_{npr}$. We define $f_A : \mathbf{R}^{\Omega} \longrightarrow \mathbf{R}$ by formula

$$f_A(j^r_{(0,e_n)}) = A_{\mathbf{R}^n}(L)(0, e_n, e_1, \ldots, e_p)$$

for every $L \in F_0(T\mathbf{R}^n)$.

Let $f \in E_{npr}$. For every n-dimensional manifold M we define $(A_f)_M : F_0(TM) \longrightarrow F_p(TM)$ by the formula

$$(A_f)_M(L)(Q) = f\left(j^r_{(0,e_n)}\left(L \circ T(\varphi^{-1} \circ \psi_{(T\times\ldots\times T)T\varphi(Q)})\right)\right)$$

for every $L \in F_0(TM)$ and every $Q \in (T \times \ldots \times T)TM$, where $\varphi : U \longrightarrow \mathbf{R}^n$ is an arbitrary chart on M such that $Q \in (T \times \ldots \times T)TU$ and where the map $\psi_{(X,V,Y_1,W_1,\ldots,Y_p,W_p)} : \mathbf{R}^n \longrightarrow \mathbf{R}^n$ is given by

$$\begin{aligned}\psi_{(X,V,Y_1,W_1,\ldots,Y_p,W_p)} &= X + Y_1x^1 + \ldots + Y_px^p + e_{p+1}x^{p+1} + \ldots + e_{n-1}x^{n-1} + Vx^n \\ &\quad + W_1x^1x^n + \ldots + W_px^px^n\end{aligned}$$

for every $(X, V, Y_1, W_1, \ldots, Y_p, W_p) \in (T \times \ldots \times T)T\mathbf{R}^n$.

It is possible to check that both definitions make sense, that the maps

$$\begin{aligned}N_{n0r} \ni A &\longrightarrow f_A \in E_{n0r}, \\ E_{n0r} \ni f &\longrightarrow A_f \in N_{n0r}\end{aligned}$$

are isomorphisms of rings and they are inverse to each other. Hence we may identify the rings E_{n0r} and E_{n0r} and we can prove that the maps

$$\begin{aligned}N_{npr} \ni A &\longrightarrow f_A \in E_{npr}, \\ E_{npr} \ni f &\longrightarrow A_f \in N_{npr}\end{aligned}$$

are isomorphisms of modules and they are inverse to each other.

4. THE MAIN RESULTS

4.1. The Case $n \geq p+2$

The following theorem gives a complete classification of natural transformations in the case $n \geq p+2$.

Theorem 1 – *Let $n \geq p+2$ and let Ψ_{pr} denote the set consisting of all elements of the form $(a, b, c, \alpha, \beta, \gamma)$, where a, b, c are non-negative integers such that $a+b+2c = p$, the maps*

$$\alpha : \{1, \ldots, a\} \longrightarrow \{0, \ldots, r-1\},$$
$$\beta : \{1, \ldots, b\} \longrightarrow \{0, \ldots, r-1\}$$

are strictly increasing and the map

$$\gamma : \{1, \ldots, c\} \longrightarrow \{0, \ldots, r-2\}$$

is increasing.

Then for every natural transformation A of order r of Lagrangians into p-forms on the tangent bundle over n-dimensional manifolds, there exist uniquely determined smooth functions

$$q_{(a,b,c,\alpha,\beta,\gamma)} : \mathbf{R}^{r+1} \longrightarrow \mathbf{R} \tag{4.1}$$

for $(a, b, c, \alpha, \beta, \gamma) \in \Psi_{pr}$, such that, for every n-dimensional manifold M and for every $L \in F_0(TM)$,

$$A_M(L) = \sum_{(a,b,c,\alpha,\beta,\gamma)\in\Psi_{pr}} q_{(a,b,c,\alpha,\beta,\gamma)} \circ \Big(L, C_M(L), \ldots, C_M^r(L)\Big) \tag{4.2}$$
$$\cdot \bigwedge_{i=1}^{a} d\Big(C_M^{\alpha(i)}(L)\Big) \wedge \bigwedge_{i=1}^{b} \Big(d\Big(C_M^{\beta(i)}(L)\Big) \circ J_M\Big) \wedge \bigwedge_{i=1}^{c} d\Big(d\Big(C_M^{\gamma(i)}(L)\Big) \circ J_M\Big).$$

On the other hand, it is evident that for every system of smooth functions (4.1) the family of maps A given by formula (4.2) is a natural transformation of order r of Lagrangians into p-forms on the tangent bundle.

4.2. The Case $n = p+1$

Now we assume that $n = p+1$. For every strictly increasing map $\varepsilon : \{1, \ldots, p+2\} \longrightarrow \{0, \ldots, r-1\}$ and every $L \in F_0(T\mathbf{R}^n)$, we denote by $d_\varepsilon(j^r_{(0,e_n)}L)$ the determinant of the following matrix

$$\begin{bmatrix} \dfrac{\partial\Big(C_{\mathbf{R}^n}^{\varepsilon(1)}(L)\Big)}{\partial v^1} & \cdots & \dfrac{\partial\Big(C_{\mathbf{R}^n}^{\varepsilon(1)}(L)\Big)}{\partial v^p} & C_{\mathbf{R}^n}^{\varepsilon(1)+1}(L) & \dfrac{\partial\Big(C_{\mathbf{R}^n}^{\varepsilon(1)}(L)\Big)}{\partial x^n} \\ \vdots & & \vdots & \vdots & \vdots \\ \dfrac{\partial\Big(C_{\mathbf{R}^n}^{\varepsilon(n+1)}(L)\Big)}{\partial v^1} & \cdots & \dfrac{\partial\Big(C_{\mathbf{R}^n}^{\varepsilon(n+1)}(L)\Big)}{\partial v^p} & C_{\mathbf{R}^n}^{\varepsilon(1)+1}(L) & \dfrac{\partial\Big(C_{\mathbf{R}^n}^{\varepsilon(n+1)}(L)\Big)}{\partial x^n} \end{bmatrix},$$

where the values of functions are evaluated at the point $(0, e_n)$. A trivial verification shows that $d_\varepsilon \in E_{npr}$for every strictly increasing map $\varepsilon : \{1, \ldots, p+2\} \longrightarrow \{0, \ldots, r-1\}$. ¿From this we deduce that, if we set $D_\varepsilon = A_{d_\varepsilon}$, then $D_\varepsilon \in N_{npr}$.

We can now formulate our main result for $p \geq 1$ and $n = p+1$.

Theorem 2 – *Let $p \geq 1$ and $n = p+1$. Let Ψ_{pr} be as that in Theorem 1 and let Φ_{pr} denote the set consisting of all strictly increasing maps $\varepsilon : \{1,\dots,p+2\} \longrightarrow \{0,\dots,r-1\}$.*

Then for every natural transformation A of order r of Lagrangians into p-forms on the tangent bundle over n-dimensional manifolds, there exist smooth functions

$$s_{(a,b,c,\alpha,\beta,\gamma)} : \mathbf{R}^{r+1} \longrightarrow \mathbf{R}$$

for $(a,b,c,\alpha,\beta,\gamma) \in \Psi_{pr}$ and

$$t_\varepsilon : \mathbf{R}^{r+1} \longrightarrow \mathbf{R} \tag{4.3}$$

for $\varepsilon \in \Phi_{pr}$, such that for every n-dimensional manifold M and for every $L \in F_0(TM)$

$$\begin{aligned} A_M(L) \;=\; & \sum_{(a,b,c,\alpha,\beta,\gamma)\in\Psi_{pr}} s_{(a,b,c,\alpha,\beta,\gamma)} \circ \Big(L, C_M(L),\dots,C_M^r(L)\Big) \\ & \cdot \bigwedge_{i=1}^{a} d\Big(C_M^{\alpha(i)}(L)\Big) \wedge \bigwedge_{i=1}^{b} \Big(d\Big(C_M^{\beta(i)}(L)\Big)\circ J_M\Big) \wedge \bigwedge_{i=1}^{c} d\Big(d\Big(C_M^{\gamma(i)}(L)\Big)\circ J_M\Big) \\ & + \sum_{\varepsilon\in\Phi_{pr}} \Big(t_\varepsilon \circ \Big(L, C_M(L),\dots,C_M^r(L)\Big)\Big)\cdot (D_\varepsilon)_M(L). \end{aligned}$$

Notice that the functions (4.3) are in general not uniquely determined. In algebraic language, Theorem 2 describes generators of the module N_{npr} for $p \geq 1$ and $n = p+1$, but this theorem yields no information about bases of the module N_{npr}. In fact, it is possible to prove that, for $p \geq 1$, $n = p+1$ and $r = p+3$, the module N_{npr} is not free.

It is also worth pointing out that the assumption $p \geq 1$ in Theorem 2 cannot be dropped. Actually, it is possible to construct new examples of natural transformations in the case $p = 0$ and $n = 1$.

References

1. J. Dębecki, Natural transformations of Lagrangians, *Proc. of the Winter School at Zdikov*, Czechoslovakia (1993)

2. J. Dębecki, J. Gancarzewicz, M. de León, W. Mikulski, Invariants of Lagrangians and their classifications, *J. Math. Phys.* 35:4568-4593 (1994)

3. J. Gancarzewicz, W. Mikulski, Z. Pogoda, Natural bundles and natural liftings. Prolongations of geometric structures, *in*: "Proc. of the Conf. on Diff. Geometry and Its Applications, 24-29 September 1992", pp.281-320; Silesian University, Opava (1993)

4. S. Ishihara, K.Yano, "Tangent and Cotangent Bundle: Differential Geometry", Marcel Dekker, New York (1973)

5. I. Kolář, Natural operators related with the variation calculus, Prolongations of geometric structures, *in*: "Proc. of the Conf. on Diff. Geometry and Its Applications, 24-29 September 1992", Silesian University, Opava (1993)

6. I. Kolář, P. Michor, J. Slovák, "Natural Transformations in Differential Geometry", Springer-Verlag, Berlin (1993)

7. R. S. Palais, C.-L. Terng, Natural bundles have finite order, *Topology* 16:271-277 (1977).

$SL(2, \mathbb{R})$-COHERENT STATES AND INTEGRABLE SYSTEMS IN CLASSICAL AND QUANTUM PHYSICS

Jean-Pierre Gazeau

LPTM, Université Paris 7 – Denis Diderot
2, Place Jussieu, F - 75251 Paris Cedex 05, France

1. INTRODUCTION

The group $SL(2, \mathbb{R}) \cong SU(1,1) \cong Sp(1, \mathbb{R})$ appears in various domains of physics as being the *raison d'être* of the integrability of the considered system. We do not here pretend to give an exhaustive list of such systems. We just want to stress the simplifying role played by the complex structure and the related coherent states which naturally appear for some of them. We also intend to clarify the relation existing between those different physical models, their respective complex structures and between two different types of coherent states, namely the $SU(1,1)$ Perelomov coherent states and the $SU(1,1)$ Barut-Girardello ones.

After this conference text was written, the author learned from J-P. Antoine that similar results on integral transforms between different U.I.R realisations of $SL(2, \mathbb{R})$ have recently been presented by D.Basu.[1]

2. TWO IDENTICAL PARTICLES IN ONE DIMENSION

Leinaas and Myrheim exhibit the underlying $SL(2, \mathbb{R})$ symmetry of this system in the following way.

In appropriate units, the relative coordinate and momentum

$$x = x_{(1)} - x_{(2)}, \qquad p = \frac{1}{2}\left(p_{(1)} - p_{(2)}\right) \tag{2.1}$$

of the two particles satisfy the canonical commutation relations, either classical and quantum mechanical

$$\{x, p\} = 1, \qquad [x, p] = i. \tag{2.2}$$

If the particles are identical, x and p do not exist as observables since they are antisymmetric under exchange of particle indices. Observables should be of higher degree and the minimal choice for a basic set is the following:

$$A = \frac{1}{4}\left(p^2 + x^2\right), \qquad B = \frac{1}{4}\left(x^2 - p^2\right), \qquad C = -\frac{1}{4}(xp + px). \tag{2.3}$$

The Poisson brackets

$$\{A, B\} = C \qquad \{A, C\} = -B, \qquad \{B, C\} = -A \tag{2.4}$$

Quantization, Coherent States, and Complex Structures
Edited by J.-P. Antoine *et al.*, Plenum Press, New York, 1995

define the Lie algebra $sl(2,\mathbb{R})$. Note the constraint equation on the classical level

$$\Gamma = A^2 - B^2 - C^2 = 0. \tag{2.5}$$

Quantization of the system in agreement with (2.2) leads to the commutation relations:

$$[A,B] = iC \qquad [A,C] = -iB, \qquad [B,C] = -iA. \tag{2.6}$$

Now the Casimir operator assumes the value

$$\Gamma = A^2 - B^2 - C^2 = -\frac{3}{16}. \tag{2.7}$$

This number lies in the range of values assumed by Γ when we are in presence of the discrete series of representations of $sl(2,\mathbb{R})$. If we denote by $\{|\eta,n\rangle,\ n \in \mathbb{N}\}$ the orthonormal basis of the representation space for a given representation $\mathcal{U}_\eta$, we have the characteristic ladder formulas:

$$\Gamma|\eta,n\rangle = \eta\,(\eta-1)\,|\eta,n\rangle, \tag{2.8}$$

$$\begin{aligned} A|\eta,n\rangle &= (\eta+n)\,|\eta,n\rangle, \\ B_+|\eta,n\rangle &= \sqrt{(n+1)\,(2\eta+n)}|\eta,n+1\rangle, \\ B_-|\eta,n\rangle &= \sqrt{n\,(2\eta+n-1)}|\eta,n-1\rangle, \end{aligned} \tag{2.9}$$

where $B_\pm = B \pm iC$ are raising and lowering operators for the eigenvalues of A.

Discrete series representations correspond to the range $\eta > 0$ and the value $-\frac{3}{16}$ assumed by Γ in the harmonic oscillator representation (2.3) corresponds to the two cases

$$\eta = \frac{1}{4} \qquad \text{or} \qquad \eta = \frac{3}{4}. \tag{2.10}$$

Barut-Girardello coherent states[1] are defined as eigenvectors of B_- with complex eigenvalue β:

$$B_-|\eta,\beta\rangle = \beta|\eta,\beta\rangle, \tag{2.11}$$

$$|\eta,\beta\rangle = \nu_{\eta,\beta}\sum_{n=0}^{\infty}\frac{\beta^n}{\sqrt{n!\Gamma\,(n+2\eta)}}|\eta,\beta\rangle, \tag{2.12}$$

where $\nu_{\eta,\beta}$ is a normalisation factor. Those coherent states saturate the $sl(2,\mathbb{R})$ Heisenberg inequality

$$\Delta B\,\Delta C \geq \frac{1}{2}\langle A\rangle. \tag{2.13}$$

The question to be addressed now is the physical interpretation of the parameter η and of these coherent states in terms of quantum statistics, since we deal with classically identical particles. The departure points are the special harmonic oscillator values $\eta = \frac{1}{4}$ and $\frac{3}{4}$. They are interpreted in this way because A is the Hamiltonian of a harmonic oscillator. In the coordinate representation, the spectrum $\left\{\frac{1}{4}+n,\ n \in \mathbb{N}\right\}$ of $2A$ corresponds to symmetric wave functions, whereas the spectrum $\left\{\frac{3}{4}+n,\ n \in \mathbb{N}\right\}$ corresponds to antisymmetric wave functions. Therefore we are allowed to give the interpretation of the boson (resp. fermion) representation to the case $\eta = \frac{1}{4}$ (resp. $\eta = \frac{3}{4}$). In addition, we get a continuum of intermediate statistics cases corresponding to other positive values of η. We shall clarify this physical interpretation of η in the next sections, as well as the relevance of the Barut-Girardello states in this context.

3. $SL(2,\mathbb{R})$ AND THE 2-BODY CALOGERO MODEL

The one-dimensional harmonic oscillator is still integrable if we add a x^{-2} potential:

$$H_{cal} = -\frac{1}{2}\frac{d^2}{dx^2} + \frac{1}{2}x^2 + \frac{\lambda}{x^2} = a^+a + \frac{1}{2} + \frac{\lambda}{x^2} \tag{3.1}$$

where $a = \frac{1}{\sqrt{2}}(x + ip)$.

Indeed we recover the algebra $sl(2,\mathbb{R})$ (2.6) if we now identify[10]

$$\begin{aligned} A &= \frac{1}{2}H_{cal} \\ B_+ &= \frac{1}{2}\left(a^+\right)^2 - \frac{\lambda}{2x^2} \\ B_- &= \frac{1}{2}\left(a\right)^2 - \frac{\lambda}{2x^2}. \end{aligned} \tag{3.2}$$

The Casimir operator $\Gamma = A^2 - B^2 - C^2 = A^2 - \frac{1}{2}(B_+B_- + B_-B_+)$ then assumes the value

$$\Gamma = \frac{1}{4}\left(\lambda - \frac{3}{4}\right). \tag{3.3}$$

If we compare with (2.8), we obtain the relation between λ and η:

$$\lambda = 4\left(\eta - \frac{1}{4}\right)\left(\eta - \frac{3}{4}\right), \tag{3.4}$$

which clearly expresses the meaning of the two values $\eta = \frac{1}{4}$ and $\eta = \frac{3}{4}$. Thus the harmonic oscillator whith the additional $\frac{1}{x^2}$ potential can be interpreted as describing a two-particle system of identical particles whith fractional statistics in one space dimension. Note that the corresponding N-body problem given by

$$H_{cal} = \sum_{i=1}^{N}\left(a^+_{(i)}a_{(i)}\right) + \sum_{i\neq j}\frac{\lambda}{\left(x_{(i)} - x_{(j)}\right)^2} \tag{3.5}$$

was exactly solved by Calogero.[5]

4. $SL(2,\mathbb{R})$ AND TWO ANYONS IN THE LOWEST LANDAU LEVEL

Here we consider identical charged particles in two-dimensional space. They are submitted to a magnetic field strong enough to restrict the problem to the lowest Landau level. Translational invariance allows us to separate out the center-of-mass motion and to deal with the following relative Hamiltonian

$$H = \frac{1}{m}[\mathbf{p} - e\mathbf{A}(\mathbf{r})]^2 \tag{4.1}$$

where the vector potential, in the radial gauge, is related to the magnetic flux density B_0 by $\mathbf{A} = B_0\,{}^*\mathbf{r}/4$ (using the notation ${}^*\mathbf{r} = (-y, x)$).

Let us consider the components (X, Y) of the constant of motion

$$\mathbf{R} = \mathbf{r} - \frac{m}{eB_0}{}^*\mathbf{v} = \frac{1}{2}\mathbf{r} - \frac{2}{eB_0}{}^*\mathbf{p} \tag{4.2}$$

(in the radial gauge). They obey canonical commution rules in quantum mechanics:

$$[X,Y] = -\frac{2i}{eB_0} \tag{4.3}$$

and this allows us to introduce standard annihilation and creation operators

$$a = \sqrt{\frac{eB_0}{4}}\,(X - iY)\,, \qquad a^+ = \sqrt{\frac{eB_0}{4}}\,(X + iY)\,. \tag{4.4}$$

The lowest Landau level is spanned by the complete set

$$\psi_l = N_l \mathbf{z}^l e^{-\frac{\mathbf{z}\bar{\mathbf{z}}}{2}}, \qquad N_l = \left[\pi\Gamma\left(l+1\right)\right]^{-\frac{1}{2}}, \tag{4.5}$$

with $l = 0, 1, 2, \ldots$. Here we have introduced the complex variable

$$\mathbf{z} = \sqrt{\frac{eB_0}{4}}\,(x + iy)\,, \tag{4.6}$$

in terms of which the algebra (4.4) is represented by

$$a = \frac{\partial}{\partial \mathbf{z}} + \frac{\bar{\mathbf{z}}}{2} \equiv D, \qquad a^+ = \mathbf{z}. \tag{4.7}$$

The integer l is the eigenvalue of the angular momentum operator

$$L = \frac{1}{2}\left(\frac{eB_0}{2}R^2 - 1\right) = a^+a = \mathbf{z}D. \tag{4.8}$$

We see in the expression (4.3) in what sense we can consider this problem as one dimensional and in fact equivalent to the two-identical particle problem in one space dimension when we deal with anyons. Indeed, in the latter case, there is a supplementary symmetry constraint on the wave function related to the interchange of particles $\mathbf{z} \to -\mathbf{z}$:

$$\psi\left(e^{i\pi}\mathbf{z}\right) = e^{i\theta}\psi\left(\mathbf{z}\right). \tag{4.9}$$

Since X and Y are not observable in this context we follow the same procedure as in Section 2.

We introduce the quadratic observables

$$A = \frac{1}{4}\left(a^+a + aa^+\right) = \frac{1}{2}\left(L + \frac{1}{2}\right) = \frac{1}{2}\left(\mathbf{z}D + \frac{1}{2}\right), \tag{4.10}$$

$$B = \frac{1}{4}\left(a^2 + a^{+2}\right), \qquad C = \frac{1}{4}i\left(a^2 - a^{+2}\right), \tag{4.11}$$

which satisfy the $sl(2,\mathbb{R})$ algebra (2.6). In the discrete series representation (2.9) we see from (4.10) that bosons correspond to $\eta = \frac{1}{4}$ with $l = 0, 2, 4, \ldots$, while $\eta = \frac{3}{4}$ for fermions with $l = 0, 3, 5, \ldots$. Again we interpolate those limiting cases by introducing "anyonic" wave functions[10]

$$\psi_n^\nu = N_{2n+\nu}\mathbf{z}^{2n+\nu}e^{-\frac{\mathbf{z}\bar{\mathbf{z}}}{2}}, \tag{4.12}$$

where ν can be restricted to the interval $0 \le \nu < 2$ with n taking on the values $0, 1, 2, \ldots$. The parameter ν is related to the statistics parameter of the anyons by

$$\theta = \pi\nu. \tag{4.13}$$

The energy eigenvalues of the functions (4.12) are independent of ν, which only produces a shift in the eigenvalues of L. The relations between ν, η, and the Calogero coupling λ are the following

$$\eta = \frac{\nu}{2} + \frac{1}{4}, \qquad \lambda = \nu(\nu - 1). \tag{4.14}$$

Note that bosons (resp. fermions) are described by $\nu = 0$ (resp. $\nu = 1$).

The algebra $sl(2,\mathbb{R})$ is now represented by the following operators

$$A = \frac{1}{2}\left(\mathbf{z}D + \frac{1}{2}\right), \qquad B_+ = \frac{1}{2}\mathbf{z}^2 K, \qquad B_- = \frac{1}{2}KD^2, \tag{4.15}$$

where K is defined as the square root of the operator

$$K^2 = 1 - \frac{\nu(\nu-1)}{(L+1)(L+2)} \equiv 1 - \nu(\nu-1)\mathbf{z}^{-2}D^{-2}. \tag{4.16}$$

They act on the Hilbert space generated by the states $\psi_n^\nu \equiv |\eta, n\rangle$ on which Γ consistently assumes the value

$$\Gamma = \frac{1}{4}\left(-\frac{3}{4} + \nu(\nu-1)\right) = \eta(\eta - 1). \tag{4.17}$$

Note the restriction on the values of ν constrains $\eta = \frac{\theta}{2\pi} + \frac{1}{4} \,(\text{mod } 1)$ to lie within the interval

$$\frac{1}{4} \leq \eta < \frac{5}{4}. \tag{4.18}$$

The Barut-Girardello coherent states (2.12) describe a maximally localized pair of anyons in the lowest Landau level. This can be inferred from the particular boson and fermion cases where they reduce to symmetric (resp. antisymmetric) linear combinations of Weyl-Heisenberg (standard) coherent states.[9,12] The latter are eigenvectors of the operator $a = \sqrt{\frac{eB_0}{4}}\,(X - iY)$

$$a|\alpha\rangle = \alpha|\alpha\rangle, \tag{4.19}$$

and in the $\mathbf{z}$-representation are gaussian (up to a phase) centered around the position $\mathbf{R}_\alpha$ corresponding to the complex number α :

$$\alpha = \sqrt{\frac{eB_0}{4}}(X_\alpha - iY_\alpha) \equiv \sqrt{\frac{eB_0}{4}}R_\alpha e^{-i\Phi_\alpha}, \tag{4.20}$$

$$\langle \overline{\mathbf{z}}\,|\alpha\rangle = \mu \exp\left(\frac{1}{2}(\alpha\mathbf{z} - \overline{\alpha\mathbf{z}})\right) \exp\left(-\frac{1}{2}|\mathbf{z} - \overline{\alpha}|^2\right), \tag{4.21}$$

where μ is a normalisation factor. Particularizing the expansion (2.12) to $\eta = \frac{1}{4}$ and $\frac{3}{4}$, and using the expression (4.12) for ψ_n^ν, we obtain:

$$\left|\frac{1}{4}, \beta\right\rangle(\mathbf{z}) = \nu_{\frac{1}{4},\beta}\frac{e^{-\frac{|\mathbf{z}|^2}{2}}}{\sqrt{\pi}}\sum_{n=0}^{+\infty}\frac{\left(\sqrt{2\beta}\mathbf{z}\right)^{2n}}{2n!}, \tag{4.22}$$

$$\left|\frac{3}{4}, \beta\right\rangle(\mathbf{z}) = \nu_{\frac{3}{4},\beta}\frac{e^{-\frac{|\mathbf{z}|^2}{2}}}{\sqrt{\pi\beta}}\sum_{n=0}^{+\infty}\frac{\left(\sqrt{2\beta}\mathbf{z}\right)^{2n+1}}{(2n+1)!}. \tag{4.23}$$

Taking $2\beta = \alpha^2$ and appropriate normalisation factors, one can compare (4.22) and (4.23) with (4.21):

$$\left|\frac{1}{4}, \frac{\alpha^2}{2}\right\rangle \propto (|\alpha\rangle + |-\alpha\rangle),$$
$$\left|\frac{3}{4}, \frac{\alpha^2}{2}\right\rangle \propto (|\alpha\rangle - |-\alpha\rangle). \tag{4.24}$$

This relation allows one to calculate the Berry connection corresponding to the polar angle Φ_α in Eq. (4.20) in the general case of intermediate statistics. With the following expression for the normalisation factor in terms of a modified Bessel function,

$$\nu_{\eta,\beta} = \sqrt{\frac{|\beta|^{2\eta-1}}{I_{2\eta-1}(2|\beta|)}}, \tag{4.25}$$

Hansson, Leinaas and Myrheim have obtained,

$$\begin{aligned} A_{\Phi_\alpha}^{(\eta)} &= \left\langle \eta, \frac{\alpha^2}{2} \middle| i\partial_{\Phi_\alpha} \middle| \eta, \frac{\alpha^2}{2} \right\rangle = 2\langle \eta, \beta | \beta\partial_\beta - \overline{\beta}\partial_{\overline{\beta}} | \eta, \beta \rangle \\ &= 2|\beta| I_{2\eta}(2|\beta|) / I_{2\eta-1}(|\beta|) \end{aligned} \tag{4.26}$$

The limit $R_\alpha \gg \frac{1}{eB_0}$ makes apparent the statistical interpretation of the parameter η:

$$A_{\Phi_\alpha}^{(\eta)} \simeq \frac{eB_0}{4} R_\alpha^2 - \left(2\eta - \frac{1}{2}\right). \tag{4.27}$$

The first term is due to the magnetic field. The second term is a geometrical phase which arises when the two particles are interchanged.

5. ELEMENTARY PARTICLE IN TWO-DIMENSIONAL ANTI-DE SITTER SPACE-TIME

The two-dimensional Anti-de-Sitter space time is a constant curvature space-time with kinematical group $SO_0(1,2)$. It is actually the coset $SO_0(1,2)/SO(1,1)$, where $SO(1,1)$ corresponds to Lorentz boosts. A free particle in such a universe has the following phase space

$$SO_0(1,2)/SO(2) \cong SU(1,1)/U(1), \tag{5.1}$$

where $SO(2)$ corresponds to time translations.

The bounded version of (5.1) is the complex unit disk

$$\mathcal{D} = \{\xi \in \mathbb{C}, |\xi| < 1\}. \tag{5.2}$$

Let us give an example of (q,p)-parametrization of this phase space

$$\xi = \frac{\sinh \kappa q - i(\cosh \kappa q)\frac{p}{mc}}{1 + (\cosh \kappa q)\frac{p_0}{mc}} \tag{5.3}$$

where $p_0 = (m^2c^2 + p^2)^{\frac{1}{2}}$. κ is the curvature or inverse fundamental length, c is a fundamental speed, and m is the Poincaré mass of the particle in the flat space limit $\kappa = 0$.

The Hilbert space of quantum states of the particle can be described in this context as a Fock-Bargmann space of analytic functions. More precisely, this space is

$$\mathcal{F}_\eta = \left\{ f; f \text{ holomorphic in } \mathcal{D}, f \in L^2(\mathcal{D}, dP_\eta) \right\} \tag{5.4}$$

where $\eta = 1, \frac{3}{2}, 2, \frac{5}{2}, \ldots$, and

$$dP_\eta = \frac{(2\eta - 1)}{\pi} \left(1 - \xi\bar{\xi}\right)^{2\eta - 2} d^2\xi, \tag{5.5}$$

where $d^2\xi$ means $d(Re\xi)\, d(Im\xi)$.

This measure is such that $f(z) = 1$ is normalized: $||f|| = 1$. The space $\mathcal{F}_\eta$ carries the discrete series representation $\mathcal{U}_\eta$ of $SU(1,1)$ introduced at the infinitesimal level in Section 2: $g = \begin{pmatrix} \alpha, & \beta \\ \bar{\beta}, & \bar{\alpha} \end{pmatrix}$, $|\alpha|^2 - |\beta|^2 = 1$, acts on $\mathcal{F}_\eta$ as follows:

$$\left(\mathcal{U}_\eta(g) f\right)(\xi) = \left(-\bar{\beta}\xi + \alpha\right)^{-2\eta} f\left(\frac{\bar{\alpha}\xi - \beta}{-\bar{\beta}\xi + \alpha}\right). \tag{5.6}$$

The three basic infinitesimal generators for this representation are given by

$$\begin{aligned} K_0 &= \xi\frac{\partial}{\partial\xi} + \eta, \\ K_1 &= -\frac{1}{2}i\left(1 - \xi^2\right)\frac{\partial}{\partial\xi} + i\eta\xi, \\ K_2 &= \frac{1}{2}\left(1 + \xi^2\right)\frac{\partial}{\partial\xi} + \eta\xi. \end{aligned} \tag{5.7}$$

K_0 is the hamiltonian of the particle and corresponds to the operator A. Its spectrum on B_η is $\eta, \eta + 1, \ldots, \eta + n, \ldots$, and the corresponding normalized eigenstates are

$$u_n(\xi) = \left[\frac{\Gamma(2\eta + n)}{\Gamma(2\eta)\, n!}\right]^{\frac{1}{2}} \xi^n, \qquad n \in \mathbb{N}. \tag{5.8}$$

K_1 corresponds to $-C$ and K_2 corresponds to B. Raising and lowering operators have the following expressions

$$\begin{aligned} K_+ &= K_2 - iK_1 = \xi^2\frac{\partial}{\partial\xi} + 2\eta\xi, \\ K_- &= K_2 + iK_1 = \frac{\partial}{\partial\xi}. \end{aligned} \tag{5.9}$$

In this context, the physical interpretation of the parameter η is given in terms of the quantity $\frac{\hbar\kappa}{mc}$, the unique dimensionless combination of the four physical constants. The free antidesitterian particle is both a deformation of the one-dimensional harmonic oscillator of mass m and frequency $\omega = \kappa c$ and a deformation of the free relativistic particle of mass m. This leads to the expression[8]

$$\eta = \frac{mc}{\hbar\kappa} + \frac{1}{2} + O(\kappa), \tag{5.10}$$

i.e., the lowest energy of the particle is given by

$$\hbar\omega\eta = mc^2 + \frac{1}{2}\hbar\omega + O(\kappa). \tag{5.11}$$

This formula is a sort of illustration of the de Broglie wave-particle duality.

Quantization of the classical mechanics for this antidesitterian system is best achieved with the Berezin prescription[3,7]

$$\mathcal{D} \ni \xi \rightarrow \left|\overline{\xi}\right\rangle\left\langle\overline{\xi}\right| \in \mathcal{L}\left(\mathcal{F}_{\eta}\right). \tag{5.12}$$

The states $\left|\overline{\xi}\right\rangle$ are coherent states in the Perelomov sense[12]

$$\left|\overline{\xi}\right\rangle(\xi') = \left\langle\overline{\xi}'\left|\overline{\xi}\right.\right\rangle = \left(1 - \xi'\overline{\xi}\right)^{-2\eta}. \tag{5.13}$$

They resolve the identity with respect to the measure dP_{η}:

$$I_d = \int_{\mathcal{D}} \left|\overline{\xi}\right\rangle\left\langle\overline{\xi}\right| dP_{\eta}(\xi) \tag{5.14}$$

If we now consider the Barut-Girardello coherent states, they assume their simplest form on the unit disk. Indeed, as eigenvectors of $K_- = \partial_{\xi}$, they read with an appropriate choice of normalisation

$$\begin{aligned} \partial_{\xi}|\beta\rangle_b &= \beta|\beta\rangle_b, \\ |\beta\rangle_b(\xi) &= \frac{1}{\sqrt{\pi}}e^{\beta\xi}. \end{aligned} \tag{5.15}$$

They also resolve the identity with respect to a certain measure to be given later.

6. THREE ANALYTIC REALIZATIONS OF THE DISCRETE SERIES OF $SL(2,\mathbb{R})$

In the previous section, we have presented the Fock-Bargmann realization on the unit disk $\mathcal{D}$ of the discrete series of representations of $SL(2,\mathbb{R})$. This realization is limited to the values $\eta = 1, \frac{3}{2}, 2, \frac{5}{2}, \ldots$ taken by the parameter η, but can be extended to the half line

$$\eta \in \left(\frac{1}{2}, +\infty\right) \tag{6.1}$$

for the universal covering or for the Lie algebra $sl(2,\mathbb{R})$. In Section 4 we have presented the "2-anyon Landau level" realization on the complex plane. Let us now make things more precise after including the nonanalytic factor in (4.12) into the related scalar-product measure of the underlying Hilbert space. We now define the latter as follows

$$\mathcal{L}_{\nu} = \left\{\psi, \psi \text{ entire in } \mathbb{C},\ \psi \text{ even},\ \psi \in L^2\left(\mathbb{C}, dL_{\nu}\right)\right\}, \tag{6.2}$$

where

$$dL_{\nu}(\mathbf{z}) = |\mathbf{z}|^{2\nu} e^{-\mathbf{z}\overline{\mathbf{z}}} d^2\mathbf{z}, \qquad \nu = 2\eta - \frac{1}{2}. \tag{6.3}$$

The $sl(2,\mathbb{R})$ algebra is then represented as follows

$$A = \frac{1}{2}\left(\mathbf{z}D_{\mathbf{z}} + \frac{1}{2}\right), \qquad \text{with} \quad D_{\mathbf{z}} = \partial_{\mathbf{z}} + \frac{\nu}{\mathbf{z}}, \tag{6.4}$$

$$B_+ = \frac{1}{2}\mathbf{z}^2 K_{\mathbf{z}}, \qquad B_- = \frac{1}{2} K D_{\mathbf{z}}^2 \tag{6.5}$$

where $K_{\mathbf{z}}^2 = 1 - \nu(\nu - 1)\mathbf{z}^{-2} D_{\mathbf{z}}^{-2}$.

The normalized eigenstates for A are given by:

$$\psi_n^{\nu}(\mathbf{z}) = N_{2n+\nu}\mathbf{z}^{2n}, \qquad n \in \mathbb{N}, \tag{6.6}$$

where $N_l = [\pi\Gamma(l+1)]^{-\frac{1}{2}}$.

The Barut-Girardello coherent states $|\beta\rangle_b$ read in this realisation:

$$|\beta\rangle_b = \sum_{n=0}^{+\infty}\left[\frac{\Gamma\left(\nu+\frac{1}{2}\right)}{\pi n!\Gamma\left(\nu+n+\frac{1}{2}\right)}\right]^{\frac{1}{2}}\beta^n\psi_n^{\nu}. \tag{6.7}$$

Note that their scalar product

$${}_b\langle\beta'|\beta\rangle_b = \sum_{n=0}^{+\infty}\frac{\Gamma\left(\nu+\frac{1}{2}\right)}{\pi n!\Gamma\left(\nu+n+\frac{1}{2}\right)}\left(\overline{\beta}'\beta\right)^n \tag{6.8}$$

is, up to a factor $\left(\overline{\beta}'\beta\right)^{\frac{1}{2}\nu}$, a modified Bessel function.

It is clear that the representation (6.4), (6.5) of $sl(2,\mathbb{R})$ is still not satisfactory, since it involves inverse of differential operators. Another representation was proposed by Brink et al.,[4] which is in fact the original Barut-Girardello realization. The aim of those authors was to display the equivalence between the N-body Calogero model and the N-anyon model in the lowest Landau level. Let us summarize their procedure in the two-body case relevant to the present context. The Hamiltonian H_{cal} given in (3.2) in terms of the relative coordinate $x = x_{(1)} - x_{(2)}$ is transformed into the following one acting on $\Phi^{\pm}$:

$$H = \frac{1}{2}\left(-D^2 + x^2\right), \tag{6.9}$$

$$D = \frac{\partial}{\partial x} + \frac{\nu}{x}(1-\Pi), \tag{6.10}$$

$$\Pi x = -x\Pi \qquad \text{(Klein operator)}, \tag{6.11}$$

if we make the Ansatz on the original wave function

$$\psi^{\pm} = x^{\nu}\Phi^{\pm}. \tag{6.12}$$

Here Φ^+ is even and Φ^- is odd, and the relation between λ and ν reads

$$\lambda = \nu(\nu \mp 1) \tag{6.13}$$

according to the even or odd parity of $\Phi^{\pm}$.

Next one introduces two operators

$$a^{(\mp)} = \frac{1}{\sqrt{2}}(x \pm D) \tag{6.14}$$

which satisfy

$$[a^{(-)}, a^{(+)}] = 1 + 2\nu\Pi. \tag{6.15}$$

The Hamiltonian (6.9) then reads

$$H = \frac{1}{2}\left\{a^{(-)}, a^{(+)}\right\}. \tag{6.16}$$

A complex realization of the algebra (6.15) is given by

$$\begin{aligned}\tilde{a}^{(-)} &= \frac{\partial}{\partial \mathbf{z}} + \frac{\nu}{\mathbf{z}}(1-\Pi) \equiv \widetilde{D},\\ \tilde{a}^{(+)} &= \mathbf{z},\end{aligned} \tag{6.17}$$

and the Hamiltonian (6.16) generalizes to

$$\widetilde{H} = \frac{1}{2}\left\{\tilde{a}^{(-)}, \tilde{a}^{(+)}\right\} = \mathbf{z}\frac{\partial}{\partial \mathbf{z}} + \nu + \frac{1}{2}. \tag{6.18}$$

$\widetilde{H}$ describes a systems of two anyons. It enters in the following realization of the discrete series of $sl(2, \mathbb{R})$

$$\tilde{A} = \frac{1}{2}\widetilde{H} = \frac{1}{2}\left(\mathbf{z}\frac{\partial}{\partial \mathbf{z}} + \nu + \frac{1}{2}\right) \tag{6.19}$$

$$\tilde{B}_+ = \frac{1}{2}\mathbf{z}^2 \tag{6.20}$$

$$\tilde{B}_- = \frac{1}{2}\widetilde{D}^2 = \frac{1}{2}\frac{\partial^2}{\partial \mathbf{z}^2} + \frac{2\nu}{\mathbf{z}}\frac{\partial}{\partial \mathbf{z}} - \frac{\nu}{\mathbf{z}^2}(1-\Pi) \tag{6.21}$$

$$\Pi\mathbf{z} = -\mathbf{z}\Pi. \tag{6.22}$$

This realization considerably simplifies the previous one (6.4)-(6.5). Let us choose the even parity. Then $(1-\Pi)$ gives zero in (6.22) and the carrier Hilbert space is generated by the eigenstates:

$$\tilde{\psi}_n^{\nu} = \widetilde{N}_n \mathbf{z}^{2n}, \qquad n = 0, 1, 2, \ldots \tag{6.23}$$

$$\widetilde{N}_n = \frac{1}{2^n}\left[\frac{\Gamma\left(\nu + \frac{1}{2}\right)}{\Gamma\left(\nu + n + \frac{1}{2}\right) n!}\right]^{\frac{1}{2}}. \tag{6.24}$$

The above normalisation requires a new measure $dB_\nu(\mathbf{z})$ in the definition of the Hilbert space:

$$\mathcal{B}_\nu = \left\{\tilde{\psi}, \tilde{\psi} \text{ entire in } \mathbb{C},\ \text{even},\ \int \left|\tilde{\psi}(\mathbf{z})\right|^2 d\mathcal{B}_\nu(\mathbf{z}) < \infty\right\}, \tag{6.25}$$

$$d\mathcal{B}_\nu(\mathbf{z}) = \frac{2^{\frac{1}{2}-\nu}|\mathbf{z}|^{2\nu+1}}{\pi\Gamma\left(\nu+\frac{1}{2}\right)} K_{\frac{1}{4}-\frac{1}{2}\nu}\left(|\mathbf{z}|^2\right) d^2\mathbf{z}. \tag{6.26}$$

$K_\nu(\xi)$ is a modified Bessel function of the third kind:

$$\begin{aligned}K_\nu(\xi) &= \frac{\pi}{2}\frac{I_{-\nu}(\xi) - I_\nu(\xi)}{\sin \pi\nu}\\ I_\nu(\xi) &= \sum_{n=0}^{+\infty} \frac{\left(\frac{1}{2}\xi\right)^{\nu+2n}}{n!\Gamma(\nu+n+1)}.\end{aligned} \tag{6.27}$$

The measure dB_ν was derived by Barut and Girardello.[1] It also is the measure with respect to which the Barut-Girardello coherent states $|\beta\rangle_b$ with $\beta = \frac{1}{2}\mathbf{z}^2$ solve the identity:

$$\pi \int_{\mathbb{C}} \left|\frac{\mathbf{z}^2}{2}\right\rangle_b {}_b\!\left\langle\frac{\mathbf{z}^2}{2}\right| d\mathcal{B}_\nu(\mathbf{z}) = I_d. \tag{6.28}$$

Note that the parametrization $\beta = \frac{1}{2}\mathbf{z}^2$ is in agreement with the special boson and fermion cases (4.24).

Finally let us give the realizations of the two types of coherent states in the Barut-Girardello Hilbert space $\mathcal{B}_\nu$.

The Perelomov coherent states assume their simplest expression, independent from η:

$$\begin{aligned} \widetilde{\left|\overline{\xi}\right\rangle}(\mathbf{z}) &= e^{\frac{\mathbf{z}^2}{2}\overline{\xi}} \\ &= \sqrt{\pi}\left|\frac{\mathbf{z}^2}{2}\right\rangle_b\left(\overline{\xi}\right). \end{aligned} \tag{6.29}$$

We shall note here the duality between the two types of coherent states versus the two types of realizations: Fock-Bargmann and Barut-Girardello.

On the other hand, the Barut-Girardello coherent states assume their original expression (in agreement with (6.8)):

$$\begin{aligned} \widetilde{\left|\frac{\overline{\mathbf{z}}^2}{2}\right\rangle_b}(\mathbf{z}') &= \sqrt{\pi}\,{}_b\!\left\langle\frac{\overline{\mathbf{z}'}^2}{2}\middle|\frac{\overline{\mathbf{z}}^2}{2}\right\rangle_b = \frac{1}{\sqrt{\pi}}\,{}_0F_1\left(\nu+\frac{1}{2};\frac{\overline{\mathbf{z}}^2\mathbf{z}'^2}{2}\right) \\ &= \frac{1}{\sqrt{\pi}}\Gamma\left(\nu+\frac{1}{2}\right)\left(i\frac{\overline{\mathbf{z}}\mathbf{z}'}{\sqrt{2}}\right)^{\frac{1}{2}-\nu} J_{\nu-\frac{1}{2}}\left(\sqrt{2}i\overline{\mathbf{z}}\mathbf{z}'\right) \end{aligned} \tag{6.30}$$

where

$${}_0F_1(c;\xi) = \sum_{n=0}^{+\infty}\frac{\Gamma(c)}{\Gamma(c+n)}\frac{\xi^n}{n!}.$$

We now come to the unitary maps between pairs of these three realisations. They are defined by integral kernels as follows.

(i) *Unitary map between the Fock-Bargmann realization and the "2-anyon Landau level" realization*

It is defined by the integral transforms:

$$\psi(\mathbf{z}) = \int_D K_{fl}^{\nu}\left(\mathbf{z},\overline{\xi}\right) f(\xi)\, dP_\eta(\xi) \tag{6.31}$$

$$f(\mathbf{z}) = \int_{\mathbb{C}} K_{fl}^{\nu}(\overline{\mathbf{z}},\xi)\,\psi(\mathbf{z})\, dL_\nu(\xi) \tag{6.32}$$

for $f \in \mathcal{F}_\eta$, $\psi \in \mathcal{L}_\nu$, and $\eta = \frac{1}{2}\nu + \frac{1}{4} > \frac{1}{2} \Rightarrow \nu > \frac{1}{2}$.

The kernel K_{fl} is given by:

$$K_{fl}^{\nu}\left(\mathbf{z},\overline{\xi}\right) = \sum_{n=0}^{+\infty}\psi_n^{\nu}(\mathbf{z})\, u_n\left(\overline{\xi}\right) \tag{6.33}$$

where u_n and ψ_n^ν are defined in Eq. (5.8) and (6.6) respectively. Actually K_{fl} is just the Perelomov coherent state in the "2-anyon Landau level realisation"

$$K_{fl}\left(\mathbf{z},\overline{\xi}\right) = \left|\overline{\xi}\right\rangle(\mathbf{z}) \tag{6.34}$$

This fact is simply inferred from the reproducing property of the kernel $\left\langle\overline{\xi}'\middle|\overline{\xi}\right\rangle$ in the Fock-Bargmann realization.

Note the particular boson and fermion cases

$$K_{fl}^{0}\left(\mathbf{z},\overline{\xi}\right) = K_{fl}^{1}\left(\mathbf{z},\overline{\xi}\right) = e^{\frac{\mathbf{z}^2}{2}\overline{\xi}}, \tag{6.35}$$

but we must recall that the unitary map is not defined for $\nu = 0$, since the Fock-Bargmann realisation has no meaning for $\nu < \frac{1}{2}$. On the other hand, the unitary map is perfectly defined for $\nu = 1$ and this provides a way for fully reaching the "harmonic oscillator" representation of the double covering of $SU(1,1)$, when one first works on the unit disk and next go onto the two-fold complex plane via the unitary map (6.31).

(ii) *Unitary map between the Fock-Bargmann realization and the Barut-Girardello realization*

The corresponding kernel is now independent of ν:

$$\begin{aligned} K^{\nu}_{fb}\left(\mathbf{z},\overline{\xi}\right) &= \sum_{n=0}^{+\infty} \widetilde{\psi}^{\nu}_{n}(\mathbf{z})\, u_n\left(\overline{\xi}\right) = e^{\frac{\mathbf{z}^2}{2}\overline{\xi}} = \\ &= \sqrt{\pi}\left|\frac{\mathbf{z}^2}{2}\right\rangle_b \left(\overline{\xi}\right) = \widetilde{\left|\overline{\xi}\right\rangle}(\mathbf{z}), \end{aligned} \tag{6.36}$$

and we can see again the crucial role played by the reproducing property of both types of coherent states.

(iii) *Unitary map between the "2-anyon Landau level" realization and the Barut-Girardello realization*

The corresponding kernel is now just the Barut-Girardello coherent state:

$$K^{\nu}_{lb}\left(\mathbf{z}', \overline{\mathbf{z}}\right) = \widetilde{\left|\frac{\overline{\mathbf{z}}^2}{2}\right\rangle_b}\left(\mathbf{z}'\right) \tag{6.37}$$

given by (6.30).

References

1. D. Basu, *J. Math. Phys.* 35:3612 (1994)
2. A.O. Barut and L. Girardello, *Commun. Math. Phys.* 21:41 (1971)
3. F.A. Berezin, *Commun. Math. Phys.* 40:153 (1975)
4. L. Brink, J.H. Hansson, S. Konstein, and M.A. Vasiliev, The Calogero model – Anyonic representation, fermionic extension, and supersymmetry, preprint USITP-92-14 (1993)
5. F. Calogero, *J. Math. Phys.* 16:419 (1971)
6. J-P. Gazeau, On two analytic elementary systems in quantum mechanics, *in* "Colloque de Géométrie analytique, Paris, Juillet 1992", Hermann, Paris (1995) (to appear)
7. J-P. Gazeau and J .Renaud, *Ann. Phys. (N.Y.)* 222:86 (1993)
8. J-P. Gazeau and J. Renaud, *Phys. Lett. A* 179:67 (1993)
9. R.J. Glauber, *Phys. Rev.* 131:2766 (1963)
10. J.H. Hansson, J.H. Leinaas, and J. Myrheim, *Nucl. Phys. B* 384:559 (1992)
11. J.H. Leinaas and J. Myrheim, *Int. J. Mod. Phys. A* 8:3649 (1993)
12. A. Perelomov, "Generalized Coherent States and Their Applications", Springer, Berlin, Heidelberg (1986).

SYMPLECTIC AND LAGRANGIAN REALIZATION OF POISSON MANIFOLDS

M. Giordano,[1] G. Marmo[1,2] and A. Simoni[1,2]

[1] Dipartimento di Scienze Fisiche, Università Federico II
Napoli, Italy
[2] I.N.F.N.
Mostra D'Oltremare, Pad 19
80125 Napoli, Italy

1. INTRODUCTION

An action principle is usually the starting point to describe physical systems, both particles and fields. A preliminary study usually considers fields as given, i.e. as external fields while particles are thought of as test particles, therefore only the point particle dynamics is dealt with. In this approach the Lagrangian function usually is the sum of three terms: a kinematic term which is quadratic in the velocities, a current-potential coupling term which is linear in the velocities, and a term which depends only on the positions, for instance the electrostatic potential.

The transition to the Hamiltonian description in the symplectic or Poisson formalism allows to absorb the magnetic field in a change of coordinates, so that the momentum $p = \partial\mathcal{L}/\partial\dot{q}$ is replaced by $p + eA$ where A is the vector potential. The electrostatic or other effective potentials appears in the Hamiltonian function.

There are, however, other situations, like the electron-monopole system, where the magnetic field cannot be absorbed in a change of coordinates, this has to do with the fact that the symplectic form in this case is closed but not exact. What we learn from this case is that some external fields, like the magnetic field, will modify the symplectic structure while some other fields will modify the Hamiltonian function. Moreover, to incorporate the magnetic field of the monopole in the Lagrangian picture, one has to add fictitious degrees of freedom, i.e. degrees of freedom which do not carry a dynamical evolution.[1,2]

From the symplectic structure one usually defines Poisson brackets, so that the correspondence with quantum systems is more transparent. As a matter of fact, Poisson brackets arise in the classical limit of quantum systems also for those variables which carry a first order dynamics rather than a second order one, like those arising in the Lagrangian picture, e.g., spin, isospin, color, or other inner variables carried by particles.

We seem to be facing the following situation: when classical mechanics is thought of as limit of quantum mechanics, it seems more natural to deal with the Poisson formalism. If, viceversa, we start with a classical system and try to quantize it, it seems more natural to start with a Lagrangian function to be used either in a Feynman

Quantization, Coherent States, and Complex Structures
Edited by J.-P. Antoine *et al.*, Plenum Press, New York, 1995

path-integral approach to construct the propagator, or to define canonical coordinates $p = \partial\mathcal{L}/\partial\dot{q}$ and q and replace them with operators like in the Dirac procedure.

It seems therefore quite natural to investigate if there is some systematic way to compare the Poisson formalism with the Lagrangian one. More likely, it was in this spirit that Feynman tried, as reported by Dyson,[3] to start with a Poisson structure and look for those forces which could be described in this formalism to investigate if they were more general than those coming from a Lagrangian description.

On the other hand, if one would find that the formalism is more general than the Lagrangian one, then it is important to develop an intuition on how to add interactions, i.e. external fields for point particles, directly in the Poisson bracket. This would replace our present way of composing systems and adding interactions in the Lagrangian formalism.

After this long introduction, we are going to expose briefly what we may call Feynman's problem. Then we introduce an appropriate way to compare symplectic and Poisson formalism and then we move to the comparison of Lagrangian and Poisson formalism. We also give some examples to show what is our present experience in comparing these various formalisms.

2. THE FEYNMAN PROBLEM

As reported by Dyson,[3] Feynman was able to derive the Lorentz force for a point particle by postulating some specific Poisson brackets on $T\mathbb{R}^3$.

We start with variables $(x^i, v^i), i = 1,2,3$, x^i denoting position and v^i velocity. The postulated form of brackets is

$$\left\{x^i, x^j\right\} = 0 \tag{2.1}$$

$$m\left\{x^i, v^j\right\} = \delta_{ij} \tag{2.2}$$

$$\left\{v^i, v^j\right\} = f_{ij}, \tag{2.3}$$

with $f_{i,j}$ to be determined by the Jacobi identity and the requirement that there exists a second order dynamics described by

$$\frac{dx^i}{dt} = \left\{x^i, H\right\} = v^i \tag{2.4}$$

$$\frac{dv^i}{dt} = \left\{v^i, H\right\} = \frac{1}{m}F^i(\vec{x}, \vec{v}). \tag{2.5}$$

Thus f_{ij} is to be determined and this will determine the most general field and forces which are compatible with the description in terms of a function H via the determined Poisson brackets.

Feynman found that there is a unique solution giving the coupling to electromagnetic fields. We show now how one proceeds.

One first defines the magnetic field B^i by setting $f_{ij} = \frac{1}{m^2}\varepsilon^{ijk}B_k$, or

$$m^2\left\{v^i, v^j\right\} = \varepsilon^{ijk}B_k. \tag{2.6}$$

From the Jacobi identity

$$\left\{x^i, \left\{v^i, v^k\right\}\right\} + \left\{v^i, \left\{v^k, x^i\right\}\right\} + \left\{v^k, \left\{x^i, v^i\right\}\right\} = 0, \tag{2.7}$$

together with (2.1)-(2.2), it follows $\{x^i, B_k\} = 0$. Thus B_k must be independent of the particle velocity v^i. By using the Jacobi identity involving v^i, v^j and v^k, one finds that B_k has zero divergence, recovering the homogeneous Maxwell equations. Next one takes the total time derivatives of the Poisson brackets, i.e. one requires the dynamical evolution to be a canonical transformation. One gets

$$\left\{v^i, v^j\right\} + \left\{x^i, \dot{v}^j\right\} = 0 \tag{2.8}$$

or

$$\left\{x^i, F^j\right\} = -\frac{1}{m}\epsilon^{ijk} B_k(x), \tag{2.9}$$

from which it follows that F^i is at most linear in the velocities

$$F^i = E^i(x) + \epsilon^{ijk} v_j B_k(x). \tag{2.10}$$

Equation (2.10) is the Lorentz force law (where the electric charge is set equal to one, and the electric field emerges from the fact that (2.9) determines F^j up to a vector depending on $\vec{x}$. From the time derivative of (2.6) one finds

$$m\epsilon^{ijk}\left\{F^i, v^j\right\} = \dot{B}_k. \tag{2.11}$$

Using the divergenceless condition on B_k, we get

$$m\epsilon^{ijk}\left\{E^i, v^j\right\} = \dot{B}_k - m\left\{B_k, v^i\right\} v^j \quad . \tag{2.12}$$

Because of the time independence assumption on B_k we get

$$\epsilon^{ijk}\left\{E^i, v^j\right\} = 0, \tag{2.13}$$

i.e. the vector $\vec{E}$ has zero curl, as in the case for time independent magnetic fields.

If one repeats this calculation for time dependent fields, one instead recovers the Faraday law and thus both homogeneous Maxwell equations.

In this case both the velocity v^i and the force F^i are allowed to have explicit dependence on t, so that equation (2.5) is generalized to

$$\frac{d}{dt}v^i = \left\{v^i, H\right\} + \frac{\partial v^i}{\partial t} = \frac{1}{m}F^i(\vec{x}, \vec{v}, t). \tag{2.14}$$

The analysis then proceeds as above with the static fields $E^i(\vec{x})$ and $B^i(\vec{x})$ replaced by $E^i(\vec{x}, t)$ and $B^i(\vec{x}, t)$. The Hamiltonian function for the above system is

$$H = \frac{m}{2}v^2 + \Phi(\vec{x}, t), \tag{2.15}$$

where $\Phi(\vec{x}, t)$ is the standard scalar potential. From this one reconstructs the electric field according to

$$\vec{E} = -\vec{\nabla}\Phi + m\frac{\partial \vec{v}}{\partial t}. \tag{2.16}$$

It is now clear that one has not gone beyond the Lagrangian formalism. We could have obtained the same field of forces starting with a Lagrangian function.

We shall now consider how to relate Poisson structures with symplectic structures and Lagrangian functions. To this aim we first review some Poisson geometry.

3. REVIEW OF POISSON MANIFOLDS

We recall[4,5] that a Poisson structure on a manifold M is a skew-symmetric R-bilinear map, denoted

$$\{.,.\} : \mathcal{F}(M) \times \mathcal{F}(M) \to \mathcal{F}(M),$$

such that

1. $(\mathcal{F}(M), \{.,.\})$ is a real Lie algebra, i.e. it satisfies the Jacobi identity

$$\{F, \{G, H\}\} + \{G, \{H, F\}\} + \{H, \{F, G\}\} = 0. \tag{3.1}$$

2. The map $F \mapsto X_F = \{., F\}$ is a Lie algebra homomorphism from $(\mathcal{F}(M), \{.,.\})$ to the Lie algebra of derivations of the associative algebra $\mathcal{F}$ of functions on M, i.e. it satisfies the Leibnitz rule

$$\{F, GH\} = G\{F, H\} + \{F, G\}\, H. \tag{3.2}$$

The pair $(M, \{.,.\})$ is called a Poisson manifold. The Poisson structure is said to be nondegenerate or regular, if for any open submanifold $U \subset M, f \in \mathcal{F}(U), g \in \mathcal{F}(U), \{f, g\} = 0$ for any $g \in \mathcal{F}(U)$ implies that f is constant on U.

If there exist globally defined functions C such that $\{C, G\} = 0 \ \forall G \in \mathcal{F}(U)$, they will be called Casimir functions.

The best known example of Poisson manifold is R^{2n} with standard brackets

$$\{F, G\} = \frac{\partial F}{\partial x_i}\frac{\partial G}{\partial p^i} - \frac{\partial F}{\partial p^i}\frac{\partial G}{\partial x_i} \tag{3.3}$$

where $(x_i, p^i), i \in \{1, 2, \ldots, n\}$ are canonical position and momentum variables used to parametrize R^{2n}.

A vector field X_F defined by

$$X_F \cdot G = \{G, F\}$$

is said to be Hamiltonian with Hamiltonian function F. The differential equation associated with any such vector field is given by

$$\frac{d}{dt}\xi^a = \{\xi^a, F\}. \tag{3.4}$$

Using coordinates $(\xi^a), a \in \{1, 2, \ldots, n\}$, we get

$$\{F, G\} = \left\{\xi^a, \xi^b\right\}\frac{\partial F}{\partial \xi^a}\frac{\partial G}{\partial \xi^b} = \Lambda^{ab}(\xi)\frac{\partial F}{\partial \xi^a}\frac{\partial G}{\partial \xi^b}, \tag{3.5}$$

which allows to define a bivector field, i.e. a twice-contravariant skew-symmetric tensor,

$$\Lambda = \Lambda^{ab}\frac{\partial}{\partial \xi^a} \wedge \frac{\partial}{\partial \xi^b}. \tag{3.6}$$

This tensor is usually called a Poisson tensor. In terms of Λ, it is possible to write the Poisson bracket in the form

$$\Lambda(dF, dG) = \{F, G\} \tag{3.7}$$

and Hamiltonian vector fields

$$X_F = -\Lambda(dF). \tag{3.8}$$

When Λ is invertible, the inverse tensor is a two-form which turns out to be closed, i.e. a symplectic form.[6] If in addition the symplectic form is defined on the tangent bundle TQ of a configuration space Q and satisfies some additional conditions, it can be associated with a Lagrangian function, at least locally. In this way one recovers the usual Lagrangian formalism on the tangent bundle TQ of some configuration manifold Q.

4. REVIEW OF THE LAGRANGIAN FORMALISM

The geometrical approach to time independent Lagrangian formalism starts with a differential manifold Q and its tangent bundle $\tau : TQ \to Q$.[7,8] This vector bundle carries a canonical (1,1)-tensor field S called the vertical endomorphism. In natural coordinates (q^i, v^i) of the tangent bundle TQ, S is given by

$$S = dq^i \otimes \frac{\partial}{\partial v^i}. \tag{4.1}$$

The action on one-forms on TQ is given by $S(dq^i) = 0$, $S(dv^i) = dq^i$ and extended by linearity. On vector fields we get

$$S(\frac{\partial}{\partial q^i}) = \frac{\partial}{\partial v^i}, \quad S(\frac{\partial}{\partial v^i}) = 0, \tag{4.2}$$

giving

$$S(a^i\frac{\partial}{\partial q^i} + b^i\frac{\partial}{\partial v^i}) = a^i\frac{\partial}{\partial v^i}, \tag{4.3}$$

which is called the vertical lift X^v of the vector field $a^i\frac{\partial}{\partial q^i} + b^i\frac{\partial}{\partial v^i}$.

In addition to S, on the tangent bundle TQ, like on any other vector bundle, there is a vector field Δ, the infinitesimal generator of dilations on the fibers. In natural local coordinates for TQ we have

$$\Delta = v^i\frac{\partial}{\partial v^i}. \tag{4.4}$$

A second order vector field is any vector field Γ on TQ such that

$$S(\Gamma) = \Delta. \tag{4.5}$$

Given a Lagrangian function $\mathcal{L}$ on TQ, we associate with it a two form $\omega_{\mathcal{L}}$ by setting

$$\omega_{\mathcal{L}} = -dS(d\mathcal{L}). \tag{4.6}$$

Often we shall denote by $d_S f$ the one form $S(df)$. With this notation we find $\omega_{\mathcal{L}} = -dd_S\mathcal{L}$. It is possible to show that $dd_S + d_S d = 0$.

The (1,1)-tensor can be made to act also on higher forms, for instance

$$S(\alpha \wedge \beta) = (S\alpha) \wedge \beta + \alpha \wedge S\beta$$

and then extend by linearity. We find that $SdSdf = 0$ for any function f.

The following theorem characterize two-forms ω which are of the Lagrangian type:
Theorem. – Let ω be a closed form on a tangent bundle TQ. If $S\omega = 0$, there exists, locally, a function $\mathcal{L}$ such that $\omega = \omega_{\mathcal{L}} = -dS(d\mathcal{L})$.

We do not prove this theorem here and refer to Morandi et al.[8]

With any Lagrangian function $\mathcal{L}$ we can associate a Poisson bracket by solving the following equations:[9]

$$\left\{\frac{\partial \mathcal{L}}{\partial \dot{q}^i}, \frac{\partial \mathcal{L}}{\partial \dot{q}^j}\right\} = 0, \tag{4.7}$$

which can be expanded into

$$0 = \frac{\partial^2 \mathcal{L}}{\partial \dot{q}^i \partial \dot{q}^k}\left\{\dot{q}^k, \dot{q}^r\right\}\frac{\partial^2 \mathcal{L}}{\partial \dot{q}^j \partial \dot{q}^r} + \frac{\partial^2 \mathcal{L}}{\partial \dot{q}^i \partial q^k}\left\{q^k, \dot{q}^r\right\}\frac{\partial^2 \mathcal{L}}{\partial \dot{q}^j \partial \dot{q}^r} + \frac{\partial^2 \mathcal{L}}{\partial \dot{q}^i \partial \dot{q}^k}\left\{\dot{q}^k, q^r\right\}\frac{\partial^2 \mathcal{L}}{\partial \dot{q}^j \partial q^r}. \tag{4.8}$$

We also have

$$\delta_{ij} = \left\{\frac{\partial \mathcal{L}}{\partial \dot{q}^i}, q^j\right\} = \frac{\partial^2 \mathcal{L}}{\partial \dot{q}^i \partial \dot{q}^k}\left\{\dot{q}^k q^j\right\}, \tag{4.9}$$

with the assumption $\{q^i, q^j\} = 0$.

It is clear that if

$$\det |\frac{\partial^2 \mathcal{L}}{\partial \dot{q}^i \partial \dot{q}^k}| \neq 0,$$

we have a unique Poisson bracket by solving the above algebraic equations.

If $\mathcal{L}$ is degenerate, i.e. $\det ||\frac{\partial^2 \mathcal{L}}{\partial \dot{q}^i \partial \dot{q}^k}|| = 0$, we might find several solutions or none. Usually the situation is handled via the so-called Dirac-Bergman constraint procedure;[10,11] for a different approach see Dubrovin et al.[12]

We are now ready to compare the Poisson formalism with the symplectic and Lagrangian formalism.

5. SYMPLECTIC AND LAGRANGIAN REALIZATION OF POISSON MANIFOLDS

A map $\Phi : M_1 \to M_2$ between Poisson manifolds $(M_1, \{.,.\}_1)$ and $(M_2, \{.,.\}_2)$ is said to be a Poisson map when

$$\{\Phi^* F, \Phi^* G\}_1 = \Phi^*(\{F, G\}_2) \tag{5.1}$$

for any pair of functions $F, G \in \mathcal{F}(M_2)$.[5,13]

When $\{.,.\}_1$ is invertible, making M_1 into a symplectic manifold, and Φ is a submersion we say that Φ is a symplectic realization of the Poisson manifold $(M_2, \{.,.\}_2)$.

Remark : One might put different requirements on Φ as a map, for instance require only that Φ is smooth, or that Φ is a submersion onto and so on. What seems useful from the dynamical point of view is that the dynamics we are describing on M_2 can be fully recovered by projecting integral curves of a related dynamical system on M_1. It seems therefore natural to have the minimal requirement $\dim M_1 \geq \dim M_2$. When dealing with complete vector fields we may put additional restriction on Φ.

When $M_1 = TQ$ for some configuration space Q, and $\{.,.\}_1$ is the Poisson bracket associated with a regular Lagrangian function $\mathcal{L}$, we say that Φ is a Lagrangian realization. When the bracket $\{.,.\}_1$ is associated with a degenerate Lagrangian, we say that Φ is a weak Lagrangian realization of $(M_2, \{.,.\}_2)$.

The above definitions clarify the framework in which we have to compare the various formalisms. In the same spirit we would say that we have a (weak) Lagrangian realization of a symplectic manifold (M_2, ω_2) if we replace the symplectic structure ω_2

with the associated Poisson bracket and we have a (weak) Lagrangian realization of the corresponding Poisson manifold. Within this broader approach we have to investigate for necessary and sufficient conditions for a Poisson manifold to admit a Lagrangian realization. Again, if we would find that any "reasonable" Poisson bracket can be given a Lagrangian realization, we would conclude that the Poisson formalism does not take us beyond the Lagrangian formalism. There are few papers dealing with symplectic realization of Poisson manifolds,[13,14] a recent one deals with Lagrangian realization.[15]

In the remaining part of this paper, we shall give examples already available in the litterature of Lagrangian realizations of a special class of Poisson brackets. We shall also give few examples of symplectic realizations of exotic Poisson brackets. These examples might be a starting point to formulate reasonable sufficient conditions for the existence of Lagrangian realizations.

6. SPINNING PARTICLES

We shall ignore the spatial degrees of freedom of the particle and study only the spin variables.

We denote the spin variables by $S_i, i = 1,2,3$. We have $\vec{S} \in \mathbb{R}^3$. If we specialize to the dynamical system of a spin interacting with a magnetic field $\vec{B}$, the standard equations of motion for such a system gives the precession of the spin S_i,

$$\dot{S}_i = \mu\varepsilon_{ijk}S_jB_k, \tag{6.1}$$

where μ denotes the magnetic moment. In this particular situation we are able[16] to exhibit all Poisson brackets in R^3. indeed if we set

$$\{S_i, S_j\} = \varepsilon_{ijk}F^k(S) \tag{6.2}$$

and introduce the one-form $\alpha = F^k dS^k$, we find that the Poisson bracket (6.1) satisfies the Jacobi identity iff $d\alpha \wedge \alpha = 0$. Therefore, locally, α can be written as $\alpha = f_0 df_1$, where f_0 and f_1 are functions of S_i and we have

$$\{S_i, S_j\} = f_0(S)\varepsilon_{ijk}\frac{\partial f_1}{\partial S_k}. \tag{6.3}$$

As a result, f_1 has zero Poisson bracket with all S_i i.e. defines a Casimir function.

If we now require that the dynamics be such that there is a Hamiltonian function $H = H(S)$, then f_1 is also a constant of the motion and therefore

$$\dot{S}_i\frac{\partial f_1}{\partial S_i} = 0 \tag{6.4}$$

By using $\dot{S}_i = \mu\varepsilon_{ijk}S_jB_k$ we get $f_1 = f_1(|\vec{S}|^2, \vec{S}\cdot\vec{B})$. Of course H also must have the same form, because it will be a constant of the motion, $H = H(|\vec{S}|^2, \vec{S}\cdot\vec{B})$. Therefore all possible Hamiltonian descriptions for our dynamics are given by

$$\dot{S}_i = \mu\varepsilon_{ijk}S_jB_k = f_0(S)\varepsilon_{ijk}\frac{\partial f_i}{\partial S_k}\frac{\partial H}{\partial S_j}. \tag{6.5}$$

The standard canonical formalism for a spinning particle is recovered when we write

$$f_0 = \frac{1}{2}, \quad f_1 = |\vec{S}|^2, \quad H = -\mu\vec{S}\cdot\vec{B}. \tag{6.6}$$

For this choice the Poisson bracket algebra corresponds to the $su(2)$ Lie algebra. Reducible representations of $SU(2)$ then arise upon performing a canonical quantization of this system. Alternatively, to obtain irreducible representations for $SU(2)$, one must impose a constraint on the variables S_i. Namely, one must require that the classical Casimir function f_1 takes only certain constant values (and as a result S_i will span S^2).[2]

Alternative brackets of some relevance can be given. These will give rise to alternative Hamiltonian formulations for the same dynamics, but canonically inequivalent to the standard one. One such formulation results from the choice

$$f_0 = \frac{1}{2}, \quad f_1 = S_1^2 + S_2^2 + \frac{1}{2\lambda}\left[\frac{\cosh 2\lambda S_3}{\sinh\lambda} - \frac{1}{\lambda}\right] \tag{6.7}$$

and

$$H = -\mu\lambda S_3, \tag{6.8}$$

where, with no loss of generality, we have taken the magnetic field to be in the third direction. Here λ is a deformation parameter, the standard formalism is recovered when $\lambda \to 0$. For nonzero λ, we get the Poisson brackets

$$\{S_1, S_2\} = \frac{1}{2}\frac{\sinh(2\lambda S_3)}{\sinh\lambda},$$

$$\{S_2, S_3\} = S_1, \quad \{S_3, S_1\} = S_2 \tag{6.9}$$

These brackets are a classical realization of the quantum commutation relations for generators of the $U_q(sl(2))$ Hopf algebra. If we denote the latter generators by J_+, J_- and J_0, their commutation relations are

$$\begin{aligned} [J_0, J_\pm] &= \pm J_\pm \\ [J_+, J_-] &= \frac{q^{2J_0} - q^{-2J_0}}{q - q^{-1}}. \end{aligned} \tag{6.10}$$

These commutation relations reduce to the $su(2)$ Lie algebra relations in the limit $q \to 1$. To obtain the classical system above, one can replace the quantum operators J_+, J_- and J_0 by $S_1 + iS_2$, $S_1 - iS_2$ and S_3 respectively, q by $\exp\lambda$, and the commutation relations by i times Poisson brackets.

There is another choice for f_0 and f_1 which is known to correspond to the classical limit of the $U_q(sl(2))$ Hopf algebra. It is

$$F_0 = \frac{\lambda}{4}S_3, \quad f_1 = {S_1}^2 + {S_2}^2 + {S_3}^2 + {S_3}^{-2}. \tag{6.11}$$

It leads to the following brackets for S_i:

$$\{S_2, S_3\} = \frac{\lambda}{2}S_1S_3, \quad \{S_3, S_{s-1}\} = \frac{\lambda}{2}S_2S_3, \tag{6.12}$$

$$\{S_1, S_2\} = \frac{\lambda}{2}((S_3)^2 - (S_3)^{-2}). \tag{6.13}$$

Now, in order to obtain the equations of motion for the spin, one can choose the Hamiltonian

$$H = \frac{2\mu B}{\lambda}\ln S_3, \tag{6.14}$$

where again we have chosen the magnetic field along the third direction.

It is clear that other choices will give rise to other interesting descriptions of the spin precession. Now we would like to comment on the possible symplectic or Lagrangian realizations of previous Poisson stuctures.

We shall give a regular Lagrangian realization (therefore also a symplectic realization) and a weak Lagrangian realization for the standard Poisson bracket giving rise to the $su(2)$ algebra. As the procedure works for any Lie algebra realized in terms of Poisson brackets, we shall describe this more general procedure.

7. SYMPLECTIC AND LAGRANGIAN REALIZATION OF LINEAR POISSON MANIFOLDS

On $\mathbb{R}^n$ with coordinates $x_i, i = 1, 2, \cdots, n$ we consider the Poisson bracket

$$\{x_i, x_j\} = c_{ij}^k x_k. \tag{7.1}$$

From the Jacobi identity of the given Poisson brackets, we get that c_{ij}^k are the structure constants of a Lie algebra, say g. We can consider any Lie group G whose Lie algebra is g and start with G as a configuration space. First we need to introduce some notation and facts about the geometry of Lie groups and their tangent bundles.

Let L or R denote the action of G on itself by left or right translations respectively, i.e., $L : G \times G \to G, L(g,h) = L_g(h) = gh, R(g,h) = R_g(h) = hg$, for all $g, h \in G$. A vector field X on G is called left or right invariant if $TL_G X(h) = X(gh)$ or $TR_g X(h) = X(hg)$. If X is left or right invariant then $X(g) = TL_g(e)$ or $X(g) = TR_g(e)$. The Lie algebra g of G is the Lie algebra of left invariant or right invariant vector fields on G. These invariant vector fields can be identified with their values at the identity element of the group. Fixing a basis E_i of T_eG, we define the corresponding basis of invariant vector fields $X_i^L(g) = TL_g E_i, X_i^R(g) = TR_g E_i$. From now on we identify g with T_eG and we fix a basis E_i of g. We next define left and right invariant forms on G. We shall denote a basis of one-forms by θ_L and θ_R respectively, they satisfy $\theta_L^i(X_i^L) = \delta_j^i$ $\theta_R^i(X_i^R) = \delta_j^i$. The canonical Maurer-Cartan g-valued one-forms Θ_L and Θ_R on G can be written as

$$\Theta_L = E_i \otimes \theta_L^i, \Theta_R = E_i \otimes \theta_R^i \tag{7.2}$$

The structure constants c_{ij}^k are defined by

$$\left[X_i^L, X_j^L\right] = c_{ij}^k X_k^L \tag{7.3}$$

or from the corresponding commutator in terms of right invariant vector fields. The vector bundle TG is trivial and can be identified with $G \times g$, using left or right translations. Coordinates on TG obtained using left translations will be called body coordinates, and coordinates defined by right translations will be called spatial coordinates. To be definite we shall now use only body coordinates, therefore we suppress the indices L.

It is clear that the canonical(1-1)-tensor S on TG has the following form when written in body coordinates:

$$S = \frac{\partial}{\partial v^i} \otimes \theta^i. \tag{7.4}$$

Second order vector fields will be written as

$$\Gamma = v^i X_i + f^i(g,v)\frac{\partial}{\partial v^i}. \tag{7.5}$$

If $\mathcal{L}$ is any Lagrangian function on TG, we have a Cartan one-form

$$\theta_{\mathcal{L}} = Sd\mathcal{L} = \frac{\partial \mathcal{L}}{\partial v^i}\theta^i. \tag{7.6}$$

The associated Cartan two-form is given by

$$\omega_{\mathcal{L}} = -d\theta_{\mathcal{L}}. \tag{7.7}$$

When $\omega_{\mathcal{L}}$ is a symplectic two-form, i.e. $\mathcal{L}$ is a regular form, we say that $\theta_{\mathcal{L}}$ is a Liouville form. We denote by $\mathcal{X}(\theta_{\mathcal{L}})$ the Lie algebra of all projectable vector fields on TG which preserve the Liouville form $\theta_{\mathcal{L}}$. A right inverse to the projection

$$\mathcal{X}(\theta_{\mathcal{L}}) \to \mathcal{X}(G)$$

is called the Liouville lift

$$\mathcal{X}(G) \to \mathcal{X}(\theta_{\mathcal{L}}).$$

It assigns to a vector field $X \in \mathcal{X}(G)$ the Hamiltonian vector field $X^{\mathcal{L}}$ on TG for the function $\theta_{\mathcal{L}}(X \circ \tau)$, where $\tau : TG \to G$ is the projection. If we consider the basis of left invariant vector fields X_i, we get $P_L = \theta_{\mathcal{L}}(X_i \circ \tau)$ and they define a map

$$TG \to g^* \equiv \mathbb{R}^n, \tag{7.8}$$

which provides a Lagrangian realization of the bracket, since we can show that $\{P_i, P_j\} = C_{ij}^k P_k$.

Had we started with the canonical lifting of the left or right action of G T^*G, the standard momentum map associated with these lifts would again provide a symplectic realization of the initial Poisson bracket.

In our Lagrangian realization on TG we have obtained indeed several Lagrangian realizations because $\mathcal{L}$ is arbitrary, the only requirement being the regularity. In particular any metric or pseudometric tensor on G, independently of its transformation properties, will give rise to a Lagrangian realization of the starting Poisson manifold. In particular when G is compact we can use a biinvariant metric tensor, when G is semisimple we can use the biinvariant one associated with the Cartan-Killing form on T_eG. This freedom in the realization can be exploited to accomodate particular dynamical requirements. For instance for a given Poisson dynamics on $\mathbb{R}^n$ we can use this freedom to find a related vector field on TG which is second order.

Remark : Any map $f : TG \to g^*$ which is a submersion can be realized as a momentum map by using the following construction. We consider either of the Maurer-Cartan forms Θ_L or Θ_R. Both of them are Lie algebra valued, therefore it is possible to define one forms

$$\Theta_L^f = f_i\theta_L^i, \quad \Theta_R^f = f_i\theta_R^i \tag{7.9}$$

We have two-forms $\omega_L^f = -d\Theta_L^f$ and $\omega_R^f = -d\Theta_R^f$:

$$\omega_L^f = -df_i \wedge \theta_L^i + f_i\frac{1}{2}C_{jk}^i\theta_L^j \wedge \theta_L^k,$$
$$\omega_R^f = -df_i \wedge \theta_R^i - f_i\frac{1}{2}C_{jk}^i\theta_R^j \wedge \theta_R^k. \tag{7.10}$$

Because f is a submersion,we get symplectic structures and the one-forms we have defined are Liouville one-forms. Of course, they will be associated with a Lagrangian function $\mathcal{L}$ if we have

$$Sd\mathcal{L} = \theta_L^f \quad \text{or} \quad Sd\mathcal{L} = \theta_R^f. \tag{7.11}$$

For this to be the case, we need

$$Sd\theta_L^f = 0, \tag{7.12}$$

i.e

$$(Sdf_i) \wedge \theta_L^i = 0 \quad \text{or} \quad (Sdf_i) \wedge \theta_R^i = 0. \tag{7.13}$$

Indeed these are necessary and sufficient conditions.

It should be noticed that we have started with a map $f : TG \to g^*$ which was not required to be a Poisson map. However if ω_L^f or ω_R^f are symplectic, we can use the one-forms as Liouville forms and associate with them momentum maps which become a Poisson map.

In the special case when f is a constant map, we can associate with it a degenerate Lagrangian function $\mathcal{L}_f$ which turns out to be linear along the fibers. The image of TG via the momentum map associated with $\mathcal{L}_f$ will be the coadjoint orbit in g^* going through the constant value of f.

For additional details and more general treatment of Poisson structures on the tangent and cotangent bundle of a Lie group and their reduction we refer to Alekseevsky et al.[17]

8. SYMPLECTIC REALIZATION OF POISSON MANIFOLDS FOR SPINNING PARTICLES

Going back to our Poisson brackets for spinning particles, while in the previous section we have considered the Lagrangian realization for the linear case, here we shall consider the Poisson brackets

$$\{S_1, S_2\} = \frac{1}{2}\frac{\sinh(2\lambda S_3)}{\sinh\lambda},$$

$$\{S_2, S_3\} = S_1, \quad \{S_3, S_1\} = S_2. \tag{8.1}$$

We first notice that these brackets may be obtained from

$$\{x_1, x_2\} = x_3, \quad \{x_2, x_3\} = x_1, \quad \{x_3, x_1\} = x_2 \tag{8.2}$$

by setting

$$\begin{aligned} S_a &= \sqrt{\frac{\sinh\lambda(J+x_3)\sinh\lambda(J-x_3)}{(J+x_3)(J-x_3)\sinh^2\lambda}}\sqrt{\frac{\sinh\lambda}{\lambda}}x_a, \quad a = 1, 2 \\ S_3 &= x_3. \end{aligned} \tag{8.3}$$

If we compare this map with our previous symplectic or Lagrangian realization of the linear bracket we find a similar realization for this alternative bracket for the spinning particle.

Remark: Using results in Refs. 18-19 and our general construction in Section 7, it is possible to exhibit a Lagrangian and weak Lagrangian realization of the following algebras in $\mathbb{R}^3$:

1. Deformation of the Euclidean algebra:

$$\{x, y\} = 0 \quad , \quad \{J, P_x\} = P_y \quad , \quad \{J, P_y\} = -\frac{1}{\lambda}\sinh(\lambda P_x).$$

2. Deformation of Heisemberg-Weyl algebra:

$$\{P_x, P_y\} = \frac{1}{\lambda}\sinh(\lambda w) \quad , \quad \{w, x\} = 0 \quad , \quad \{w, y\} = 0.$$

3. Deformation of $sl(2, \mathbb{R})$

$$\{J_\pm, J_0\} = \mp J_\pm \quad , \quad \{J_+, J_-\} = \frac{\sinh 2\lambda J_0}{\sinh \lambda}.$$

9. CONCLUSIONS

It seems that a large family of Poisson structures can be given a (weak) Lagrangian description the way we have discussed. The general case is however completely open. It would be interesting to investigate the same problem when the procedure is required to be equivariant with respect to the "canonical" action of some group G. In particular it would be interesting to extend the present analysis to relativistic field theory.

10. ACKNOWLEDGMENTS

The material in this note borrows heavily from the work done in collaboration with many friends as will be evident from the references. We thanks these friends for this collaboration over the years. G.M. thanks the organizers of the workshop on Geometrical Methods in Physics in Białowieza for their warm hospitality and for giving him an opportunity to present the material in this note.

References

1. A.P. Balachandran, S. Borchardt and A. Stern, Lagrangian and Hamiltonian descriptions of Yang-Mills particles, *Phys. Rev.*, D 17:3247 (1978)

2. A. P. Balachandran, G. Marmo, B. Skagerstam and A. Stern, Gauge symmetries and fibre bundles – Applications to particle dynamics, *Lecture Notes in Physics* 188, Springer (1982)

3. F.J. Dyson , Feynman's proof of the Maxwell equations, *Amer. J. Phys.*, 58:209 (1990)

4. A. Lichnerowicz, Les variétés de Poisson et leurs algèbres de Lie associées, *J. Diff. Geom.* 17:253 (1977)

5. A. Weinstein, On the local structure of Poisson manifolds, *J. Diff. Geom.* 18:523 (1983)

6. R. Jost, Poisson brackets (an unpedagogical lecture), *Rev. Mod. Phys.* 36:572 (1964)

7. M. Crampin, Tangent bundle geometry for Lagrangian dynamics, *J. Phys. A: Math. Gen.* 16:3775 (1983)

8. G. Morandi, C. Ferrario, G. Lo Vecchio, G. Marmo and C. Rubano, The inverse problem in the calculus of variations and the geometry of the tangent bundle, *Phys. Reports* 188 (1990)

9. E.C.G. Sudarshan and N. Mukunda, "Classical Dynamics: a Modern Perspective", Wiley, New York (1974)

10. P.G. Bergman and I. Goldberg, Dirac bracket transformation in phase space, *Phys. Rev.* 98:531 (1955)

11. P.A.M. Dirac, "Lectures on Quantum Mechanics", Belfer Graduate School of Science, Yeshiva University, New York (1964)

12. B. A. Dubrovin, M. Giordano, G.Marmo and A. Simoni, Poisson brackets on presymplectic manifolds, *Int. J. Mod. Phys. A* 8:3747 (1993)

13. M. V. Karasev, Analogues of objects of the theory of Lie groups for nonlinear Poisson brackets, *Math. USSR Izvestiya* 28:497 (1987)

14. P. Xu, Morita equivalence and symplectic realizations of Poisson manifolds, *Ann. Scient. Ec. Norm. Sup.* 25:307 (1992)

15. J. F. Cariñena, L. A. Ibort, G. Marmo and A. Stern, The classical Feynman's problem the invers problem for Poisson dynamics, *Physics Reports* (to appear)

16. J. Grabowski, G. Marmo and A. Perelomov, Poisson structures: towards a classification, *Mod. Phys. Lett. A* 8:1719 (1993)

17. D. Alekseevsky, G. Grabowski, G. Marmo and P. W. Michor, Poisson structures on the cotangent bundle of a Lie group or a principal bundle and their reduction, *J. Math. Phys.* 35:4909 (1994)

18. V.I. Man'ko, G. Marmo, P. Vitale and F. Zaccaria, A generalization of the Jordan-Schwinger map: the classical version and its q-deformation, *Int. J. Mod. Phys. A* 9:5541 (1994)

19. A. Stern and I. Yakushin, Deformed Wong particles, *Phys. Rev. D* 48:4974 (1993)

FROM THE POINCARÉ–CARTAN FORM TO A GERSTENHABER ALGEBRA OF POISSON BRACKETS IN FIELD THEORY

Igor V. Kanatchikov

Institut für Theoretische Physik
RWTH Aachen
D-52056 Aachen, Germany

Abstract

We consider the generalization of the basic structures of classical analytical mechanics to field theory within the framework of the De Donder-Weyl (DW) covariant canonical theory. We start from the Poincaré-Cartan form and construct the analogue of the symplectic form – the polysymplectic form of degree $(n+1)$, n is the dimension of the space-time. The dynamical variables are represented by differential forms and the polysymplectic form leads to a natural definition of the Poisson brackets on forms. The Poisson brackets equip the exterior algebra of dynamical variables with the structure of a "higher-order" Gerstenhaber algebra. We also briefly discuss a possible approach to field quantization which proceeds from the DW Hamiltonian formalism and the Poisson brackets of forms.

1. INTRODUCTION

In this communication I discuss the canonical structure underlying the so-called De Donder–Weyl (DW) Hamiltonian formulation in field theory and its possible application to a quantization of fields. The abovementioned structure was found in a recent paper of mine,[1] to which I refer both for further references and for additional details. In particular, I am going to show that the relationships between the Poincaré-Cartan form, the symplectic structure and the Poisson structure, which are well known in the mathematical formalism of classical mechanics, have their natural counterparts also in field theory within the framework of the DW canonical theory. This leads to the analogue of the symplectic structure, which I call polysymplectic, and to the analogue of the Poisson brackets which are defined on differential forms.

Recall that the Euler-Lagrange field equations may be written in the following form (see for instance Refs. 2-4)

$$\frac{\partial p^i_a}{\partial x^i} = -\frac{\partial H}{\partial y^a}, \qquad \frac{\partial y^a}{\partial x^i} = \frac{\partial H}{\partial p^i_a} \tag{1.1}$$

in terms of the variables

$$p^i_a := \frac{\partial L}{\partial(\partial_i y^a)}, \tag{1.2}$$

$$H := p^i_a \partial_i y^a - L \tag{1.3}$$

which are to be refered to as the DW momenta and the DW Hamiltonian function respectively. Here $L = L(y^a, \partial_i y^a, x^i)$ is the Lagrangian density, $x^i, i = 1, ..., n$ are space-time coordinates and $y^a, a = 1, ..., m$ are field variables. Eqs. (1) are reminiscent of Hamilton's canonical equations in mechanics and, therefore, may be thought of as a specific covariant Hamiltonian formulation of field equations. We call Eqs. (1) the DW Hamiltonian field equations and the formulation of field theory in terms of the variables p^i_a and H above the DW Hamiltonian formulation. The formulation above originates from the work of De Donder and Weyl (1935) on the variational calculus of multiple integrals.

The mathematical structures underlying this formulation of field theory were considered earlier by several authors in the context of the so-called multisymplectic formalism[5] which was recently studied in detail in Refs. 6-8. However, the possible analogues of the symplectic structure and the Poisson brackets, which are known to be so fruitful in the canonical formulation of classical mechanics, still are not properly understood within the DW canonical theory.

Our interest to this subject is motivated by the explicit covariance of the formulation above, in the sense that the space and time variables are not discriminated as usual, and its finite dimensionality, in the sense that the formulation refers to the finite dimensional analogue of the phase space namely, the space of variables (y^a, p^i_a), as well as by the attempts to understand if or how it is possible to construct a formulation of quantum field theory which would be based on the DW Hamiltonian formulation. Clearly, the answer to the latter question requires the analogue of the Poisson brackets and the bracket representation of the equations of motion corresponding to the DW formulation.

The canonical formalism in classical mechanics is related to the variational principle of least action and it may be derived from the fundamental object of the calculus of variations – the Poincaré-Cartan (P-C) form (see e.g. Ref. 9). The corresponding construction leads to structures which are known to be important for quantization. Conventional generalization to field theory implies setting off the time dimension from other space-time dimensions and leads to the infinite dimensional functional version of the abovementioned construction. Here we are interested in the field theoretical generalization of these structures within the space-time symmetric DW formulation.

2. POINCARÉ–CARTAN FORM, CLASSICAL EXTREMALS AND THE POLYSYMPLECTIC FORM

In field theory, which is related to the variational problems with several independent variables, the analogue of the P-C form written in terms of the DW Hamiltonian variables (1.2), (1.3) reads[6–8]

$$\Theta = p^i_a dy^a \wedge \omega_i - H\omega, \tag{2.1}$$

where $\omega := dx^1 \wedge ... \wedge dx^n$ and $\omega_i := \partial_i \rfloor \omega$. The equations of motion in the DW Hamiltonian form, Eqs. (1.1), may be shown to follow from the statement that the classical extremals are the integral hypersurfaces of the multivector field of degree n, $\overset{n}{X}$,

$$\overset{n}{X} := \tfrac{1}{n!} X^{M_1 ... M_n}(z)\, \partial_{M_1 ... M_n}, \tag{2.2}$$

where $\partial_{M_1 ... M_n} := \partial_{M_1} \wedge ... \wedge \partial_{M_n}$, which annihilates the canonical $(n+1)$-form

$$\Omega_{DW} := d\Theta, \tag{2.3}$$

that is

$$\overset{n}{X} \,\lrcorner\, \Omega_{DW} = 0. \tag{2.4}$$

The integral hypersurfaces of $\overset{n}{X}$ are defined as the solutions of the equations

$$\overset{n}{X}{}^{M_1 \ldots M_n}(z) = \mathcal{N} \frac{\partial(z^{M_1}, \ldots, z^{M_n})}{\partial(x^1, \ldots, x^n)} \tag{2.5}$$

where a multiplier $\mathcal{N}$ depends on the chosen parametrization of a hypersurface and $z^M := (x^i, y^a, p^i_a)$. The component calculations show that Eq. (2.4) specifies only a part of the components of $\overset{n}{X}$ and that the DW canonical equations (1.1) actually follow from the "vertical" components $X^{v i_1 \ldots i_{n-1}}$. We call vertical the field and the DW momenta variables $z^v := (y^a, p^i_a)$ and horizontal the space-time (independent) variables x^i.

Introducing the notions of the vertical multivector field of degree p:

$$\overset{p}{X}{}^V := \frac{1}{(p-1)!} X^{v i_1 \ldots i_{p-1}} \partial_{v i_1 \ldots i_{p-1}}, \tag{2.6}$$

the vertical exterior differential, d^V, $d^V \ldots := dz^v \wedge \partial_v \ldots$, and the form

$$\Omega := -dy^a \wedge dp^i_a \wedge \omega_i, \tag{2.7}$$

one may check that (2.4) is equivalent to

$$\overset{n}{X}{}^V \,\lrcorner\, \Omega = (-)^n d^V H, \tag{2.8}$$

if the parametrization in (2.5) is chosen such that

$$\frac{1}{n!} \overset{n}{X}{}^{i_1 \ldots i_n} \partial_{i_1 \ldots i_n} \,\lrcorner\, \omega = 1.$$

The form Ω in (2.7) is to be referred to as *polysymplectic*.

The appearance of the DW field equations in the form of (2.8) suggests (cf. with mechanics!) that the polysymplectic form is a field theoretical analogue of the symplectic form, so that its properties should be taken seriously as a starting point for the canonical formalism.

As a generalization of (2.8), it is easy to see that the polysymplectic form maps in general the horizontal q-forms, $\overset{q}{F}$,

$$\overset{q}{F} := \frac{1}{q!} F_{i_1 \ldots i_q}(z) dx^{i_1 \ldots i_q}, \tag{2.9}$$

where

$$dx^{i_1 \ldots i_q} := dx^{i_1} \wedge \ldots \wedge dx^{i_q},$$

to the vertical multivectors of degree $(n-q)$:

$$\overset{n-q}{X} \,\lrcorner\, \Omega = d^V \overset{q}{F} \tag{2.10}$$

for all $q = 0, \ldots, n-1$. Evidently, the horizontal forms play a role of dynamical variables within the present formalism. Henceforth we omit the superscripts ^{V} labelling the vertical multivectors.

The hierarchy of maps (2.10) may be viewed as a local consequence of the hierarchy of "graded canonical symmetries"

$$\pounds_{\overset{p}{X}} \Omega = 0, \tag{2.11}$$

$p = 1, ..., n$, which are formulated in terms of the generalized Lie derivatives with respect to the vertical multivector fields. By definition,[10] for any form μ

$$\pounds_{\overset{p}{X}} \mu := \overset{p}{X} \lrcorner\, d^V \mu - (-1)^p\, d^V(\overset{p}{X} \lrcorner\, \mu). \tag{2.12}$$

Now, by analogy with the terminology known from mechanics, I call the vertical multivector fields fulfilling (2.11) *locally Hamiltonian* and those fulfilling (2.10) (globally) *Hamiltonian*. Correspondingly, the forms to which the Hamiltonian multivector fields can be associated through the map (2.10) are referred to as the *Hamiltonian forms*.

The notion of a Hamiltonian form implies certain restriction on the dependence of its components on the DW momenta. For example, the components of the vector field $X_F := X^a \partial_a + X^i_a \partial^a_i$ associated through the map $X_F \lrcorner \Omega = d^V F$ with the $(n-1)$–form $F := F^i \omega_i$ are given by

$$X^i_a = \partial_a F^i, \quad -X^a \delta^i_j = \partial^a_j F^i. \tag{2.13}$$

The latter relation restricts the admissible $(n-1)$–forms to those which have a simple dependence on the DW momenta namely, $F^i(y, p, x) = f^a(y, x)\, p^i_a + g^i(y, x)$.

Note also that the Hamiltonian multivector field associated with a form through the map (2.10) is actually defined up to an addition of *primitive* fields which annihilate the polysymplectic form

$$\overset{p}{X}_0 \lrcorner\, \Omega = 0. \tag{2.14}$$

Therefore, the image of a Hamiltonian form under the map (2.10) given by the polysymplectic form is rather the equivalence class of Hamiltonian multivector fields of corresponding degree modulo an addition of primitive fields.

3. THE POISSON BRACKETS ON FORMS AND A GERSTENHABER ALGEBRA

It is natural to define the bracket of two locally Hamiltonian multivector fields as follows:

$$[\overset{p}{X}_1, \overset{q}{X}_2] \lrcorner\, \Omega := \pounds_{\overset{p}{X}_1} (\overset{q}{X}_2 \lrcorner\, \Omega). \tag{3.1}$$

From the definition it follows that

$$deg([\overset{p}{X}_1, \overset{q}{X}_2]) = p + q - 1, \tag{3.2}$$

$$[\overset{p}{X}_1, \overset{q}{X}_2] = -(-1)^{(p-1)(q-1)} [\overset{q}{X}_2, \overset{p}{X}_1], \tag{3.3}$$

$$\begin{aligned}(-1)^{g_1 g_3} [\overset{p}{X}, [\overset{q}{X}, \overset{r}{X}]] &+ (-1)^{g_1 g_2} [\overset{q}{X}, [\overset{r}{X}, \overset{p}{X}]] \\ &+ (-1)^{g_2 g_3} [\overset{r}{X}, [\overset{p}{X}, \overset{q}{X}]] = 0,\end{aligned} \tag{3.4}$$

where $g_1 = p - 1$, $g_2 = q - 1$ and $g_3 = r - 1$.

These properties allow us to identify the bracket in (3.1) with the vertical (i.e. taken w.r.t. the vertical variables) Schouten–Nijenhuis (SN) bracket of multivector fields and to conclude that the space of LH fields is a graded Lie algebra with respect to the (vertical) SN bracket.

For two Hamiltonian multivector fields one obtains

$$
\begin{aligned}
[\overset{p}{X}_1\,, \overset{q}{X}_2] \rfloor \Omega &= \pounds_{\overset{p}{X}_1} d^V \overset{s}{F}_2 \\
&= (-1)^{p+1} d^V (\overset{p}{X}_1 \rfloor d^V \overset{s}{F}_2) \qquad (3.5)\\
&=: -d^V \{\![\overset{r}{F}_1, \overset{s}{F}_2]\!\}, \qquad (3.6)
\end{aligned}
$$

where $r = n - p$ and $s = n - q$. From (3.5) it follows that the SN bracket of two Hamiltonian fields is a Hamiltonian field (as in mechanics). In (3.6) one defines the bracket operation on Hamiltonian forms which is induced by the vertical SN bracket of multivector fields associated with them.

From the definition in (3.6) it follows

$$
\{\![\overset{r}{F}_1, \overset{s}{F}_2]\!\} = (-1)^{(n-r)} X_1 \rfloor d^V \overset{s}{F}_2 = (-1)^{(n-r)} X_1 \rfloor X_2 \rfloor \Omega \qquad (3.7)
$$

and

$$
\begin{aligned}
deg \{\![\overset{r}{F}_1, \overset{s}{F}_2]\!\} &= r + s - n + 1, \qquad (3.8)\\
\{\![\overset{p}{F}_1, \overset{q}{F}_2]\!\} &= -(-1)^{g_1 g_2} \{\![\overset{q}{F}_2, \overset{p}{F}_1]\!\}, \qquad (3.9)\\
(-1)^{g_1 g_3} \{\![\overset{p}{F}, \{\![\overset{q}{F}, \overset{r}{F}]\!\}]\!\} &+ \\
(-1)^{g_1 g_2} \{\![\overset{q}{F}, \{\![\overset{r}{F}, \overset{p}{F}]\!\}]\!\} &+ (-1)^{g_2 g_3} \{\![\overset{r}{F}, \{\![\overset{p}{F}, \overset{q}{F}]\!\}]\!\} = 0, \qquad (3.10)\\
\{\![\overset{p}{F}, \overset{q}{F} \wedge \overset{r}{F}]\!\} &= \{\![\overset{p}{F}, \overset{q}{F}]\!\} \wedge \overset{r}{F} + (-1)^{q(n-p-1)} \overset{q}{F} \wedge \{\![\overset{p}{F}, \overset{r}{F}]\!\} \\
&+ \textit{higher-order corrections}, \qquad (3.11)
\end{aligned}
$$

where $g_1 = n - p - 1$, $g_2 = n - q - 1$ and $g_3 = n - r - 1$.

The algebraic construction which satisfies the axioms (3.9), (3.10) and (3.11) without higher-order corrections, together with the familiar properties of the exterior product, is known as the Gerstenhaber algebra.[11] Higher-order corrections in Eq. (3.11) are composed of the terms like

$$
\frac{1}{(n-p-1)!} X^{v i_1 \ldots i_{n-p-1}} (\partial_{v i_1 \ldots i_s} \rfloor d^V \overset{q}{F}) \wedge \partial_{i_{s+1} i_{s+2} \ldots i_{n-p-1}} \rfloor \overset{r}{F},
$$

with $s = 1, \ldots, n-p-1$, and are similar to the last term in the "Leibniz rule" for, say, the second derivative: $(fg)'' = f''g + fg'' + 2f'g'$. They appear due to the fact that the multivector field $\overset{n-p}{X}$ associated with the form $\overset{p}{F}$ does not act on exterior forms as a graded derivation, but rather as a graded differential operator of order $-(n-p)$ which is composed of the subsequent actions of graded derivations of order -1. The latter are given by the vector fields ∂_v and ∂_{i_s}, $s = 1, \ldots, n-p-1$, which constitute the vertical multivector $\overset{n-p}{X}$. The algebraic structure given by Eqs. (3.9)–(3.11) may be called the higher-order Gerstenhaber algebra, but in the following we will continue to refer to it as a Gerstenhaber algebra, for short.

Remark: Strictly speaking, the space of Hamiltonian forms is not closed with respect to the exterior product, so that the full justification of the Leibniz rule (3.11) requires a generalization of the above construction which admits arbitrary horizontal forms as the dynamical variables (see Ref. 12). Higher-order corrections in (3.11) were overlooked in previous communications (cf. Refs. 1, 12).

4. EQUATIONS OF MOTION IN THE BRACKET FORM

By analogy with mechanics, one can expect that the equations of motion are given by the bracket with the DW Hamiltonian function. Indeed, for the bracket of H with the $(n-1)$–form $F := F^i\omega_i$ one obtains

$$\begin{aligned}\{\![H, F]\!\} &= X_F \lrcorner\, d^V H = X_F{}^a \partial_a H + X_F{}^i_a \partial^a_i H \\ &= \partial_i p^j_a \partial^a_j F^i + \partial_i y^a \partial_a F^i,\end{aligned}$$

where we have used (2.3) and (1.1). Introducing the *total* (i.e. evaluated on extremals) exterior differential $\boldsymbol{d}$ of a horizontal form of degree p, $\overset{p}{F}$:

$$\boldsymbol{d}\overset{p}{F} := \partial_i z^M dx^i \wedge \partial_M \overset{p}{F} = \partial_i z^v dx^i \wedge \partial_v \overset{p}{F} + dx^i \wedge \partial_i \overset{p}{F} = \boldsymbol{d}^V F + d^{hor} F,$$

one can write the equation of motion of Hamiltonian $(n-1)$–form F as (by definition, $*^{-1}\omega := 1$)

$$*^{-1}\,\boldsymbol{d}F = \{\![H, F]\!\} + \partial_i F^i. \tag{4.1}$$

The bracket of a p–form with H vanishes for $p < n-1$. The equations of motion of arbitrary forms may be written in terms of the bracket with the n–form $H\omega$. This implies a certain extension of the construction in Section 2. Namely, we map $H\omega$ to a vector-valued form $\tilde{X} := \tilde{X}^v{}_i dx^i \otimes \partial_v$ by

$$\tilde{X} \lrcorner\, \Omega = d^V H\omega, \tag{4.2}$$

where $\tilde{X} \lrcorner\, \Omega := X^v_{\cdot k} dx^k \wedge (\partial_v \lrcorner\, \Omega)$ is the Fröhlicher-Nijenhuis inner product. From (4.2) it follows

$$\tilde{X}^a_{\cdot k} = \partial^a_k H, \quad \tilde{X}^i_{ak}\delta^k_i = -\partial_a H. \tag{4.3}$$

Substitution of the natural parametrization of $\tilde{X}$:

$$\tilde{X}^v_{\cdot k} = \frac{\partial z^v}{\partial x^k},$$

into (4.3) leads to the DW Hamiltonian equations (1.1).

Now, we define the bracket with $H\omega$ (cf. (3.7))

$$\{\![H\omega, \overset{p}{F}]\!\} := \tilde{X}_{H\omega} \lrcorner\, d^V \overset{p}{F} \tag{4.4}$$

and find that

$$\boldsymbol{d}\overset{p}{F} = \{\![H\omega, \overset{p}{F}]\!\} + d^{hor}\overset{p}{F}. \tag{4.5}$$

Thus, we have shown that the bracket with the DW Hamiltonian n–form $H\omega$ is related to the exterior differential of a form.

Remark: The bracket which is naively defined in (4.4) does not satisfy in general the axioms of a Gerstenhaber algebra. The appropriate extension of a Gerstenhaber algebra structure to n–forms is a part of the generalization of the present construction to the forms which are not Hamiltonian according to the definition in Section 2 (see Ref. 12).

5. TOWARDS A QUANTIZATION

An appropriate quantization of a Gerstenhaber algebra of exterior forms, which generalizes to field theory the Poisson algebra of dynamical variables, may in principle lead to certain quantization procedure in field theory. The purpose of this section is to discuss briefly a possible heuristic approach to such a quantization.

We start from the observation that

$$\{\![p_a, y^b]\!\} = \delta^b_a, \tag{5.1}$$

where $p_a := p^i_a \omega_i$ is the $(n-1)$-form which may be considered as the momentum variable canonically conjugate to fields y^a. Applying Dirac's quantization prescription $[\ ,\ ]_{\pm} = i\hbar \{\![\ ,\ ]\!\}_{\pm}$, one obtains the canonical commutation relation for the operators corresponding to fields and the $(n-1)$-form momenta

$$[\hat{p}_a, \hat{y}^b] = i\hbar\delta^b_a. \tag{5.2}$$

In the "y-representation" one finds the differential operator realization of $\hat{p}_a$

$$\hat{p}_a = i\hbar \frac{\partial}{\partial y^a}. \tag{5.3}$$

Based on the analogy with the quantization of classical mechanics in Schrödinger's representation and the observation made in Section 4 that the exterior differential is related to the DW Hamiltonian n-form, one can conjecture the following form of the covariant "Schrödinger equation"

$$i\hbar\, d\Psi = (H\omega)^{op}\Psi \tag{5.4}$$

for the "wave function" $\Psi = \Psi(x^i, y^a)$, which depends on the space-time and field variables which form the analogue of a configuration space within the present formulation.

In the particular example of a system of scalar fields y^a interacting through the potential $V(y)$, which is given by the Lagrangian

$$L = -\frac{1}{2}\partial_i y^a \partial^i y_a - V(y),$$

the DW Hamiltonian function takes the form

$$H = -\frac{1}{2}p^i_a p^a_i + V(y). \tag{5.5}$$

In terms of the $(n-1)$-form momenta variables p_a the $n-$form $H\omega$ may be written as

$$H\omega = \frac{1}{2} * p_a \wedge p^a + V(y)\omega, \tag{5.6}$$

where $*p_a = -p^i_a dx_i$ (the Minkowski metric in the x-space is assumed). The realization of the operator corresponding to the non-Hamiltonian one-form $*p_a$

$$\widehat{*p}_a = *\,\hat{p}_a \tag{5.7}$$

is suggested by the quantization of the bracket

$$\{\![*p_a, y^b\omega_i]\!\} = -\delta^b_a dx_i = *\{\![p_a, y^b\omega_i]\!\}, \tag{5.8}$$

which may be calculated either with the help of the Leibniz rule (3.10) or within a more general scheme,[12] where one associates arbitrary horizontal forms with the differential operators on exterior algebra which are represented by the multivector-valued forms instead of the multivectors as in the case of Hamiltonian forms.

Furthermore, the classical identity $\omega = *1$ suggests that $\hat{\omega} = *$ and, therefore, one can write

$$(H\omega)^{op} = *(-\frac{\hbar^2}{2} \triangle + V(y)) =: *H^{op}, \tag{5.9}$$

where $\triangle := \partial^a \partial_a$ is the Laplace operator in a field space. Thus, the Schrödinger equation (5.4) may also be written as

$$i\hbar *^{-1} d\Psi = H^{op}\Psi. \tag{5.10}$$

Evidently, this equation makes sense only if the wave function Ψ is a nonhomogeneous horizontal form. In the simple case of the DW Hamiltonian operator (5.9), which does not depend explicitly on the space-time coordinates, one can take into account only the zero- and $(n-1)$-form contributions, so that

$$\Psi = \psi_0(x, y) + \psi^i(x, y)\, \omega_i. \tag{5.11}$$

Substituting (5.11) into (5.10), one obtains the component form of our Schrödinger equation:

$$i\hbar\partial_i\psi^i = H^{op}\psi_0, \tag{5.12}$$

$$-i\hbar\partial_i\psi_0 = H^{op}\psi_i. \tag{5.13}$$

The integrability condition of this set of equations is

$$\delta\Psi = 0. \tag{5.14}$$

By a straightforward calculation, one can derive from (5.12) and (5.13) the following conservation law

$$\partial_i[\bar{\psi}_0\psi^i + \psi_0\bar{\psi}^i] = -\frac{i\hbar}{2}\partial_a[\bar{\psi}_0 \overleftrightarrow{\partial}_a \psi_0 - \bar{\psi}^i \overleftrightarrow{\partial}_a \psi_i]. \tag{5.15}$$

If one assumes a sufficiently rapid decay of the wave function $\Psi(x, y)$ for large values of fields $|y| \to \infty$, by Gauss' theorem one obtains

$$\partial_i \int dy\, [\bar{\psi}_0\psi^i + \psi_0\bar{\psi}^i] = 0. \tag{5.16}$$

Thus, the current

$$j^i := \int dy\, [\bar{\psi}_0\psi^i + \psi_0\bar{\psi}^i]$$

is the conserved space-time current of the theory. It suggests the inner product of nonhomogeneous forms Ψ, which one needs for the calculation of quantum theoretical expectation values.

The covariant Schrödinger equation may be solved by means of the separation of field and space-time variables. Namely, let us write

$$\Psi(x, y) = \Phi(x) f(y), \tag{5.17}$$

where $\Phi(x)$ is a nonhomogeneous form with the components depending on x:

$$\Phi(x) := \phi_0(x) + \phi^i(x)\omega_i \tag{5.18}$$

and $f(y)$ is a function of field variables. Substituting this Ansatz into the Schrödinger equation (5.10), we arrive at the eigenvalue problem for the DW Hamiltonian operator

$$H^{op} f = \kappa f, \tag{5.19}$$

and the equation on $\Phi(x)$:

$$i\hbar *^{-1} d\Phi(x) = \kappa\Phi(x). \tag{5.20}$$

From the latter equation it follows that

$$\Box\phi_0 = \frac{\kappa^2}{\hbar^2}\phi_0, \quad \phi_i = -\frac{i\hbar}{\kappa}\partial_i\phi_0. \tag{5.21}$$

The solutions of (5.19) and (5.21) provide us with a basis for decomposition of an arbitrary solution of the covariant Schrödinger equation.

Remarks:

1. The canonical bracket (5.1) belongs to the subalgebra of zero- and $(n-1)$-forms of a Gerstenhaber algebra of dynamical variables. The other canonical brackets from this subalgebra are

$$\{\![p^i_a, y^b\omega_j]\!\} = \delta^b_a\delta^i_j, \tag{5.22}$$
$$\{\![p_a, y^b\omega_i]\!\} = \delta^b_a\omega_i. \tag{5.23}$$

Quantization of these three brackets is a part of the problem of quantization of the center of a Gerstenhaber algebra, which is formed by the forms of the kind $p^i_a dx^{\cdots}$ and $y^a dx^{\cdots}$, where $dx^{\cdots}$ denotes the basis elements of Grassmann algebra of horizontal forms. The question as to which subalgebra of a Gerstenhaber algebra of dynamical variables should or can be quantized remains open and deserves the same careful study as the similar question concerning the quantizable subalgebra of the Poisson algebra of observables in mechanics. The minimal subalgebra is that of $(n-1)$-forms and the canonical bracket from this subalgebra is given by (5.23). Its quantization rather than a quantization of (5.1) gives rise to the operator realization of p_a in (5.3). The quantization of the subalgebra of zero- and $(n-1)$-forms with the canonical brackets (5.1), (5.22) and (5.23) leads to the problem of realization of the operator $\hat{p}^i_a$ which would be consistent with the realization of $\hat{p}_a$ and the requirement $\hat{p}_a = \hat{p}^i_a \circ \hat{\omega}_i$, as well as to the problem of the proper realization of the operation $\circ$ of the multiplication of quantum operators. When quantizing the (centre of the) Gerstenhaber algebra, the latter problem is that of the proper realization of the quantized wedge product, which is in this case a generalization of the Jordan symmetric product of operators in quantum mechanics.

2. The quantization of the bracket (5.1) leads to the operator realization of the $(n-1)$–form p_a which is the 0-form. In general, the form degree of the operator corresponding to a dynamical variable is different from the classical form degree of the latter. This gives rise to an additional problem of which degree should define the graded products of operators which correspond to the exterior product and the quantized Poisson bracket respectively.

3. It is interesting to note that the realization (5.3) is not consistent with the classical property $dx^i \wedge p_a = (-)^{n-1}p_a \wedge dx^i$ which one may require to be also fulfilled

on the quantum level. This may be achieved if the quantization prescription is modified in such a way that

$$[\ ,\ ]_{\pm} = \gamma\hbar \{\!\![\ ,\ ]\!\!\}_{\pm} \tag{5.24}$$

where γ denotes the imaginary unit corresponding to the Clifford algebra of the n-dimensional space-time over which a field theory under quantization is formulated. In Minkowski space-time $\gamma := \gamma_0\gamma_1\gamma_2\gamma_3$. In the case of mechanics ($n = 1$) $\gamma = i$ and the above quantization prescription reduces to that of Dirac. The quantization according to (5.24) leads to the realization

$$\hat{p}_a = \gamma\hbar\partial_a \quad \text{and} \quad \widehat{dx^i}\wedge = \gamma^i\wedge, \tag{5.25}$$

where $\wedge$ on the right hand side denotes the graded symmetrized Clifford product. Correspondingly, the wave function may be considered as taking values in the Clifford–Kähler algebra of nonhomogeneous forms which corresponds to the n-dimensional space-time (see e.g. Ref. 13 and the references quoted there). The latter reduces to complex numbers in the case of mechanics. This quantization prescription leads to the same realization of the DW Hamiltonian operator as in (5.9). However, in general, it is not clear which quantization prescription is more appropriate both physically and mathematically for the quantization of the suitable "quantizable" subalgebra of a Gerstenhaber algebra of forms–dynamical–variables in field theory.

4. The elements of quantum theory presented above possess the basic features of a quantum description of dynamics and its connections with the structures of classical mechanics. These elements are easily seen to reduce to the corresponding elements of quantum mechanics at $n = 1$. In this sense at least our formulation may be viewed as an approach to the quantum description of fields. Establishing the possible links with the known approaches and results in quantum field theory and a physical interpretation of the present formulation poses many conceptual questions and needs a further study which we hope to communicate elsewhere. In particular, it would be interesting to understand a possible relation of our nonhomogeneous form-valued wave function $\Psi(x, y)$ to the Schrödinger wave functional $\Psi(t, [y(\mathbf{x})])$ and of our covariant Schrödinger equation to the functional Schrödinger equation.

Acknowledgements. I thank Prof. A. Odzijewicz and the organizers for inviting me to present this talk. I acknowledge useful discussions with F. Cantrijn, M. Gotay, M. Modugno and J. Śniatycki during the time of the workshop. Thanks to Z. Oziewicz for several inspirating discussions on the subject of this paper and encouragement.

References

1. I. V. Kanatchikov, On the canonical structure of the De Donder-Weyl covariant Hamiltonian formulation of field theory I. Graded Poisson brackets and equations of motion, PITHA 93/41 (November 1993) and hep-th/9312162

2. H. Rund, "The Hamilton-Jacobi Theory in the Calculus of Variations", van Nostrand, Toronto (1966)

3. H. Kastrup, Canonical theories of Lagrangian dynamical systems in physics, *Phys. Rep.* 101:1 (1983)

4. E. Binz, J. Śniatycki, H. Fisher, "Geometry of Classical Fields", North-Holland, Amsterdam (1989)

5. J. Kijowski, A finite dimensional canonical formalism in the classical field theory, *Comm. Math. Phys.* 30:99 (1973);
J. Kijowski, Multiphase spaces and gauge in the calculus of variations, *Bull. de l'Acad. Polon. des Sci., Sér sci. math., astr. et Phys.* XXII:1219 (1974);
J. Kijowski and W. Szczyrba, A canonical structure for classical field theories, *Comm. Math. Phys.* 46:183 (1976)

6. M.J. Gotay, An exterior differential systems approach to the Cartan form, *in*: "Géométrie Symplectique & Physique Mathématique", P. Donato, C. Duval et al. (eds.), Birkhäuser, Boston (1991)
M. J. Gotay, A multisymplectic framework for classical field theory and the calculus of variations I. Covariant Hamiltonain formalism, *in*: "Mechanics, Analysis and Geometry: 200 Years after Lagrange", M. Francaviglia (ed.), North Holland, Amsterdam (1991)
M. J. Gotay, A multisymplectic framework for classical field theory and the calculus of variations II. Space + time decomposition, *Diff. Geom. and its Appl.* 1:375 (1991)

7. M. J. Gotay, J. Isenberg, J. E. Marsden, R. Montgomery, J. Śniatycki and Ph. B. Yasskin: Momentum maps and classical relativistic fields: The Lagrangian and Hamiltonian structure of classical field theories with constraints, preprint, Berkeley (1992)

8. J.F. Cariñena, M. Crampin, L.A. Ibort, On the multisymplectic formalism for first order field theories, *Diff. Geom. and its Appl.* 1:345 (1991)

9. R. Abraham and J.E. Marsden, "Foundations of Mechanics", 2nd ed., Benjamin and Cummings, N.Y. (1978)

10. W.M. Tulczyjew, The graded Lie algebra of multivector fields and the generalized Lie derivative of forms, *Bull. de l'Acad. Polon. sci., Sér sci. math., astr. et phys.* XXII:937 (1974)

11. M. Gerstenhaber, The cohomology structure of an associative ring, Ann. Math. 78:267 (1963);
M. Gerstenhaber and S. D. Schack, Algebraic cohomology and deformation theory, *in*: "Deformation Theory of Algebras and Structures and Applications", M. Hazewinkel and M. Gerstenhaber (eds.), Kluwer, Dordrecht (1988);
B.H. Lian and G.J. Zuckerman, New perspectives of the BRST-algebraic structure of string theory, *Commun. Math. Phys.* 154:613 (1993)

12. I.V. Kanatchikov, Basic structures of the covariant canonical formalism for fields based on the De Donder-Weyl theory, preprint PITHA 94/17 and hep-th/9410238;
I.V. Kanatchikov, On the finite dimensional covariant Hamiltonian formalism in field theory, *in*: "New Frontiers in Gravitation", R. Santilli and G. Sardanashvily (eds.), Hadronic Press, Palm Harbor (1995) (to appear)

13. P. Becher and H. Joos, The Dirac-Kähler equation and fermions on the lattice, *Z. Phys. C* 15:343 (1982),
I.M. Benn and R.W. Tucker, Fermions without spinors, *Comm. Math. Phys.* 89:341 (1983),
N.A. Salingaros, G.P. Wene, The Clifford algebra of differential forms, *Acta Appl. Math.* 4:271 (1985).

GEOMETRIC COHERENT STATES, MEMBRANES, AND STAR PRODUCTS

Mikhail Karasev *

Department of Applied Mathematics
Moscow Institute of Electronics & Mathematics
B.Vuzovsky per., 3/12, Moscow 109028, Russia

Abstract

An exact star product on symplectic-Kähler manifolds is constructed via quadrangle and hexagon membrane amplitudes. Coherent states and realization of the Dirac axioms over lagrangian submanifolds are described by triangle and pentagon membrane amplitudes. Relations between the star-product quantization and Dirac-type quantization are found.

1. INTRODUCTION

There is an old idea: to quantize classical observables (i.e., functions f on a phase space $\mathcal{X}$) by the integral

$$\mathrm{Op}(f) = \int_{\mathcal{X}} f(x)\mathbf{T}(x)\,dl(x),$$

where dl is a measure on $\mathcal{X}$, and $\mathbf{T}$ is an appropriate family of operators. The nature of this family depends on a concrete situation, system or the type of the quantization theory. The family $\mathbf{T}$ is known under many different names: a Fourier transform of a Lie group representation,[1] a coherent state,[2-4] a reproducing kernel or an overcomplete system,[5-7] a frame,[8] a quantizer.[9] It is important that $\mathbf{T}$ controls a noncommutative "star" product $\circledast$ in the space of classical observables:

$$\mathrm{Op}(f)\cdot\mathrm{Op}(g) = \mathrm{Op}(f \circledast g).$$

The problem is to describe such products $\circledast$ explicitly. An axiomatization of this problem and a search of formulas for $\circledast$ in terms of formal power series in a semiclassical parameter $\hbar$ has led to the famous theory of deformation quantization due to F. Bayen, M. Flato, C. Fronsdal, A. Lichnerowicz, D. Sternheimer[10] (for the existence of such formal products, see Refs.11,12).

There is a crucial open question: how to construct a star product exactly (informally) and geometrically. It would be natural to expect that the global geometry of $\mathcal{X}$

*This work was supported in part by the International Science Foundation and the Russian Foundation for Fundamental Research.

and intrinsic geometric structures inside $\mathcal{X}$ are reflected in the quantum multiplication, as well as in the quantum spaces of Connes.[13]

Here our viewpoint is the following: the more "quantum dispersion" to be included into $\mathbf{T}$, the more classically and geometrically the product $\circledast$ looks. Quantum fluctuations should be maximally extracted from symbols f and injected into the model family of operators $\mathbf{T}$. In other words, we try to maximize the first quantum correlator

$$\operatorname{tr}(\mathbf{T}(x)\mathbf{T}(y)), \qquad x, y \in \mathcal{X},$$

between the points of the classical space in order to free the geometry from quantum uncertainty.

We aim to represent quantum wave functions using classical geometric objects inside of $\mathcal{X}$, e.g., orbits of classical Hamilton flows or invariant submanifolds. This should be an exact (not approximate) representation for solutions of quantum equations. In the semiclassical approximation it will look especially simple and explicit.

The well-known method of characteristics for solving differential equations, as well as the orbit method for constructing irreducible representations of Lie groups are naturally interpreted in this way.

This area can be called "noncommutative characteristics" if we think about solutions of differential equations, or the "membrane and string quantization" if we think about irreducible representations and star products. Below we follow and generalize the results of a number of papers.[14–22]

2. STAR PRODUCTS VIA INTEGRALS OF SYMPLECTIC AND RICCI FORMS OVER MEMBRANES

2.1. Terminology around the Notion "Quantization"

Let $\mathcal{X}$ be a set, and $\widehat{\mathcal{A}}$ be an associative algebra with trivial center and with a certain convergence (the interesting particular cases: von Neumann, C^* or Hilbert algebras).

Definition 2.1. An injective map $\mathcal{X} \mapsto \widehat{\mathcal{A}}$ is called a *quantization of* $\mathcal{X}$ *in* $\widehat{\mathcal{A}}$ if the linear envelope of its image is dense. A quantization of $\mathcal{X}$ is called *complete* with respect to a measure dl on $\mathcal{X}$ if it is measurable and its total (weak) integral over $\mathcal{X}$ approximates the unit. A quantization in a Hilbert algebra is called *pure* if its image consists of pure states.

Pure quantization of $\mathcal{X}$ can be always realized as a map $\mathbf{P}$ from $\mathcal{X}$ into the set of orthogonal one-dimensional projectors on a certain Hilbert space $\mathcal{L}$, such that the envelope of all projectors $\{\mathbf{P}(x) \mid x \in \mathcal{X}\}$ is dense in the algebra $\widehat{\mathcal{A}}$ of Hilbert–Schmidt operators on $\mathcal{L}$. We shall say that the quantization $\mathbf{P}$ is *represented* in $\mathcal{L}$.

In this operator representation the completeness of $\mathbf{P}$ means that

$$\int_{\mathcal{X}} \mathbf{P}(x)\, dl(x) = I \qquad \text{(weak integral)},$$

where I is the unit operator in $\mathcal{L}$.

Note that the total volume

$$N = \int_{\mathcal{X}} dl \tag{2.0}$$

is finite, and hence, the number N is an integer, if and only if the Hilbert space $\mathcal{L}$ is N-dimensional.

Definition 2.2. Let $\mathbf{P}$ be a pure quantization. Its first quantum correlator

$$p(x,y) = \operatorname{tr}(\mathbf{P}(x)\mathbf{P}(y)) \tag{2.1}$$

will be called the *probability function*. The second correlator

$$b(x,y,w) = \operatorname{tr}(\mathbf{P}(x)\mathbf{P}(y)\mathbf{P}(w)) \tag{2.2}$$

will be called the *holonomy function*.

We shall use these objects below in Section 2.4 when we reconstruct an algebra $\mathcal{A}$ of functions over $\mathcal{X}$ isomorphic to the quantum algebra $\widehat{\mathcal{A}}$.

Now let us fix a linear space $\mathcal{F}$ of functions f over $\mathcal{X}$ for which the operator distribution

$$f \longmapsto \hat{f} \stackrel{\text{def}}{=} \int_{\mathcal{X}} f(x)\mathbf{P}(x)\,dl(x) \tag{2.3}$$

is well-defined in the weak sense on an invariant dense subspace in $\mathcal{L}$. In particular, the Dirac δ-functions δ_x localized at points $x \in \mathcal{X}$ belong to $\mathcal{F}$, and we get $\hat{\delta}_x = \mathbf{P}(x)$. After the identification $x \longleftrightarrow \mathbf{P}(x)$, the image $\widehat{\mathcal{F}}$ of the map (2.3) can be considered simply as a linear envelope of points from $\mathcal{X}$ completed in a certain topology. Our original algebra $\widehat{\mathcal{A}}$ is a subset in $\widehat{\mathcal{F}}$. So, we see that the map $f \longmapsto \hat{f}$ is an extension of the quantization $\mathbf{P}$ onto the space $\mathcal{F}$; this map will be also called a *quantization*.

2.2. Quantization of Symplectic-Kähler Manifolds

Suppose now that $\mathcal{X}$ is a $2n$-dimensional connected and simply connected manifold with complex structure J. We shall denote by $z = (z^1,\dots,z^n)$ the local complex coordinates and by $\partial = \partial_x$ the holomorphic differential at the point $x \in \mathcal{X}$. Let $\Pi \sim \{\partial/\partial\bar{z}\}$ be the polarization over $\mathcal{X}$, i.e., the distribution of n-dimensional eigenspaces of J corresponding to the eigenvalue i.

The manifold $\mathcal{X}$ will be called *symplectic-Kähler* if it is equipped with a symplectic form ω compatible with the complex structure, and also esuipped with a metric g which is Kählerian with respect to this complex structure.

Definition 2.3. A smooth pure quantization $\mathbf{P}$ of a complex manifold $\mathcal{X}$ is called *holomorphic*, if the one-form $(I-\mathbf{P})\,d\mathbf{P}$ is of 1-type, i.e., $(I-\mathbf{P})\bar{\partial}\mathbf{P} = 0$. The two-form

$$\partial_x\bar{\partial}_x \ln p(x,y) = \sum_{j,k=1}^{n} \frac{\partial^2 p(x,y)}{\partial\bar{z}^j(x)\partial z^k(x)}\,dz^k(x)\wedge d\bar{z}^j(x), \tag{2.4}$$

determined over $\mathcal{X} \ni x$ by means of the probability function p given in (2.1) and really independent of y, is called a *fundamental form* of the quantization $\mathbf{P}$.

Definition 2.4. Let $\hbar > 0$. We shall say that a quantization $\mathbf{P}$ is an *$\hbar$-quantization of a symplectic-Kähler manifold* $(\mathcal{X},\omega,g)$ if $\mathbf{P}$ is holomorphic and its fundamental form coincides with

$$\frac{1}{i}\Big(\frac{\omega}{\hbar} + \rho\Big), \qquad \text{where} \quad \rho \stackrel{\text{def}}{=} \frac{1}{2}\text{Ricci} \equiv i\bar{\partial}\partial \ln\det g. \tag{2.5}$$

Note that the de Rham class of the fundamental form (2.4) divided by 2π is integer-valued. So, for any $\hbar$-quantized symplectic-Kähler manifold $(\mathcal{X},\omega,g)$ the following quantization condition automatically holds:

$$-\frac{1}{2\pi\hbar}[\omega] + \frac{1}{2}c_1 \in H^2(\mathcal{X},\mathbf{Z}). \tag{2.6}$$

Here c_1 is the first Chern class of $\mathcal{X}$ generated by J.

2.3. Complex Groupoid and Membranes

Let $\mathcal{X}$ be a manifold with complex structure J. Denote by $\mathcal{X}^{(-)}$ the manifold $\mathcal{X}$ with complex structure $(-J)$. Then the direct product $\mathcal{X}^{\#} = \mathcal{X} \times \mathcal{X}^{(-)}$ equipped with the natural groupoid structure will be called the *complex groupoid corresponding to the complex manifold* $\mathcal{X}$. This means that $\mathcal{X}$ is imbedded into $\mathcal{X}^{\#}$ as the submanifold of units $\operatorname{diag}(\mathcal{X} \times \mathcal{X})$, and the projections $\pi_1 : \mathcal{X}^{\#} \to \mathcal{X}$, $\pi_2 : \mathcal{X}^{\#} \to \mathcal{X}$ on the first and second components are, respectively, the morphism and antimorphism of complex structures. The fibers of these projections can be considered as integral leaves of the polarizations Π and $\overline{\Pi}$.

Now, following Refs.19,20, we show how to describe the probability and holonomy functions in terms of integrals over certain membranes in the groupoid $\mathcal{X}^{\#}$. First of all, note that the form $i\bar{\partial}\partial \ln p$ provides an extension of the form $\frac{\omega}{\hbar} + \rho$ from $\mathcal{X}$ to $\mathcal{X}^{\#}$, except for the singular subset

$$\mathcal{X}_0^{\#} = \{(x,y) \mid p(x,y) = 0\}. \tag{2.7}$$

The complement $\mathcal{X}^{\#} \setminus \mathcal{X}_0^{\#}$ is dense in $\mathcal{X}^{\#}$ and includes a neighbourhood of the diagonal $\operatorname{diag}(\mathcal{X} \times \mathcal{X}) \approx \mathcal{X}$. We shall denote the extended form also by $\frac{\omega}{\hbar} + \rho$. This form vanishes on the leaves of Π and $\overline{\Pi}$.

For each $x \in \mathcal{X}$, we denote by $\Pi(x) \subset \mathcal{X}^{\#}$ the leaf of Π (i.e., the fiber of π_1) intersecting the submanifold $\operatorname{diag}(\mathcal{X} \times \mathcal{X}) \approx \mathcal{X}$ at the point $(x,x) \approx x$. For any two different points $x, y \in \mathcal{X}$, we have a unique point $x|y \in \overline{\Pi}(x) \cap \Pi(y)$.

Let us fix three points $x, y, w \in \mathcal{X}$ and consider another three point set $x|y, y|w, w|x \in \mathcal{X}^{\#}$. We connect these six points by paths going along the leaves of Π or $\overline{\Pi}$ in turn and get a cycle:

$$x \longleftarrow x|y \longleftarrow y \longleftarrow y|w \longleftarrow w \longleftarrow w|x \longleftarrow x. \tag{2.8}$$

Denote by *hexagon*(w,y,x) a membrane (i.e., a two-dimensional oriented immersed surface) in $\mathcal{X}^{\#}$ whose boundary coincides with the circle (2.8).

If we take $w = y$, then the hexagon turns into a quadrangle membrane with boundary

$$x \longleftarrow x|y \longleftarrow y \longleftarrow y|x \longleftarrow x,$$

whose paths go along the leaves $\Pi(x)$, $\overline{\Pi}(y)$, $\Pi(y)$, $\overline{\Pi}(x)$ respectively. Denote this membrane by *quadrangle*(y,x).

Theorem 2.1. *The probability and holonomy functions corresponding to a quantization of a symplectic-Kähler manifold* $(\mathcal{X}, \omega, g)$ *are given by the following formulas*

$$p(y,x) = \exp\{-\operatorname{Im} \textstyle\int_{quadrangle(y,x)} (\frac{\omega}{\hbar} + \rho)\}, \tag{2.9}$$

$$b(w,y,x) = \exp\{i \textstyle\int_{hexagon(w,y,x)} (\frac{\omega}{\hbar} + \rho)\}, \tag{2.10}$$

where ρ *is one half of the Ricci form (see (2.5)). The integrals on the right-hand sides are well-defined outside of the singular subset* $\mathcal{X}_0^{\#}$ *(2.7), but the exponential functions are continued globally as smooth functions on* $\mathcal{X} \times \mathcal{X}$ *and* $\mathcal{X} \times \mathcal{X} \times \mathcal{X}$.

If the quantization is complete with respect to a measure dl*, then*

$$\int_{\mathcal{X}} \exp\left\{i \textstyle\int_{hexagon(w,y,x)} (\frac{\omega}{\hbar} + \rho)\right\} dl(w) = \exp\{-\operatorname{Im} \textstyle\int_{quadrangle(y,x)} (\frac{\omega}{\hbar} + \rho)\}.$$

Note that the last identity looks very intriguing and, perhaps, correlates with the well-known Duistermaat-Heckman formula.[23]

Definition 2.5. A function of the type $\exp\{i\int_{membrane}(\frac{\omega}{\hbar}+\rho)\}$ will be called a *membrane amplitude.*

2.4. Integral Equation for Star Product via Probability and Holonomy Functions

Now we are able to describe explicitly an algebra $\mathcal{A}$ of functions over $\mathcal{X}$ isomorphic to our initial Hilbert algebra $\widehat{\mathcal{A}}$ from Definition 2.1.

Let us introduce integral operators $\mathcal{P}$, $\mathcal{B}$ by the following formulas:

$$\begin{aligned}\mathcal{P}(f)(x) &\stackrel{\text{def}}{=} \int_{\mathcal{X}} p(x,y)f(y)\,dl(y),\\ \mathcal{B}(f\otimes g)(w) &\stackrel{\text{def}}{=} \int_{\mathcal{X}}\int_{\mathcal{X}} b(w,y,x)f(y)g(x)\,dl(y)dl(x).\end{aligned} \tag{2.11}$$

Denote by $L_0^2=\operatorname{Ran}_{L^2}\mathcal{P}$ the range of the operator $\mathcal{P}: L^2\to L^2$, i.e., the orthogonal complement to the null-kernel $\operatorname{Ker}_{L^2}\mathcal{P}$ in the space $L^2=L^2(\mathcal{X},dl)$. Then one gets well-defined operators $\mathcal{P}: L_0^2\to L_0^2$, $\mathcal{B}: L_0^2\times L_0^2\to L_0^2$, and $\mathcal{P}$ is injective.

Taking two functions $f,g\in L_0^2$, we determine the product $f*g$ by the integral equation

$$\int_{\mathcal{X}} p(w,y)(f*g)(y)\,dl(y)=\int_{\mathcal{X}}\int_{\mathcal{X}} b(w,y,x)f(y)g(x)\,dl(y)dl(x) \tag{2.12}$$

or, in other words, by the formula

$$f*g=\mathcal{P}^{-1}\mathcal{B}(f\otimes g). \tag{2.13}$$

Theorem 2.2. *Let* $\mathbf{P}$ *be a pure complete quantization of a set* $\mathcal{X}$. *Then on the space* $L_0^2=\operatorname{Ran}_{L^2}\mathcal{P}$ *there is an associative multiplication* (2.13) *defined by the probability and holonomy functions* (2.1), (2.2). *The completion* $\mathcal{A}\stackrel{\text{def}}{=}[L_0^2]$ *with respect to the Hilbert norm*

$$\|f\|=(f,f)^{1/2},\qquad (f,g)\stackrel{\text{def}}{=}(\mathcal{P}f,g)_{L^2},$$

is a Hilbert algebra of functions over $\mathcal{X}$, *i.e.*

$$\begin{aligned}(f,g) &= (\bar g,\bar f), & \|f*\bar f\| &\le \|f\|^2,\\ \overline{f*g} &= \bar g*\bar f, & (f*g,k) &= (g,\bar f*k).\end{aligned}$$

Moreover, the following property holds:

$$(f,g)=\int_{\mathcal{X}} f*\bar g\,dl=\int_{\mathcal{X}}\bar g*f\,dl. \tag{2.14}$$

The map $f\mapsto\hat f$ (2.3) *is an isomorphism of* $\mathcal{A}$ *onto* $\widehat{\mathcal{A}}$.

If the total volume N *of* $\mathcal{X}$ (2.0) *is finite (for example, if* $\mathcal{X}$ *is compact), then* $\mathcal{A}$ *is* N^2*-dimensional algebra with unit* 1, *and* $\|1\|=\sqrt{N}$.

Remark 2.1. The bilinear operator $*$ (2.13) is well-defined on a space larger than L_0^2 (or $\mathcal{A}$). For example, it may be extended to $L^2 \times L^2 \to L_0^2$, but such an algebra L^2 will have a nontrivial center, unless $\mathrm{Ker}_{L^2} \mathcal{P}$ becomes trivial. A similar extension $L^1 \times L^1 \to L_0^1$ is possible, and so on. Below we simply assume that $*$ is extended to a certain space $\mathcal{F}$ of functions over $\mathcal{X}$ (see the end of Section 2.1), and do not define this space more exactly.

Remark 2.2. Abstract $*$-products for which the second equation (2.14) holds were considered by A. Connes, M. Flato, D. Sternheimer[24] under the name "closed products".

Remark 2.3. In the original terminology of Berezin,[6] the function f is called a contravariant symbol of the operator $\hat{f}$, and the function $\mathcal{P}(f) \equiv \mathrm{tr}(\hat{f}\mathbf{P})$ is called a covariant symbol. For general $\mathcal{X}$, Berezin was interested in a *multiplication of covariant symbols*, not contravariant ones. The main idea was that it is possible to obtain an explicit formula for the multiplication of covariant symbols using their continuation onto $\mathcal{X} \times \mathcal{X}$ as functions with singularities. The advanced version of such singular covariant multiplication was given in Ref.25.

A similar covariant multiplication in terms of Kostant-Souriau geometric quantization was described by J. Rawnsley, M. Cahen, S. Gutt.[26]

All these constructions did not consider the basic equation (2.12)

$$\mathcal{P}(f * g) = \mathcal{B}(f \otimes g), \tag{2.12a}$$

which, in fact, not only defines the product $*$, but also controls how many functions should be excluded from $C^\infty(\mathcal{X})$ to obtain a quantized algebra. The crucial observation for us at this point (see Ref. 21) was the result by M. Rieffel[27] and H. Omori[28] on the nonexistence of $*$-product on the whole space $C^\infty(\mathcal{X})$ for $\mathcal{X} = \mathbf{S}^2$. From the viewpoint of (2.12a), the reason of this result is evident: the large null-kernel of $\mathcal{P}$ makes the "right" quantized algebra $C^\infty(\mathcal{X}) / \mathrm{Ker}\, \mathcal{P}$ very small (finite-dimensional for compact $\mathcal{X}$).

2.5. Geometric Star Products over Symplectic-Kähler Manifolds

Now suppose that $\mathcal{X}$ is a symplectic-Kähler manifold. Then the probability and holonomy functions can be represented as membrane amplitudes (2.9), (2.10) through the geometric data ω, ρ, dl. So, in this case the star product $*$ (2.12), (2.13) can be also expressed in terms of ω, ρ, dl. In fact, these geometric formulas for the star product via membranes work independently of the presence of a quantization map $\mathbf{P}$ (see Ref. 21). In particular, if ω, ρ, and $\hbar^{\dim \mathcal{X}/2} dl$ depend regularly on $\hbar$ as $\hbar \to 0$, then formulas (2.9), (2.10), and (2.12) expanded in formal power series in $\hbar$ give an explicit global and absolutely geometric expression for a formal star product over $\mathcal{X}$ (for details, see Ref. 21). Of course, the correspondence principle $f * g = fg + O(\hbar)$, $f * g - g * f = -i\hbar\{f, g\} + O(\hbar^2)$ holds under the additional assumption that

$$dl = \frac{\text{Liouville measure} + O(\hbar)}{(2\pi\hbar)^{\dim \mathcal{X}/2}}.$$

Note that the expansion of the probability operator $\mathcal{P}$ (2.11), as $\hbar \to 0$, looks as follows: $\mathcal{P} = I + \hbar \mathcal{P}^{(1)} + \hbar^2 \mathcal{P}^{(2)} + \ldots$, so the inverse $\mathcal{P}^{-1}$ always exists in the sense of formal series in $\hbar$, and the problem of a nontrivial null-kernel $\mathrm{Ker}\, \mathcal{P}$ does not arise. Hence, we see that formal series for the star products do not feel some essential properties of the global topology of the space $\mathcal{X}$. In particular, for compact $\mathcal{X}$ "right" finite-dimensional algebras of functions never appear in such a formal approach.

Remark 2.4. We can define a large family of star products over $\mathcal{X}$ by the formula

$$f *_\mu g \stackrel{\text{def}}{=} \mathcal{P}^\mu(\mathcal{P}^{-\mu} f * \mathcal{P}^{-\mu} g), \qquad \forall \mu \in \mathbf{C}.$$

In particular, if $\mu = 0$, then one gets the initial "anti-Wick" product (2.12); if $\mu = 1$, we get the "Wick" product, and if $\mu = 1/2$, then we have the "Weyl" product. Explicitly, all these products are determined on the space L_0 (where $\mathcal{P} > 0$ and the inverse $\mathcal{P}^{-1}$ exists), but as formal $\hbar$-power series these products are well-defined on any $f, g \in C^\infty(\mathcal{X})$. For the operator $\mathcal{P}$ and the multiplication $*$, we know formal $\hbar$-expansions with geometrically invariant coefficients, so we shall get similar ones for each product $*_\mu$. This means, for example, that the formal "Weyl" product over any symplectic–Kähler manifold $\mathcal{X}$ can be defined in a geometric invariant way, and moreover, by explicit formulas, using only metric, symplectic, and complex structures.

We conclude this section by a statement that relates in a precise way the star product (2.13) and the Poisson brackets on $\mathcal{X}$.

Let $\text{ad}(f)$ be the Hamiltonian field corresponding to a function f. Denote by $\text{ad}^+(f)$ the $\overline{\Pi}$-component of this field.

Consider the local expression for the measure dl in the complex coordinates:

$$dl = \varkappa \, d\bar{z} \, dz, \qquad \varkappa \text{ is the local density,} \tag{2.15}$$

and introduce on $\mathcal{X}$ a new global measure $dl^\#$ with the following local density

$$dl^\# = \frac{\varkappa^2}{\det g} \, d\bar{z} \, dz. \tag{2.16}$$

Denote by $^\#\text{div}\,(f)$ the divergence of the field $\text{ad}^+(f)$ with respect to the measure (2.16).

Denote by $C^\infty(\mathcal{X}, \Pi)$ the space of smooth functions on $\mathcal{X}$ whose Hamiltonian flows preserve the polarization Π. The map $f \mapsto f_\hbar$, defined on $C^\infty(\mathcal{X}, \Pi)$ by the formula

$$f_\hbar \stackrel{\text{def}}{=} f - \frac{i\hbar}{2} \, {}^\#\text{div}(f), \tag{2.17}$$

will be called a *quantum deformation.*

Note that $C^\infty(\mathcal{X}, \Pi)$ is a Lie algebra with respect to the Poisson brackets. Denote by $\mathcal{F}(\mathcal{X}, \Pi) \subset C^\infty(\mathcal{X}, \Pi)$ the Lie subalgebra of functions whose quantum deformations belong to the space $\mathcal{F}$ (see Remark 2.1).

Theorem 2.3. *For functions* $f \in \mathcal{F}(\mathcal{X}, \Pi)$, *the Hamiltonian field* $\text{ad}(f)$, *the quantum deformation* $f_\hbar$, *and the left multiplication* $f_\hbar *$ *in the sense of operation* (2.13), *are related as follows:*

$$f_\hbar * = f_\hbar - i\hbar \, \text{ad}^+(f).$$

In particular, the quantum deformation generates a Lie algebra homomorphism:

$$f_\hbar * g_\hbar - g_\hbar * f_\hbar = -i\hbar(\{f, g\})_\hbar, \qquad \forall f, g \in \mathcal{F}(\mathcal{X}, \Pi).$$

3. CONNECTIONS WITH ZERO CURVATURE OVER LAGRANGIAN SUBMANIFOLDS AND GEOMETRIC COHERENT STATES

For the case $\mathcal{X} = \mathbf{R}^{2n}$, it was shown in Ref. 14 how to use the natural quantum connection with zero curvature over lagrangian submanifolds $\Lambda \subset \mathbf{R}^{2n}$ in order to define coherent states geometrically by parallel translation. It is possible to do the same for general phase spaces $\mathcal{X}$, even for compact ones. Below we show how to obtain geometric Λ-coherent states which are defined not only locally (as sections of a bundle), but are smooth global functions over Λ with values in a given Hilbert space.

3.1. Fundamental 1-Form and Two-Point Transition Amplitude over a Lagrangian Submanifold

First we recall certain definitions from Refs. 14,15,19. Let $(\mathcal{X}, \omega, g)$ be a symplectic-Kähler manifold, $\Lambda \subset \mathcal{X}$ be a lagrangian submanifold, i.e.

$$\omega\big|_{\Lambda} = 0, \qquad \dim \Lambda = \frac{1}{2} \dim \mathcal{X}.$$

Let $d\sigma$ be a smooth measure on Λ. Consider the local Jacobian

$$j = \det g^{1/2} \frac{\mathcal{D}z}{\mathcal{D}\sigma}, \tag{3.1}$$

the local 1-form on Λ (the *Arnold form*):

$$\mu = -\frac{1}{2\pi i} d \ln(j/\bar{j}),$$

and the local connection form, generated by the metric:

$$\zeta = \frac{i}{2}(\partial \ln \det g - \bar{\partial} \ln \det g)\big|_{\Lambda}.$$

Definition 3.1. The global 1-form on Λ defined by

$$\nu_{\Lambda} \stackrel{\text{def}}{=} \frac{1}{2}(\zeta + \pi\mu) \tag{3.2}$$

is called a *fundamental form* on Λ.

Note that $d\nu_{\Lambda} = \rho\big|_{\Lambda}$ (see (2.5)).

Lemma 3.1. *For any two-dimensional membrane $\Sigma \subset \mathcal{X}$ with boundary $\partial\Sigma \subset \Lambda$ the number*

$$m_{\Lambda}(\Sigma) \stackrel{\text{def}}{=} \frac{2}{\pi}\Big(\oint_{\partial\Sigma} \nu_{\Lambda} - \int_{\Sigma} \rho\Big)$$

is an integer and coincides with the index[29] *of the membrane Σ. For closed ($\partial\Sigma = \varnothing$) membranes the index is equal to twice the Chern class: $m_{\Lambda}(\Sigma) = 2c_1(\Sigma)$.*

Definition 3.2. A lagrangian submanifold $\Lambda \subset \mathcal{X}$ will be called an *$\hbar$-submanifold* if

$$\frac{1}{2\pi\hbar}\int_{\Sigma} \omega - \frac{1}{4} m_{\Lambda}(\Sigma) \in \mathbf{Z}$$

for any membrane $\Sigma \subset \mathcal{X}$ with boundary $\partial\Sigma \subset \Lambda$.

In particular, this condition for closed membranes includes (2.6).

Let us take two points $\alpha, \beta \in \Lambda$ and consider a path $\Gamma(\beta|\alpha)$ going from β to α along Λ. Let *triangle*$(\beta|\alpha) \subset \mathcal{X}^{\#}$ be a complex triangle membrane with boundary

$$\alpha \overset{\Gamma}{\longleftarrow} \beta \longleftarrow \beta|\alpha \longleftarrow \alpha.$$

The first and the second paths here are going along $\Pi(\alpha)$ and $\overline{\Pi}(\beta)$ respectively.

Let us define a function

$$S_\Lambda(\beta|\alpha) = \int_{triangle(\beta|\alpha)} \Big(\frac{\omega}{\hbar} + \rho\Big) - \int_{\Gamma(\beta|\alpha)} \nu_\Lambda. \tag{3.3}$$

Lemma 3.2. *If Λ is an $\hbar$-submanifold, then the function* $\exp\{iS_\Lambda\}$ *does not depend on the choice of a membrane in* (3.3) *and is globally defined on* $\Lambda \times \Lambda$.

Definition 3.3. The globally defined function $\Delta_\Lambda = |j|$ (see (3.1)) is called a *modular function* over Λ and the function

$$a_\Lambda(\beta|\alpha) \overset{\text{def}}{=} \exp\{iS_\Lambda(\beta|\alpha)\}\Delta_\Lambda(\alpha)^{1/2}\Delta_\Lambda(\beta)^{1/2} \tag{3.4}$$

is called a *two-point transition amplitude* on Λ (from β to α).

3.2. Geometric Inner Product and Realization of the Dirac Quantization by First Order Operators over $\hbar$-Submanifolds

For any $\varphi_1, \varphi_2 \in C_0^\infty(\lambda)$ we define

$$(\varphi_1, \varphi_2)_\Lambda \overset{\text{def}}{=} \int_\Lambda\int_\Lambda a_\Lambda(\beta|\alpha)\,\varphi_1(\alpha)\overline{\varphi_2(\beta)}\,d\sigma(\alpha)d\sigma(\beta). \tag{3.5}$$

The $L^2(\Lambda, d\sigma)$-orthogonal complement to the subspace $\{\varphi|(\varphi,\varphi)_\Lambda = 0\}$ completed with respect to the norm $\|\varphi\|_\Lambda = (\varphi,\varphi)_\Lambda^{1/2}$ will be denoted by $\mathcal{L}_\Lambda$. This is a Hilbert space with respect to the inner product (3.5).

For any $H \in C^\infty(\mathcal{X})$ we define[15,19] a first order differential operator on Λ:

$$\breve{H} \overset{\text{def}}{=} H\big|_\Lambda - i\hbar\Big(v(H) + \frac{1}{2}{}^\sigma\mathrm{div}\, v(H)\Big), \tag{3.6}$$

where $v(H) = \mathrm{pr}_\Lambda^\Pi(\mathrm{ad}(H))$ is the projection of the Hamiltonian field on Λ along the polarization Π, and ${}^\sigma\mathrm{div}$ means the divergence with respect to the measure $d\sigma$.

Theorem 3.1. *Let Λ be an $\hbar$-submanifold in the $\hbar$-quantized symplectic-Kähler manifold $\mathcal{X}$. For $H \in C^\infty(\mathcal{X}, \Pi)$ the operator* (3.6) *is well-defined in the Hilbert space $\mathcal{L}_\Lambda$ with inner product* (3.5) *generated by the two-point transition amplitude* (3.4). *If* $\mathrm{Im}\, H = 0$, *then $\breve{H}$ is formally selfadjoint, and moreover, if Λ is closed, then $\breve{H}$ is essentially selfadjoint. The map $H \mapsto \frac{i}{\hbar}\breve{H}$ is a Lie algebra homomorphism, i.e., the Dirac axiom holds*:

$$[\breve{H}_1, \breve{H}_2] = -i\hbar\breve{\{H_1, H_2\}}, \qquad H_1, H_2 \in C^\infty(\mathcal{X}, \Pi).$$

Moreover, it is evident that $\breve{1} = I$ *(the identity operator in* $\mathcal{L}_\Lambda$).

For details and examples concerning this realization of the Dirac quantization axioms, see Refs. 19,20,22.

Now we shall answer the question: how to link this Lie algebra quantization with the associative algebra quantization defined by (2.3) and by (2.13).

3.3. Creation-Annihilation Forms and Parallel Transport over Lagrangian Submanifolds

Let a quantization $\mathbf{P}$ of a complex manifold $\mathcal{X}$ be represented in a Hilbert space $\mathcal{L}$. Let us consider the following 1-forms on $\mathcal{X}$ with values in the space of functions over $\mathcal{X}$:

$$a_x^+(\cdot) \stackrel{\text{def}}{=} \hbar\partial_x \ln p(x,\cdot), \qquad a_x^-(\cdot) \stackrel{\text{def}}{=} \hbar\overline{\partial}_x \ln p(x,\cdot), \qquad \forall x \in \mathcal{X}, \tag{3.7}$$

where p is a probability function (2.1). Applying to a^+, a^- the quantization map (2.3), one gets 1-forms $\hat{a}^+$, $\hat{a}^-$ on $\mathcal{X}$ with values in the space of operators on $\mathcal{L}$. These forms are conjugate to each other:

$$\hat{a}^- = (\hat{a}^+)^*,$$

and satisfy the identities

$$\hbar\partial\mathbf{P} = \hat{a}^+\mathbf{P}, \qquad \hat{a}^-\mathbf{P} = 0.$$

So, they can be considered as a generalization of the notion of "creation-annihilation operators" well-known for the flat space $\mathcal{X} = \mathbf{R}^{2n}$. The following statement is a generalization of that obtained in Ref. 14.

Theorem 3.2. *Suppose an $\hbar$-quantization $\mathbf{P}$ of a symplectic-Kähler manifold $(\mathcal{X}, \omega, g)$ is represented in a Hilbert space $\mathcal{L}$, and the 1-forms $\hat{a}^+$, $\hat{a}^-$ are determined by* (3.7) *using the probability function* (2.9). *Let $\Lambda \subset \mathcal{X}$ be an $\hbar$-submanifold with measure $d\sigma$, and ν_Λ be the fundamental 1-form* (3.2) *on Λ.*

(a) *The following operator-valued 1-form on Λ*

$$\hat{\theta} \stackrel{\text{def}}{=} -\frac{1}{\hbar}\hat{a}^+\Big|_\Lambda + i\nu_\Lambda \cdot I \tag{3.8}$$

determines a connection with zero curvature and with identity global holonomy in the trivial $\mathcal{L}$-bundle over Λ (i.e., in $\mathcal{L} \times \Lambda$).

(b) *Let us take $\alpha_0 \in \Lambda$ and fix a unit vector $\mathrm{e}_{\alpha_0} \in \mathcal{L}$ such that*

$$\mathbf{P}(\alpha_0) = \mathrm{e}_{\alpha_0} \otimes \mathrm{e}^*_{\alpha_0}.$$

Denote by e_α the parallel transport of e_{α_0} along any path $\alpha_0 \to \alpha$ on Λ by means of the connection (3.8); *in other words, the family $\mathrm{e} = \{\mathrm{e}_\alpha \,|\, \alpha \in \Lambda\} \subset \mathcal{L}$ is the solution of the "Cauchy problem" over Λ :*

$$\hat{a}^+\mathrm{e} = (\hbar d + i\hbar\nu_\Lambda)\mathrm{e}, \qquad \mathrm{e}\big|_{\alpha_0} = \mathrm{e}_{\alpha_0}.$$

Then the family e *is smooth and global over Λ, consists of unit vectors, and moreover*

$$\hat{a}^-_\alpha \mathrm{e}_\alpha = 0, \qquad \mathbf{P}(\alpha) = \mathrm{e}_\alpha \otimes \mathrm{e}^*_\alpha \quad \forall \alpha \in \Lambda.$$

(c) *The following identity holds:*

$$(\mathrm{e}_\alpha, \mathrm{e}_\beta)_\mathcal{L} = \exp\{iS_\Lambda(\beta|\alpha)\}, \tag{3.9}$$

where S_Λ is given by (3.3). (d) *For any $x, y, w \in \mathcal{X}$ let us fix a path $w \leftarrow y$ from y to w, and consider a membrane* pentagon$(x|w \leftarrow y) \subset \mathcal{X}^\#$ *with boundary*

$$y \leftarrow x|y \leftarrow x \leftarrow x|w \leftarrow w \leftarrow y.$$

Here the second and fifth paths go along $\Pi(w)$ *and* $\Pi(y)$ *respectively, the third and forth paths go along* $\overline{\Pi}(x)$. *Let us introduce the corresponding membrane amplitude*

$$c_{w\leftarrow y}(x) \stackrel{\text{def}}{=} \exp\{i\textstyle\int_{pentagon(x|w\leftarrow y)}(\frac{\omega}{\hbar}+\rho)\}. \tag{3.10}$$

Then the parallel transport $e_{\alpha_0} \to e_\alpha$ *can be calculated by the formula:*

$$e_\alpha = \exp\{-i\textstyle\int_{\alpha\leftarrow\alpha_0}\nu_\Lambda\}\cdot \hat{c}_{\alpha\leftarrow\alpha_0}e_{\alpha_0}.$$

Remark 3.1. The operators $\hat{c}_{w\leftarrow y}$ are "creation operators" in the sense that the conjugate operators are acting as "annihilation operators":

$$(\overline{c}_{w\leftarrow y})^*\mathbf{P}(x) = \overline{c}_{w\leftarrow y}(x)\mathbf{P}(x), \qquad \forall x,y,w\in\mathcal{X}.$$

They mutually commute and represent the groupoid $Path(\mathcal{X})$ consisting of all paths in $\mathcal{X}$; so, one has

$$\hat{c}_{\Gamma_1}\hat{c}_{\Gamma_2} = \hat{c}_{\Gamma_1\circ\Gamma_2}, \qquad \Gamma_1,\Gamma_2,\Gamma_1\circ\Gamma_2\in Path(\mathcal{X}),$$

and moreover, on cyclic paths, i.e., on boundaries of membranes, one gets a unitary character of the groupoid $Path(\mathcal{X})$:

$$\hat{c}_{\partial\Sigma} = \exp\{i\textstyle\int_\Sigma(\frac{\omega}{\hbar}+\rho)\}\cdot I.$$

Now let us introduce the smooth family of vectors

$$u_\alpha \stackrel{\text{def}}{=} \Delta(\alpha)^{1/2}e_\alpha\in\mathcal{L}, \qquad \alpha\in\Lambda,$$

where Δ is the modular function on Λ (Definition 3.3), and let us define the map

$$U_\Lambda : \mathcal{L}_\Lambda \longrightarrow \mathcal{L}, \qquad U_\Lambda(\varphi) \stackrel{\text{def}}{=} \int_\Lambda \varphi(\alpha)u_\alpha\, d\sigma(\alpha). \tag{3.11}$$

Theorem 3.3. *The map* U_Λ (3.11) *is a unitary homomorphism or, in other words,*

$$(u_\alpha, u_\beta)_\Lambda = a_\Lambda(\beta|\alpha),$$

where a_Λ *is the two-point transition amplitude* (3.4) *over* Λ, *and the inner product* $(\cdot,\cdot)_\Lambda$ *is given by* (3.5).

If $f\in\mathcal{F}(\mathcal{X},\Pi)$, *and* $f_\hbar$ *is the quantum deformation* (2.17), *then*

$$\hat{f}_\hbar U_\Lambda(\varphi) = U_\Lambda(\check{f}\varphi), \tag{3.12}$$

where $\check{f}$ *is defined by* (3.6).

Comparing this result with Theorems 3.1 and 2.3, we see that U_Λ is the Lie algebra homomorphism intertwining the Lie algebra quantization (3.6) and the associative algebra quantization (2.3).

Following Ref. 20, the vectors u_α will be called *geometric coherent states* over Λ represented in the Hilbert space $\mathcal{L}$ (or, briefly, Λ-*coherent states* in $\mathcal{L}$). The map U_Λ (3.11) is called a *coherent intertwining map over* Λ.

Remark 3.2. Using Theorem 3.2(d), one can express the intertwining map over Λ as follows:

$$U_\Lambda(\varphi) = \hat{c}_\Lambda(\varphi)e_{\alpha_0}, \qquad \alpha_0\in\Lambda, \tag{3.13}$$

where $c_\Lambda(\varphi)\in\mathcal{F}(\mathcal{X})$,

$$c_\Lambda(\varphi)(x) \stackrel{\text{def}}{=} \int_\Lambda \varphi(\alpha)\Delta(\alpha)^{1/2}\exp\{-i\textstyle\int_{\alpha\leftarrow\alpha_0}\nu_\Lambda\}\cdot c_{\alpha\leftarrow\alpha_0}(x)\,d\sigma(\alpha),$$

and $c_{\alpha\leftarrow\alpha_0}$ is the pentagon membrane amplitude (3.10). Formula (3.13) means that the range of the intertwining map U_Λ can be obtained by applying certain "creation" operators to the "vacuum" vector $e_{\alpha_0}\in\mathcal{L}$.

3.4. Example: Coherent Intertwining Map over the Diagonal in Groupoid

If $(\mathcal{X}, \omega, g)$ is a symplectic-Kähler manifold, and $\mathcal{X}^{(-)}$ is equipped with the symplectic form $(-\omega)$, and with the same metric g, then $\mathcal{X}^{\#} = \mathcal{X} \times \mathcal{X}^{(-)}$ is also a symplectic-Kähler manifold. So, one can speak about the *symplectic-Kähler groupoid* $\mathcal{X}^{\#}$ corresponding to the symplectic-Kähler manifold $\mathcal{X}$ (see Section 2.3).

Let $\mathbf{P}$ be a quantization of $\mathcal{X}$ represented in a Hilbert space $\mathcal{L}$. Then there is a natural quantization $\mathbf{P}^{\#}$ of $\mathcal{X}^{\#}$ represented in the Hilbert space $\mathcal{L}^{\#} = \widehat{\mathcal{A}}$ which is the space of all Hilbert–Schmidt operators on $\mathcal{L}$. Namely, $\mathbf{P}^{\#}(x, x')(R) \stackrel{\text{def}}{=} \mathbf{P}(x) R \mathbf{P}(x')$, $\forall R \in \widehat{\mathcal{A}}$. If $\mathbf{P}$ is an $\hbar$-quantization, then $\mathbf{P}^{\#}$ is also an $\hbar$-quantization.

The diagonal in $\mathcal{X}^{\#}$ is a lagrangian submanifold, which we identify with $\mathcal{X}$ and equip with the measure $dl^{\#}$ (2.16). Then one can ask: what are the geometric coherent states over the diagonal?

Theorem 3.4. *The geometric coherent states over the submanifold of units* $\operatorname{diag}(\mathcal{X} \times \mathcal{X})$ *in the symplectic-Kähler groupoid* $\mathcal{X}^{\#} = \mathcal{X} \times \mathcal{X}$ *are given by the formula*

$$\mathbf{u}_x = \Delta_{\text{diag}}(x)^{1/2} \cdot \mathbf{e}_x, \qquad \mathbf{e}_x = \mathbf{P}(x) \in \widehat{\mathcal{A}}. \tag{3.14}$$

Here $x \in \mathcal{X} \approx \operatorname{diag}(\mathcal{X} \times \mathcal{X})$, *and* $\Delta_{\text{diag}} = (\det g)/\varkappa$ *(see* (2.15)*). According to* (3.9) *one has*

$$\begin{aligned} (\mathbf{P}(x), \mathbf{P}(y))_{\widehat{\mathcal{A}}} &\equiv \operatorname{tr}(\mathbf{P}(x)\mathbf{P}(y)) \equiv p(x,y) = \exp\{iS_{\text{diag}}(y|x)\}, \\ S_{\text{diag}}(y|x) &= \int_{\text{quadrangle}(y,x)} \Big(\frac{\omega}{\hbar} + \rho\Big) \qquad (\textit{see } (2.9)). \end{aligned}$$

The inner product (3.5) *over* $\operatorname{diag}(\mathcal{X} \times \mathcal{X}) \approx \mathcal{X}$ *coincides with the inner product* $(\cdot, \cdot)$ *from Theorem* 2.2, *so,* $\mathcal{L}_{\text{diag}} = \mathcal{A}$.

The coherent state $\mathbf{P}$ (3.14) *over* $\mathcal{X} \approx \operatorname{diag}(\mathcal{X} \times \mathcal{X})$, *i.e., the quantization map, can be obtained by a parallel transport along* $\mathcal{X}$ *with respect to a connection with zero curvature; this transport can be expressed explicitly by the "creation" operator as in Theorem 3.2(d):*

$$\mathbf{P}(x) = \hat{c}_{x \leftarrow x_0} \mathbf{P}(x_0) \hat{c}_{x_0 \rightarrow x}.$$

For $f \in \mathcal{F}(\mathcal{X}, \Pi)$ *one has* $f \otimes 1 \in \mathcal{F}(\mathcal{X}^{\#}, \Pi^{\#})$ *and the operators* (3.6) *over* $\operatorname{diag}(\mathcal{X} \times \mathcal{X}) \approx \mathcal{X}$ *look as follows*

$$f \breve{\otimes} 1 = f_{\hbar} * = f - i\hbar\Big(\operatorname{ad}^{+}(f) + \frac{1}{2}{}^{\#}\operatorname{div}(f)\Big).$$

The coherent intertwining map U_{diag} *over* $\operatorname{diag}(\mathcal{X} \times \mathcal{X})$ *(see* (3.11)*) coincides with the quantization map* (2.3), *so that*

$$\hat{f} = U_{\text{diag}}(f).$$

The latter formula means that the quantization on the level of observables, star-products, etc., is a particular case of general integral representations by geometric coherent states over lagrangian submanifolds (see Ref. 15, Example 3).

3.5. Application: Integral Representation of Exact Eigenfunctions and Quasimodes

The integral operator (3.11) and the intertwining property (3.12) give a possibility to represent wave functions (solutions of differential equations) as a linear combination of geometric coherent states. In fact, many integral representations of special functions known under different names can be obtained in this way, and so can be understood geometrically.

Theorem 3.5. *Let the function $H \in \mathcal{F}(\mathcal{X}, \Pi)$ be real, let the compact $\hbar$-submanifold $\Lambda \subset \mathcal{X}$ belong to the energy level*

$$\Lambda \subset \{H = \lambda_0 = \text{const}\},$$

and let $\{u_\alpha | \alpha \in \Lambda\}$ be the Λ-coherent states. Let φ be an eigenfunction for the first order differential operator on Λ:

$$\Big(\operatorname{ad}(H)\Big|_\Lambda + \frac{1}{2}{}^\sigma\operatorname{div}\,(\operatorname{ad}(H))\Big)\varphi = i\lambda_1\varphi.$$

Then the vector

$$\psi \stackrel{\text{def}}{=} \int_\Lambda \varphi(\alpha)u_\alpha\, d\sigma(\alpha) = \hat{c}_\Lambda(\varphi)e_{\alpha_0}$$

is the eigenvector for the operator $\widehat{H}_\hbar$ in $\mathcal{L}$:

$$\widehat{H}_\hbar\psi = (\lambda_0 + \hbar\lambda_1)\psi.$$

In particular, if the measure $d\sigma$ on Λ is invariant with respect to the Hamilton flow γ^t_H generated by $\operatorname{ad}(H)$, then $\psi_0 \stackrel{\text{def}}{=} \int_\Lambda u_\alpha\, d\sigma(\alpha) = \hat{c}_\Lambda(1)e_{\alpha_0}$ is the eigenvector for $\widehat{H}_\hbar$:

$$\widehat{H}_\hbar\psi_0 = \lambda_0\psi_0.$$

Of course, the crucial assumption $H \in \mathcal{F}(\mathcal{X}, \Pi)$ (recall, this means that the flow γ^t_H preserves the polarization) looks very restrictive. But in this theorem we could take $H \in \mathcal{F}(\mathcal{X})$ and change the statement about exact eigenvectors into a statement about approximate quasimodes (for details, see Refs. 14,15,17,18,22,30).

Also note that the integral representation similar to (3.11) works well for solutions of the Cauchy and scattering problem.[15,17,18,30]

References

1. R. F. V. Anderson, The Weyl functional calculus, *J. Funct. Anal.* 4:240–267 (1969).
2. J. R. Klauder, Continuous representation theory. Generalized relation between quantum and classical dynamics, *J. Math. Phys.* 4:1058 (1963)
3. A. M. Perelomov, Coherent states for arbitrary Lie group, *Comm. Math. Phys.* 26:222 (1972)
4. E. Onofri, A note on coherent state representations of Lie groups, *J. Math. Phys.* 16:1087–1089 (1973)
5. S. Bergmann, The kernel functions and conformal mapping, *Amer. Math. Soc., Math. Surveys* 5 (1950)
6. F. A. Berezin, Covariant and contravariant symbols of operators, *Izvestiya Akad. Nauk SSSR, Mat.* 36:1134–1167 (1972) (Russian)

7. A. Odzijewicz, On reproducing kernels and quantization of states, *Comm. Math. Phys.* 114:577–597 (1988)

8. S. T. Ali and J.-P. Antoine, Quantum frames, quantization and dequantization, *in*: "Quantization and Infinite-Dimensional Systems, Proc. XII-th Coll. Geom. Methods in Physics, Białowieza, July 1993", J.-P. Antoine, S. T. Ali, W. Lisiecki, I. M. Mladenov, A. Odzijewicz (eds), Plenum, N.Y.& London 133–145 (1994)

9. J. M.Garcia–Bondia, Generalized Moyal quantization on homogeneous symplectic spaces, *Contemp. Math.* 134:93–114 (1992)

10. F. Bayen, M. Flato, C. Fronsdal, A. Lichnerowicz, and D. Sternheimer, Deformation theory and quantization, *Ann. Phys.* 111:61–151 (1978)

11. M. De Wilde and P. B. Lecomte, Existence of star products and of formal deformation of the Poisson Lie algebra of arbitrary symplectic manifolds, *Lett. Math. Phys.* 7:487–496 (1983)

12. H. Omori, Y. Maeda, and A. Yoshioka, Weyl manifolds and deformation quantization, *Adv. Math.* 85:224–255 (1991)

13. A. Connes, Non-commutative differential geometry, *Publ. IHES* 62:257–360 (1986)

14. M. V. Karasev, Connections over lagrangian submanifolds and certain problems in semi-classical approximation, *Zap. Nauchn. Sem. Leningrad Otdel. Mat. Inst. Steklov* (LOMI) 172:41–54 (1989) (Russian); English transl. in *J. Sov. Math.* 59:1053–1062 (1992)

15. M. V. Karasev, Simple Quantization formula, *in*: "Symplectic Geom. and Math. Phys., Actes du colloque en l'honneur de J.-M. Souriau", P. Donato, C. Duval, J. Elhadad, G.M. Tuynman (eds.), Birkhäuser, Boston 234–244 (1991)

16. M. V. Karasev, Quantization by parallel translation. Global formula for semiclassical wave functions. *in*: "Quantum Field Theory, Quantum Mechanics and Quantum Optics, Proc. XVIII Intern. Coll. Group Theor. Meth. in Phys., Moscow, 1990", Part I, Nova Sci. Publ., N.-Y. 189–192 (1991)

17. M. V. Karasev and M. B. Kozlov, Quantum and semiclassical representations over lagrangian submanifolds in $su(2)^*$, $so(4)^*$, $su(1,1)^*$, *J. Math. Phys.* 34:4986–5006 (1993)

18. M. V. Karasev and M. B. Kozlov, Representations of compact semisimple Lie algebras over lagrangian submanifolds, *Funktsional. Anal. i Prilozhen.* 28 (nr.4):16–27 (1994) (Russian); English transl. in *Funct. Anal. & Appl.*

19. M. V. Karasev, Integrals over membranes, transition amplitudes and quantization, *Russ. J. Math. Phys.* 1:523–526 (1993)

20. M. V. Karasev, Quantization by membranes and integral representations of wave functions, *in*: "Quantization and Infinite-Dimensional Systems, Proc. XII-th Workshop on Geom. Methods in Physics, Białowieza, July 1993", pp. 9–19; J.-P. Antoine, S. T. Ali, W. Lisiecki, I. M. Mladenov, A. Odzijewicz (eds.), Plenum, N. Y. & London (1994)

21. M. V. Karasev, Formulas for noncommutative products of functions in terms of membranes and strings. I, *Russ. J. Math. Phys.* 2:445–462 (1994)

22. M. V. Karasev, Quantization by means of two-dimensional surfaces (membranes). Geometrical formulas for wave functions, *Contemp. Math.* 179:83-113 (1994)

23. J. J. Duistermaat and H. J. Heckman, On the variation in cohomology of the symplectic form of the reduced phase space, *Invent. Math.* 69:259–268 (1982)

24. A. Connes, M. Flato, and D. Sternheimer, Closed star products and cyclic cohomology, *Lett. Math. Phys.* 24:1–12 (1992).

25. A. Odzijewicz, Covariant and contravariant Berezin symbols of bounded operators, *in*: "Quantization and Infinite-Dimensional Systems, Proc. XII-th Coll. Geom. Methods in Physics, Bialowieza, July 1993", pp. 99–108; J.-P. Antoine, S. T. Ali, W. Lisiecki, I. M. Mladenov, A. Odzijewicz (eds.), Plenum, N.Y.& London (1994)

26. M. Cahen, S. Gutt, and J. Rawnsley, Quantization of Kähler manifolds. Part I, *J. Geom. Phys*, 7:45–62 (1990); Part II, *Trans. Amer. Math. Soc.* 337:73–98 (1993)

27. M. Rieffel, Deformation quantization of Heisenberg manifolds, *Comm. Math. Phys.* 122:531–562 (1989)

28. H. Omori, Report at the Conference "Symplectic Geometry and Applications", Sanda, Japan, July 1993

29. M. V. Karasev, Poisson algebras of symmetries and asymptotics of spectral series, *Funktsional. Anal. i Prilozhen.* 20 (nr.1):21–32 (1986) (Russian); English transl. in *Funct. Anal. & Appl.* 20 (1986)

30. M. V. Karasev, Quantization and coherent states over lagrangian submanifolds. *Russ. J. Math. Phys.* 3:393-400 (1995).

INTEGRAL REPRESENTATION OF EIGENFUNCTIONS AND COHERENT STATES FOR THE ZEEMAN EFFECT

Mikhail Karasev and Elena Novikova*

Department of Applied Mathematics
Moscow Institute of Electronics & Mathematics
B.Vuzovsky per., 3/12, Moscow 109028, Russia

Abstract

Coherent states for the hydrogen atom in a magnetic field are constructed via the Bessel functions and Laguerre polynomials. Global integral representations for exact and semiclassical eigenfunctions are obtained using coherent states.

1. INTRODUCTION

We are interested in the following general problem: how to construct eigenfunctions (and eigenvalues) of a quantum system, using information about corresponding classical Hamiltonian system? We study the following integral Ansatz for quantum wave functions

$$\Psi = \int_{\mathcal{L}} \varphi(\beta)\chi_\beta \, d\beta. \tag{1.1}$$

Here $\{\chi_\beta\}$ is a family of vectors in the Hilbert space of a given quantum problem . This family is parametrized by points β running along a classical invariant submanifold $\mathcal{L}$ in the phase space of the problem. The amplitude φ under the integral sign is a certain new wave function. So, we try to transport our quantum problem from the initial Hilbert space to a space of functions over the classical submanifold $\mathcal{L}$. If we choose $\mathcal{L}$ and the family χ in an appropriate way, then the new quantum equation for the amplitude φ might be global over $\mathcal{L}$ and simpler than the initial problem. For instance, we can easily solve this equation in a semiclassical approximation.

Of course, this is only a general idea. For details, applications and a geometrical interpretation of the family χ and the equation for the amplitude φ, we refer to a series of works[1-3] begun in 1989 (for further developments, see Refs. 4-7).

Concerning the relations to the geometric quantization procedure, to the Bargmann representation and the coherent state theory, see Refs. 7-10.

In particular, there is an interesting case when the submanifold $\mathcal{L}$ is a Liouville torus of an integrable Hamiltonian system. In this case, there is a basis of first integrals

*This work was supported in part by the International Science Foundation and the Russian Foundation for Fundamental Research.

of motion, whose flows preserve the torus $\mathcal{L}$. Thus, we could try to construct the family χ to be invariant with respect to these flows. Then the equation for the amplitude φ will be just a first order differential equation.

It is possible and very easy to construct such an invariant family χ, provided there is a polarization over $\mathcal{L}$, transversal to $\mathcal{L}$ and preserved by the flows. Unfortunately, such an invariant polarization does not exist in general. At this point we face an important and difficult problem of quantization by means of non-invariant polarization. This problem is well known in the theory of geometric quantization.[11,12] How to construct quantization if we live in a space equipped with a non-invariant polarization?

In this paper we consider only an example: a hydrogen atom in an external magnetic field. In this example, a non-invariant polarization appears many times: real "vertical" polarization on the initial 6-dimensional phase space, Kählerian polarization on a 4-dimensional phase space after the first reduction by the angular momentum, and Kählerian polarization on a compact 2-dimensional space after the second reduction by the hydrogen hamiltonian. This short communication contains only a small part of our results concerning the coherent state quantization for Zeeman effect; see also Refs. 13,14 and forthcoming papers.

2. BESSEL COHERENT STATES

Let us consider the Schrödinger equation for a hydrogen atom in a homogeneous magnetic field:

$$\left[\left(i\frac{\partial}{\partial x}+\mathcal{A}\right)^2-\frac{1}{|x|}\right]\Psi=E\Psi. \tag{2.1}$$

Here $\mathcal{A}=\frac{\varepsilon}{2}(-x_2,x_1,0)$ is the magnetic vector-potential, the number ε is related to the strength $|\mathcal{H}|$ of the field by the following formula: $\varepsilon=\hbar^3|\mathcal{H}|/4e^3m^2c$. The problem is invariant with respect to rotations around the direction of the magnetic field (the third coordinate axis). Therefore, we consider Ψ as an eigenfunction of the third component of the angular momentum $\widehat{M}=-ix\times\partial/\partial x$, i.e.,

$$\widehat{M}_3\Psi(x)=m\Psi(x),\qquad m\in\mathbb{Z}. \tag{2.2}$$

Eigenvalues E and eigenfunctions Ψ of the problem (1.1) can be represented in the following form

$$E=-\varepsilon m-1/4\nu^2,\qquad \Psi(x)=\psi_m(q)\big|_{q=x/\nu}, \tag{2.3}$$

where the new function $\psi_m(q)$ and the new spectral parameter ν satisfy the following equation:

$$(\widehat{S}+\varepsilon^2\nu^4\widehat{B})\psi_m(q)=\nu\psi_m(q). \tag{2.4}$$

Here $\widehat{S}=|q|(-\Delta_q+1/4)$ is an "action" operator (its spectrum coincides with the set of positive integers), and the operator $\widehat{B}=|q|(q_1^2+q_2^2)/4$ can be represented in the form

$$\widehat{B}=\frac{1}{8}(\hat{b}_+ +\hat{b}_-)\cdot\hat{b}_+\cdot\hat{b}_-,\qquad \hat{b}_\pm\overset{\text{def}}{=}|q|\pm q_3.$$

The phase space $\widetilde{\mathfrak{X}}$ of the problem (2.4) after the Moser–Souriau regularization[15–17] looks like $T^*\mathbb{S}^3$. Natural complex coordinates in $\widetilde{\mathfrak{X}}$ are V_1, V_2, V_3, given by the formula

$$V=(|p|^2+1/4)q-2\langle q,p\rangle p+i|q|p,\qquad p\to -i\partial/\partial q.$$

(Note that V_j differ slightly from the Souriau coordinates[15]). Then $V_1^2+V_2^2+V_3^2=v^2$, where $v=-|q|(|p|^2-1/4)+i\langle q,p\rangle$. As shown in Ref. 18, the vectors $\{\partial/\partial\overline{V}\}$ define a Kählerian polarization Π on $\widetilde{\mathfrak{X}}=T^*\mathbb{S}^3$.

Let us consider an arbitrary lagrangian submanifold $\widetilde{\Lambda}\subset\widetilde{\mathfrak{X}}$. Denote by $\mathcal{Q}$, $\mathcal{P}$ the restriction of q, p on $\widetilde{\Lambda}$, and introduce the family $\tilde{\chi}=\{\tilde{\chi}_{\tilde{\alpha}}\mid\tilde{\alpha}\in\widetilde{\Lambda}\}$ by the formula

$$\begin{aligned}\tilde{\chi}_{\tilde{\alpha}}(q) &= \sqrt{\widetilde{J}(\tilde{\alpha})}\exp\Big\{-\int_{\tilde{\alpha}^0}^{\tilde{\alpha}}\frac{\bar{v}\,dv+\langle\overline{V},dV\rangle}{\sqrt{2(|v|^2+|V|^2)}}-v(\tilde{\alpha})\Big\}\\ &\quad\times\exp\{-|q|/2\}\cdot I_0\big(\sqrt{2(|q|v(\tilde{\alpha})+\langle q,V(\tilde{\alpha})\rangle)}\big).\end{aligned}\tag{2.5}$$

Here I_0 is the Bessel function of imaginary argument: $I_0(y)=\frac{1}{2\pi}\int_0^{2\pi}\exp(y\cos\varphi)\,d\varphi$. The function $\widetilde{J}$ is the Jacobian of the projection $T\Lambda\to\overline{\Pi}$ along Π. The integral along the path $\tilde{\alpha}^0\to\tilde{\alpha}$ is independent of the choice of the path, because $\widetilde{\Lambda}$ is a lagrangian submanifold.

The family of *Bessel coherent states* (2.5) was introduced in Ref. 3 by a reduction procedure from the usual Gaussian states in the phase space $\mathbb{R}^8$. The latter space appears in the well-known Kustaanheimo–Stiefel spinor regularization[19–21] of the hydrogen atom problem. Note that Bessel coherent states (2.5) differ from the states obtained in Refs. 18 and 22.

The family $\tilde{\chi}$ (2.5) defines an intertwining homomorphism between the space of functions on $\mathbb{R}^3_q$ and the space of functions on $\widetilde{\Lambda}$, i.e., the action of any differential operator on $\tilde{\chi}_{\tilde{\alpha}}(q)$ with respect to q is equivalent to the action of a certain operator in coordinates $\tilde{\alpha}\in\widetilde{\Lambda}$. In particular, the Hamiltonian $\widehat{S}+\varepsilon^2\nu^4\widehat{B}$ (2.4) is transformed to a certain differential operator on $\widetilde{\Lambda}$. For brevity, we do not write out this operator.

Let us consider the special case when the submanifold $\widetilde{\Lambda}$ belongs to the level $\{M_3=m=\text{const}\}$ (in our problem this is natural in view of (2.2)). This level is fibrated by closed trajectories of the Hamilton field $\mathrm{ad}(M_3)$. The base of this fibration denoted by π is a certain 4-dimensional surface $\mathfrak{X}_m$.

If a function is in involution with M_3, then it is constant along the fibers of π and, hence, is a function on the base $\mathfrak{X}_m$. In particular, this is true for complex functions:

$$z_\pm=v\pm V_3.\tag{2.6}$$

We take z_+ and z_- as complex coordinates on $\mathfrak{X}_m$, more precisely, $\{\partial/\partial\bar{z}_+,\partial/\partial\bar{z}_-\}$ is a Kählerian polarization on $\mathfrak{X}_m$. Note that this polarization, as well as the above polarization Π, is not invariant with respect to any group of symmetries.

The functions S and $b_\pm$ are expressed in terms of the complex coordinates (2.6) in the following way:

$$S=\big(\sqrt{|z_+|^2+m^2}+\sqrt{|z_-|^2+m^2}\big)/2,\qquad b_\pm=z_\pm+\bar{z}_\pm+2\sqrt{|z_\pm|^2+m^2}.$$

Lemma 2.1. *The reduced symplectic structure on $\mathfrak{X}_m$ is given by the 2-form*

$$\omega=\frac{i}{4}\Big[\frac{d\bar{z}_+\wedge dz_+}{\sqrt{|z_+|^2+m^2}}+\frac{d\bar{z}_-\wedge dz_-}{\sqrt{|z_-|^2+m^2}}\Big]=d\theta,\tag{2.7}$$

where

$$\theta=\frac{i}{2}\Big[\big(\sqrt{|z_+|^2+m^2}-|m|\big)\frac{dz_+}{z_+}+\big(\sqrt{|z_-|^2+m^2}-|m|\big)\frac{dz_-}{z_-}\Big].\tag{2.8}$$

Let us consider any Lagrangian submanifold $\Lambda \subset \mathfrak{X}_m$ and take $\tilde{\Lambda} = \pi^{-1}(\Lambda)$. By definition, $\tilde{\Lambda} \subset \{M_3 = m\}$. Denote by μ the "time" along the trajectories of the Hamiltonian field $\mathrm{ad}(M_3)$. Then $\tilde{\alpha} = (\alpha_1, \alpha_2, \mu)$, where α_1, α_2, are certain local coordinates on Λ, and $d\tilde{\alpha} = d\mu\, d\alpha$, where $d\alpha$ is a measure on Λ.

Let us apply the Ansatz (1.1) using the family of Bessel coherent states $\tilde{\chi}$ (2.5) and the lagrangian submanifold $\mathcal{L} = \pi^{-1}(\Lambda)$. If we take the amplitude $\varphi = \varphi(\alpha_1, \alpha_2)$ constant along the fibers of π, then for the eigenfunctions of the problem (2.2), (2.4), we get the following expression:

$$\psi_m(q) = \int_{\tilde{\Lambda}} \varphi(\tilde{\alpha})\tilde{\chi}_{\tilde{\alpha}}(q)\, d\tilde{\alpha} = \int_{\Lambda} \varphi(\alpha)\chi_\alpha(q)\, d\alpha, \qquad \text{where} \quad \chi_\alpha(q) \stackrel{\text{def}}{=} \int_0^{2\pi} \tilde{\chi}_{\tilde{\alpha}}(q)\, d\mu. \quad (2.9)$$

The reduced coherent states χ_α are parametrized by points $\alpha \in \Lambda$. Let us write out an explicit formula for these states.

Lemma 2.2. *The reduced Bessel coherent states over a lagrangian submanifold* $\Lambda \subset \mathfrak{X}_m$ *are given by the formula*

$$\begin{aligned}\chi_\alpha(q) \;&=\; c_1\sqrt{J(\alpha)}\exp\Big\{i\int_{\alpha^0}^{\alpha}\theta - (z_+ + z_-)/2\Big\}(z_+z_-)^{-|m|/2}\Big(\frac{q_1 + iq_2}{q_1 - iq_2}\Big)^{m/2} \\ &\times \exp\{-|q|/2\}\, I_m(\sqrt{(|q| + q_3)z_+})I_m(\sqrt{(|q| - q_3)z_-}). \end{aligned} \quad (2.10)$$

Here $I_m(y) = \frac{1}{2\pi}\int_0^{2\pi} \exp(im\varphi + y\cos\varphi)\, d\varphi$ *is the Bessel function of imaginary argument,* c_1 *is a constant,* $z_\pm = z_\pm(\alpha)$ *are complex coordinates* (2.6) *restricted on the submanifold* Λ*, the form* θ (2.8) *is integrated along a path on this submanifold, the Jacobian is given by* $J = \mathcal{D}(z_+, z_-)/\mathcal{D}(\alpha_1, \alpha_2)$.

Now assume that the following quantization rule

$$\frac{1}{2\pi}\int_\Sigma \omega - \frac{1}{4}\mu(\Sigma) \in \mathbb{Z}, \qquad \forall \Sigma, \partial\Sigma \subset \Lambda \quad (2.11)$$

holds on Λ. Here $\Sigma \subset \mathfrak{X}_m$ is an arbitrary 2-dimensional surface (membrane) with boundary $\partial\Sigma \subset \Lambda$. The number $\mu(\Sigma) \in \mathbb{Z}$ is an index of the membrane Σ (for details, see Ref. 10,23).

Under the condition (2.11), the family χ_α (2.10) is a global smooth function in $\alpha \in \Lambda$. So, after the substitution of (2.9) into the equation (2.4), we can transfer all operators acting in q onto the function $\varphi(\alpha)$:

$$\Big[\breve{S} + \frac{\varepsilon^2\nu^4}{8}(\breve{b}_+ + \breve{b}_-)\cdot \breve{b}_+ \cdot \breve{b}_-\Big]\varphi = \nu\varphi. \quad (2.12)$$

Here $\breve{S} = S|_\Lambda - (D[z_+; z_-] - D[z_-; z_+])$,

$$\breve{b}_\pm = b_\pm|_\Lambda \pm 4D[z_\pm + \sqrt{|z_\pm|^2 + m^2}] + 2(D[z_\mp; 1]\cdot D[z_\mp; z_\pm] + D[z_\mp; z_\pm]\cdot D[z_\mp; 1]);$$

the first order differential operators $D[f, g]$ for any $f, g \in C^\infty(\Lambda)$ are defined by the formula:

$$D[f, g] = \frac{g}{J}\Big[\frac{\partial f}{\partial\alpha_1}\frac{\partial}{\partial\alpha_2} - \frac{\partial f}{\partial\alpha_2}\frac{\partial}{\partial\alpha_1}\Big] + \frac{1}{2}\Big[\frac{\partial f}{\partial\alpha_1}\frac{\partial}{\partial\alpha_2}\Big(\frac{g}{J}\Big) - \frac{\partial f}{\partial\alpha_2}\frac{\partial}{\partial\alpha_1}\Big(\frac{g}{J}\Big)\Big].$$

So we have the following statement :

Theorem 2.1. *Let* $\Lambda = \{z_+ = z_+(\alpha_1,\alpha_2),\ z_- = z_-(\alpha_1,\alpha_2)\}$ *be a lagrangian submanifold in the 4-dimensional reduced space* $\mathfrak{X}_m$ *with the symplectic structure* ω (2.7). *Assume that the quantization rule* (2.11) *holds and the function* $\varphi(\alpha_1,\alpha_2)$ *satisfies equation* (2.12) *on* Λ. *Then the function*

$$\Psi(x) = \int_\Lambda \varphi(\alpha)\chi_\alpha\Big(\frac{x}{\nu}\Big)\,d\alpha$$

is an eigenfunction of the problem (2.1) *corresponding to the eigenvalue* (2.3).

Thus, instead of the initial Schrödinger equation for the hydrogen atom in a magnetic field, we obtain a new equation on an arbitrary 2-dimensional lagrangian submanifold $\Lambda \subset \mathfrak{X}_m$.

3. LAGUERRE COHERENT STATES

Now consider an important particular case when the magnetic field (the parameter ε) is sufficiently small, i.e., the second summand in the operator $\widehat{S} + \varepsilon^2\nu^4\widehat{B}$ can be considered as a perturbation. After the quantum averaging[24] $\widehat{B}$ is replaced by an operator commuting with $\widehat{S}$:

$$\widehat{S} + \varepsilon^2\nu^4\widehat{B} \quad \rightarrow \quad \widehat{S}(1 + \varepsilon^2\nu^4\widehat{C} + O(\varepsilon^2)),$$

where

$$[\widehat{C},\widehat{S}] = 0, \qquad \widehat{C} = \widehat{S}^2 + \widehat{M}_3^2 + 4\widehat{A}^2 - 5\widehat{A}_3^2 + 3,$$
$$\widehat{A} = q(-\Delta_q + 1/4) + \partial/\partial q \times (q \times \partial/\partial q) - (q \times \partial/\partial q) \times \partial/\partial q.$$

Note that $\widehat{A}$ is a "modified" Runge–Lenz vector, to obtain the usual Runge–Lenz vector, the first summand must be replaced by $q/|q|$.

The initial eigenvalues E take the form:

$$E = -1/4n^2 - \varepsilon m + \varepsilon^2 n^2\lambda/2 + O(\varepsilon^4 n^{10}), \tag{3.1}$$

where $n \in \mathbb{Z}_+$ are the eigenvalues of $\widehat{S}$, and λ are the eigenvalues of the hamiltonian $\widehat{C}$. So, our problem is reduced $\bmod O(\varepsilon^4)$ to the spectral problem for the operator $\widehat{C}$. (Compare with the results of Solov'ev[25] and with Ref. 26; our operator $\widehat{C}$ is slightly different from that of Refs. 25,26).

The function C on $\widetilde{\mathfrak{X}}$ is in involution with M_3 and S, so, in fact, C is a function on the base of the fibration of $\widetilde{\mathfrak{X}}$ by tori = {trajectory of M_3} × {trajectory of S}. Thus, it is natural to consider the second reduction ρ by trajectories of S:

$$\widetilde{\mathfrak{X}} \supset \{M_3 = m\} \xrightarrow{\pi} \mathfrak{X}_m \supset \{S = n\} \xrightarrow{\rho} \Omega_{m,n}.$$

The base $\Omega_{m,n}$ of the fibration ρ is diffeomorphic to the sphere if $m \neq 0$, or to the cylinder if $m = 0$.

Let us consider the energy levels $\Lambda^\# = \{C = \text{const}\}$ on $\Omega_{m,n}$ and take $\Lambda = \rho^{-1}(\Lambda^\#)$, $\varphi = \rho^*\varphi^\#$, where $\varphi^\# \in C^\infty(\Lambda^\#)$. Then $\alpha = (\tau, t)$, $d\alpha = d\tau\,dt$, where $\tau \in [0, 2\pi]$ is the "time" along S-trajectories, $t \in [0, T]$ is the time along C-trajectories $\Lambda^\#$.

Of course, to satisfy the rule (2.11) on Λ, we must suppose here that the quantization rule holds on $\Lambda^\#$, i.e.,

$$\frac{1}{2\pi}\int_{\Sigma^\#} \omega^\# = k + \frac{1}{2}, \qquad k \in \mathbb{Z}_+,$$

where $\Sigma^{\#}$ is the positively oriented membrane on $\Omega_{m,n}$ with boundary $\partial\Sigma^{\#} = \Lambda^{\#}$, and $\omega^{\#}$ is the reduced symplectic form.

Lemma 3.1. *A symplectic form on* $\Omega_{m,n}$ *is given by*

$$\omega^{\#} = \partial_{\bar{w}}\theta^{\#}, \qquad \theta^{\#} \stackrel{\text{def}}{=} i\varkappa(|w|^2)\bar{w}\,dw,$$

where

$$\varkappa(r) = \frac{(2n-|m|)r + |m| - \sqrt{4n^2 r + m^2(r-1)^2}}{2r(r-1)}$$

and $w = -(z_+/z_-)|_{S=n}$ *is a complex coordinate on* $\Omega_{m,n}$ *(so that* $\{\partial/\partial\bar{w}\}$ *determines a Kählerian polarization on* $\Omega_{m,n}$*). In addition we have*

$$\frac{1}{2\pi}\int\limits_{\Omega_{m,n}} \omega^{\#} = n - |m|.$$

The Hamiltonian C can be expressed in terms of the complex coordinate w:

$$C = 6n^2 - 4m^2 + 3 - 2n\frac{(2|w|^2\varkappa(|w|^2) - n + |m|)(2(w+\bar{w}) + 3(|w|^2+1))}{|w|^2 - 1}.$$

Recall that $\Lambda^{\#} = \{w = w(t) \mid w(t)$ is a C-trajectory$\}$. From (2.9) we get the following representation for the eigenfunctions:

$$\psi_m = \int\limits_{\Lambda^{\#}} \varphi^{\#}(t)\chi_t^{\#}\,dt, \qquad \text{where} \quad \chi_t^{\#} \stackrel{\text{def}}{=} \int\limits_0^{2\pi} \chi_\alpha\,d\tau. \tag{3.2}$$

Lemma 3.2. *One has*

$$\chi_t^{\#}(q) = c_2\sqrt{\bar{w}}\exp\left\{i\int\limits_{t^0}^{t}\theta^{\#}\right\}\exp\{-|q|/2\}(q_1 + i\,\mathrm{sgn}(m)q_2)^{|m|}\sum_{j=0}^{n-|m|-1} w(t)^j G_j(q),$$

Here

$$G_j(q) = g_j \cdot L_j^{|m|}\left(\frac{|q|+q_3}{2}\right)L_{n-|m|-1-j}^{|m|}\left(\frac{|q|-q_3}{2}\right), \quad g_j = \frac{(-1)^j}{(j+|m|)!(n-1-j)!},$$

where L_N^M *are the Laguerre polynomials,* c_2 *is a constant.*

After the second reduction, the coherent states $\chi_t^{\#}$ intertwine operators acting in q with operators acting in the time t of the averaged Hamiltonian C. The spectral problem (2.4) is reduced to the problem over the trajectory $\Lambda^{\#}$:

$$\check{C}\varphi^{\#} = \lambda\varphi^{\#}, \tag{3.3}$$

where

$$\begin{aligned} \check{C} = \ & C_0 - i\frac{\partial}{\partial t} - 2(D[w^2+3w+1]\cdot D[w] + D[w]\cdot D[w^2+3w+1]), \\ & C_0 = C|_{\Lambda^{\#}} - 2. \end{aligned} \tag{3.4}$$

Here by $D[f]$ we denote a differential operator

$$D[f] = \frac{f}{\dot{w}}\frac{\partial}{\partial t} + \frac{1}{2}\frac{\partial}{\partial t}\left(\frac{f}{\dot{w}}\right).$$

So, we reduce the problem (2.2), (2.4) with accuracy $O(\varepsilon^4)$ to the new quantum equation (3.3) on the closed curve $\Lambda^\# \subset \Omega_{m,n}$.

Remark. The operator $\breve{C}$ in the problem (3.3) is a second order differential operator with coefficients periodic in t. Its exact eigenfunctions $\varphi^\#$ are related to the well-known Heun polynomials.

If one notes that in (3.4) $C_0 \sim n^2$, $t \sim n^{-1}$, it will be very easy to get global semiclassical approximation of the eigenfunctions as $n \to \infty$. The crucial point here is that in the first approximation the summand $-i\partial/\partial t$ plays the leading role and the second order differential operators in (3.4) are the lower terms. So, we get the following statement:

In semiclassical approximation and in the first approximation with respect to the magnetic field, the eigenfunctions of the Schrödinger operator (2.1) *have the form* (3.2), *where the amplitude*

$$\varphi^\# = \varphi_k^\# = \exp\left\{i\frac{2\pi k}{T}t\right\}.$$

These eigenfunctions correspond to the eigenvalues (3.1), *where*

$$\lambda = \lambda_k = C_0 + \frac{2\pi k}{T}.$$

This is the global asymptotic behavior of the eigenfunctions for the hydrogen atom in a magnetic field via the Laguerre coherent states. This result differs certainly from the well-known approaches of the theory of semiclassical approximation. Note that the asymptotic behavior of the eigenfunctions can almost never be obtained by standard methods of WKB type.

So we have realized our main idea: to use the integral representation (1.1) with an appropriate family χ, and to reduce the problem to an elementary ordinary differential equation for the amplitude φ.

References

1. M. V. Karasev, Connections over lagrangian submanifolds and certain problems of semi-classical approximation, *Zap. Nauchn. Sem. Leningrad Otdel. Mat. Inst. Steklov* (LOMI) 172:41–54 (1989) (Russian); English transl. in *J. Sov. Math.* 59:1053–1062 (1992)

2. M. V. Karasev, New global asymptotics and anomalies for the problem of quantization of the adiabatic invariant, *Functional Anal. Appl.* 24:104–114 (1990)

3. M. V. Karasev, Simple Quantization formula, *in*: "Symplectic Geom. and Math. Phys.", Actes du colloque en l'honneur de J.-M. Souriau, Birkhäuser, Boston 234–244 (1991)

4. M. V. Karasev and Yu. M. Vorobjev, Integral representations over isotropic submanifolds and equations of zero curvature, *Preprint* No. AMath-QDS-92-01, Moscow Inst. Electron.& Math., 56p., Moscow (1992)

5. M. V. Karasev and Yu. M. Vorobjev, Connection and excited wavepackets over invariant isotropic torus, *in*: "Quantization and Coherent State Methods, Proc. of XI-th Workshop on Geometrical Methods in Physics, Bialowieza, Poland, July 1992", pp. 179–189; S. T. Ali, I. M. Mladenov, A. Odzijewicz (eds.), World Scientific (1993)

6. M. V. Karasev and M. B. Kozlov, Quantum and semiclassical representations over lagrangian submanifolds in $su(2)^*$, $so(4)^*$, $su(1,1)^*$, *J. Math. Phys.* 34:4986–5006 (1993)

7. M. V. Karasev and M. B. Kozlov, Representation of compact semisimple Lie algebras over lagrangian submanifolds, *Funktsional. Anal. i Prilozhen.* 28 (nr.4):16–27 (1994)

8. M. V. Karasev, Integrals over membranes, transition amplitudes and quantization, *Russ. J. Math. Phys.* 1:523–526 (1993)

9. M. V. Karasev, Quantization by Membranes. Integral representations for wave-functions, *in*: "Quantization and Infinite-Dimensional Systems, Proc. XII-th Coll. Geom. Methods in Physics, Białowieza, July 1993", pp. 9–19; J.-P. Antoine, S. T. Ali, W. Lisiecki, I. M. Mladenov, A. Odzijewicz (eds.), Plenum (1994)

10. M. V. Karasev, Quantization by means of two-dimensional surfaces (membranes). Geometrical formulas for wave-functions, *Contemp. Math.* (1994) (to appear)

11. R. Blattner, The metalinear geometry of nonreal polarizations, *Lect. Notes Math.* 570, Springer-Verlag, Berlin and New York (1977)

12. J. H. Rawnsley, A nonunitary pairing of polarizations for the Kepler problem, *Trans. Amer. Math. Soc.* 250:167–180 (1979)

13. M. V. Karasev and E. M. Novikova, Quadratic Poisson brackets for the Zeeman effect. Irreducible representations and coherent states, *Uspechi Mat. Nauk* 49 (nr.5):169–170 (1994) (Russian); English transl. in *Russian Math. Surveys*

14. M. V. Karasev and E. M. Novikova, Representation of exact and semiclassical eigenfunctions via coherent states. Hydrogen atom in a magnetic field, (to appear)

15. J.-M. Souriau, Sur la varieté de Kepler, *in*: Symposia Math. XIV, Academic Press, New York (1974)

16. J.-M. Souriau, "Structure des Systèmes Dynamiques", Dunod, Paris (1970)

17. J. Moser, Regularization of Kepler's problem and the averaging method on a manifold, *Comm. Pure Appl. Math.* 23:609–636 (1970)

18. J. H. Rawnsley, Coherent states and Kähler manifolds, *Quart. J. Math., Oxford* 28:403–415 (1977)

19. P. Kustaanheimo, Spinor regularization of Kepler motion, *Ann. Univ. Turkuensis A* 73:3–7 (1964)

20. P. Kustaanheimo and E. Stiefel, *J. Reine and Angew. Math.* 218:204–219 (1965)

21. Yu. P. Stepanovsky, Hydrogen atom in an external field as anharmonic oscillator, *Ukraine Phys. J.*, 32:1316–1321 (1987)

22. M. Horowski and A. Odzijewicz, Geometry of the Kepler system in coherent states approach, *Preprint* Inst. of Physics, Warsaw Univ. Division Bialystok (1993)

23. M. V. Karasev, Poisson algebras of symmetries and asymptotics of spectral series, *Funct. Anal. and Appl.* 20 (nr.1):21-32 (1986)

24. M. V. Karasev and V. P. Maslov, Asymptotic and geometric quantization, *Russian Math. Surveys* 39 (nr.6):133-205 (1984)

25. E. A. Solov'ev, Hydrogen atom in a weak magnetic field, *Zh. Eksper. Teoret. Fiz.* 82:1762–1771 (1982)

26. D. R. Herrick, Symmetry of the Zeeman effect for hydrogen, *Physical Review A* 26:323–329 (1982).

PART III

Q-DEFORMATIONS AND QUANTUM GROUPS, NONCOMMUTATIVE GEOMETRY

QUANTUM COHERENT STATES AND THE METHOD OF ORBITS

Branislav Jurčo[1] and Pavel Šťovíček[2]

[1] CERN, Theory Division
CH-1211 Geneva 23
Switzerland

[2] Department of Mathematics and Doppler Institute
Faculty of Nuclear Science, CTU
Trojanova 13, 120 00 Prague
Czech Republic

Abstract

A general coadjoint orbit of a compact group is quantized by introducing quantum "holomorphic" coordinate functions $\{z_{jk}\}$ on the big cell. Quantum coherent states are defined in a way quite parallel to the classical approach of Perelomov. Any irreducible representation of the deformed enveloping algebra is shown to act in a vector space of polynomials in non-commutative variables $\{z^*_{jk}\}$ according to a simple rule.

1. INTRODUCTION

This note is devoted to compact quantum groups. The goal is to show that, as in the non-deformed case, every irreducible representation τ_λ, with λ a lowest weight, admits an antiholomorphic realization. So this contribution may be considered as an attempt to find a counterpart to the classical Borel-Weil theory. A similar problem for solvable groups has been considered earlier.[1,2]

The construction we are going to describe is done in two steps. First, one has to quantize, as a complex manifold, the coset space of the corresponding group. Here we solve this problem by introducing quantum "holomorphic" coordinate functions $\{z_{jk}\}$ on the big cell. Second, the representation τ_λ is shown to act in a vector space formed by polynomials in non-commutative variables $\{z^*_{jk}\}$, according to a simple rule. In connection with the first step, recently efforts were made to describe deformations of manifolds, particularly to quantize flag and Grassmann manifolds through different approaches.[3–7] We note that our approach encompasses all types of coadjoint orbits. In the second part, the crucial point is the definition of the quantum coherent state which is suggested in a way quite parallel to the classical approach of Perelomov.[8] This may extend in an essential way the range of applications, since the earlier definitions were restricted mostly to the simplest rank-one cases.[9]

This paper highlights and comments on some of the basic steps and results, while the complete proofs will appear elsewhere.[10]

Quantization, Coherent States, and Complex Structures
Edited by J.-P. Antoine *et al.*, Plenum Press, New York, 1995

2. PRELIMINARIES

The deformation parameter is chosen as $q = e^{-h}$, $h > 0$. Let G be a simple and simply connected complex group from one of the principal series A_l, B_l, C_l and D_l, $K \subset G$ its compact form; $G = K \cdot AN$ is the Iwasawa decomposition. The corresponding Lie algebras are denoted by small bold face letters. The symbol $\mathcal{U}_h(\cdot)$ stands for a deformed enveloping algebra and $\mathcal{A}_q(\cdot)$ stands for a Hopf algebra of quantum functions. As usual, we denote the coproduct, the counit and the antipode by Δ, ε and S, respectively. The algebra $\mathcal{U}_h(\mathbf{g})$ is generated by the elements H_i and $X_i^{\pm}$, $i = 1, \ldots, l = \text{rank}$. Let T, U and Λ denote the vector corepresentations of the quantum groups G_q, K_q and AN_q, respectively. Let us remark that Λ is upper triangular with positive elements on the diagonal. Throughout the paper we use the standard R-matrix known for each of the four principal series. All basic definitions as well as the explicit form of R can be found in the seminal papers[11–13] and we do not recall them.

Nevertheless we wish to point out that $\mathcal{A}_q(G)$ is the same Hopf algebra as $\mathcal{A}_q(K)$ but the compact form is equipped, in addition, with the $*$-involution defined by $U^* = U^{-1}$. A similar relationship holds between $\mathcal{U}_h(\mathbf{g})$ and $\mathcal{U}_h(\mathbf{k})$. For the series B_l, C_l and D_l, in addition to the relation $RX_1X_2 = X_2X_1R$, with X standing for T or U or Λ, an orthogonality condition is also required, namely $CX^tC^{-1} = X^{-1}$, with C being a c-number matrix.[13] The quantized solvable group $\mathcal{A}_q(AN)$ is also a $*$-algebra with the commutation rule

$$\Lambda_1{}^* R^{-1} \Lambda_2 = \Lambda_2 R^{-1} \Lambda_1{}^*. \tag{2.1}$$

Quite important is the concept of duality; the $*$-algebras $\mathcal{A}_q(K)$ and $\mathcal{U}_h(\mathbf{k})$ are dually paired. A crucial fact exploited in the construction is that $\mathcal{A}_q(AN)$ is identical to $\mathcal{U}_h(\mathbf{k})$ as an algebra and opposite as a coalgebra. This identification plays the role of the classical momentum map. The pairing between $\mathcal{A}_q(K)$ and $\mathcal{A}_q(AN)$ follows unambiguously from

$$\langle \Lambda_1; U_2 \rangle = R_{21}^{-1}, \ \langle \Lambda_1{}^*; U_2 \rangle = R_{12}^{-1}. \tag{2.2}$$

Because of the duality one can introduce the so-called canonical element

$$\rho := \sum x_s \otimes a_s \in \mathcal{A}_q(AN) \otimes \mathcal{A}_q(K), \tag{2.3}$$

with $\{x_s\}$ and $\{a_s\}$ being mutually dual bases. We list its basic properties:

$$\rho^* = \rho^{-1} = (\mathrm{id} \otimes S)\rho, \tag{2.4}$$

$$(\Delta \otimes \mathrm{id})\rho = \rho_{23}\rho_{13}, \quad (\mathrm{id} \otimes \Delta)\rho = \rho_{12}\rho_{13}. \tag{2.5}$$

Using ρ one defines the (right) dressing transformation as a coaction

$$\mathcal{R} : \mathcal{A}_q(AN) \to \mathcal{A}_q(AN) \otimes \mathcal{A}_q(K) : u \mapsto \rho(u \otimes 1)\rho^{-1}. \tag{2.6}$$

It holds true that $\mathcal{R}(\Lambda^*\Lambda) = U^*\Lambda^*\Lambda U$, with evident embeddings of $\mathcal{A}_q(AN)$ and $\mathcal{A}_q(K)$ into $\mathcal{A}_q(AN) \otimes \mathcal{A}_q(K)$.

The basic fact from the representation theory says that the irreducible finite-dimensional $*$-representations of $\mathcal{U}_h(\mathbf{k})$ are labeled by lowest weights λ in the same way and with the same dimensions as in the classical case.[14,15] To each representation τ_λ in $\mathcal{H}_\lambda$ there corresponds a unitary corepresentation $\mathcal{T}^\lambda$ of $\mathcal{A}_q(K)$,

$$\mathcal{T}^\lambda := (\tau_\lambda \otimes \mathrm{id})\rho \in \mathrm{Lin}(\mathcal{H}_\lambda) \otimes \mathcal{A}_q(K), \tag{2.7}$$

and vice versa, $\tau_\lambda(x) = (\mathrm{id} \otimes \langle x, \cdot \rangle)\mathcal{T}^\lambda$, $\forall x \in \mathcal{U}_h(\mathbf{k})$. The symbol e_λ will stand for the normalized lowest-weight vector.

3. COSET SPACES QUANTIZED AS COMPLEX MANIFOLDS

Fix a subset Π_0 of the set of simple roots $\Pi = \{\alpha_1, \ldots, \alpha_l\}$. Denote by $\mathbf{k}_0$ the subalgebra (real) of $\mathbf{k}$ generated by H_i, $\forall i$, and by those $X_i^\pm$ such that $\alpha_i \in \Pi_0$. Similarly, $\mathbf{p}_0$ is a subalgebra (complex) of $\mathbf{g}$ generated by H_i, X_i^-, $\forall i$, and by those X_i^+ for which $\alpha_i \in \Pi_0$. K_0 and P_0 stand for the corresponding subgroups. Thus we have mutually paired Hopf algebras $\mathcal{U}_h(\mathbf{p}_0)$ and $\mathcal{A}_q(P_0)$ and $*$-Hopf algebras $\mathcal{U}_h(\mathbf{k}_0)$ and $\mathcal{A}_q(K_0)$. All coadjoint orbits of K are coset spaces of the type $\mathcal{O} = K_0\backslash K = P_0\backslash G$. So we deal with complex manifolds.

Let U_0 be the vector corepresentation of $\mathcal{A}_q(K_0)$ obeying $RU_{01}U_{02} = U_{02}U_{01}R$, $U_0^* = U_0^{-1}$, and possibly also $U_0^{-1} = CU_0^tC^{-1}$. U_0 is block diagonal, with the blocks corresponding to connected components of the Dynkin subdiagram supported on the roots from Π_0. To the roots from $\Pi - \Pi_0$ there correspond diagonal elements (1×1 blocks). The $*$-algebra morphism (the restriction morphism) $p_0 : \mathcal{A}_q(K) \to \mathcal{A}_q(K_0)$ is determined by $p_0(U) = U_0$. The algebra $\mathcal{A}_q(K_0\backslash K)$ of quantum functions on the coset space is a subalgebra in $\mathcal{A}_q(K)$ formed by K_0-invariant functions; $f \in \mathcal{A}_q(K)$ belongs to $\mathcal{A}_q(K_0\backslash K)$ iff

$$(p_0 \otimes \mathrm{id})\, \Delta f = 1 \otimes f\,. \tag{3.1}$$

Similarly one defines $\mathcal{A}_q(P_0\backslash G)$. Owing to the identification $\mathcal{A}_q(G) = \mathcal{A}_q(K)$ (without the $*$- structure) one has an embedding of $\mathcal{A}_q(P_0\backslash G)$ into $\mathcal{A}_q(K_0\backslash K)$. $\mathcal{A}_q(P_0\backslash G)$ is interpreted as the algebra of holomorphic functions on the coset space.

Denote by $\tilde{p}_0 : \mathcal{A}_q(G) \to \mathcal{A}_q(P_0)$ the restriction morphism. Decompose the vector corepresentation of $\mathcal{A}_q(G)$ as

$$T = \Lambda_{(-)} Z\,, \tag{3.2}$$

with $\Lambda_{(-)}$ being block lower triangular and Z block upper triangular and the blocks on the diagonal of Z are unit matrices. From the equalities

$$\begin{aligned}(\tilde{p}_0 \otimes \mathrm{id})\, \Delta T &= (\tilde{p}_0(T)\dot{\otimes}\Lambda_{(-)}) \cdot (I\dot{\otimes}Z), \quad \text{and} \\ (\tilde{p}_0 \otimes \mathrm{id})\, \Delta T &= (\tilde{p}_0 \otimes \mathrm{id})\, \Delta\Lambda_{(-)} \cdot (\tilde{p}_0 \otimes \mathrm{id})\, \Delta Z\,,\end{aligned}$$

one deduces that

$$(\tilde{p}_0 \otimes \mathrm{id})\, \Delta Z = I\dot{\otimes}Z\,. \tag{3.3}$$

The entries of the matrix Z are interpreted as quantum holomorphic coordinate functions living on the big cell of the coset space.

To derive the commutation relations for Z we will use the quantum Gauss decomposition. Denote by $\mathbf{b}_\pm$ the Borel subalgebras of $\mathbf{g}$ generated by H_i and X_i^+ respectively by H_i and X_i^-. The vector corepresentation $L^{(\pm)}$ of the quantum group $\mathcal{A}_q(B_\pm)$ is an upper (lower) triangular matrix. Again, $\mathcal{A}_q(B_\pm)$ is determined by the RLL-equation and, possibly, by the orthogonality condition. Furthermore, let $\mathcal{A}_q(H)$ denote the algebra of functions living on the maximal Abelian subgroup of G and let J be the corresponding vector corepresentation. In fact, this deformation is trivial and $\mathcal{A}_q(H)$ is a commutative algebra. The Gauss decomposition in this context means the isomorphism of algebras,

$$\mathcal{A}_q(B_-) \otimes \mathcal{A}_q(B_+) \simeq \mathcal{A}_q(G) \otimes_{\mathbf{twist}} \mathcal{A}_q(H)\,, \tag{3.4}$$

determined by the relations

$$T \equiv T \otimes 1 = L^{(-)}\dot{\otimes}L^{(+)}, \; J \equiv 1 \otimes J = (\mathrm{diag}\, L^{(-)})^{-1}\mathrm{diag}\, L^{(+)}\,. \tag{3.5}$$

The multiplication in the tensor product on the RHS of (3.4) is twisted according to

$$\mathrm{diag}(R)\, T_1 J_2 = J_2 T_1 \mathrm{diag}(R)\,. \tag{3.6}$$

In what follows we shall write simply L instead of $L^{(+)}$. To proceed further we will use the fact that all weights of the vector representation of G have multiplicity 1; every weight either belongs to the Weyl group orbit of the fundamental weight or is zero (only for the series B_l). Consequently one can choose a basis with vectors labeled by the weights. We use the standard ordering on the weight lattice: $\sigma > \nu$ iff $\sigma \neq \nu$ and $\sigma - \nu = \sum m_i \alpha_i$, with $m_i \in \mathbf{Z}_+$ $(0 \in \mathbf{Z}_+)$. But notice that for the vector representation, the ordering of weights is reversed as compared with the standard enumeration of basis vectors. In particular, $L_{\sigma\nu} = 0$ whenever $\sigma < \nu$. Set

$$\mathcal{W}_0 = \bigoplus_{\alpha_i \in \Pi_0} \mathbf{Z}_+ \alpha_i\,. \tag{3.7}$$

If $\Pi_0 = \emptyset$ (this is the generic case), we set $\mathcal{W}_0 = \{0\}$. Next we define matrices A and Q by putting to zero some entries of L and R, respectively:

$$A_{\sigma\nu} := \begin{cases} L_{\sigma\nu}, & \text{if } \sigma - \nu \in \mathcal{W}_0, \\ 0, & \text{otherwise}; \end{cases} \tag{3.8}$$

$$Q_{\sigma\tau,\mu\nu} := \begin{cases} R_{\sigma\tau,\mu\nu}, & \text{if } \tau - \nu = \mu - \sigma \in \mathcal{W}_0, \\ 0, & \text{otherwise}. \end{cases} \tag{3.9}$$

By comparing (3.2) and (3.5) one finds that

$$Z = A^{-1} L\,. \tag{3.10}$$

We don't present the derivation of the desired commutation relations for Z (see Ref.10), however we note that it relies heavily on the known commutation relations for L and on the following property of the R-matrix: $R_{\sigma\tau,\mu\nu} \neq 0$ implies $\sigma - \mu = \nu - \tau$, $\sigma \leq \mu$, $\tau \geq \nu$, and one of the following three possibilities occurs:
(i) $\sigma = \mu$, $\tau = \nu$, (ii) $\sigma = \nu < \tau = \mu$, (iii) $\sigma = -\tau < \mu = -\nu$.

Proposition 3.1. *The matrix Z obeys the commutation relations*

$$R\, Q_{12}^{-1} Z_1 Q_{12} Z_2 = Q_{21}^{-1} Z_2 Q_{21} Z_1 R\,. \tag{3.11}$$

In the case of series B_l, C_l and D_l, the matrix Z fulfills also

$$\mathcal{K}_{12} = Z_1 Q_{12} C_1^{-1} Z_1^t C_1 Q_{12}^{-1} \mathcal{K}_{12}\,, \tag{3.12}$$

with $\mathcal{K}_{\sigma\tau,\mu\nu} = C^t{}_{\sigma\tau} C^{-1}{}_{\mu\nu}$.

We shall denote by $\mathcal{C}$ the algebra generated by entries of Z and by $\mathcal{C}^*$ the algebra generated by entries of Z^* (and determined by the relations adjoint to (3.11), (3.12)).

4. COHERENT STATES IN THE PERELOMOV SENSE

To a lowest weight λ we relate a subset $\Pi_0 \subset \Pi$ according to the prescription:

$$\alpha_i \in \Pi_0 \quad \text{iff} \quad \tau_\lambda(X_i^+)\, e_\lambda = 0\,. \tag{4.1}$$

Let us introduce the "vacuum" functional on $\mathcal{U}_h(\mathbf{k})$, $\langle\ \rangle : \mathcal{U}_h(\mathbf{k}) \to \mathbf{C}$, by

$$\langle x \rangle := \langle e_\lambda, \tau_\lambda(x) e_\lambda \rangle\,. \tag{4.2}$$

Furthermore, we define an element $w_\lambda \in \mathcal{A}_q(K)$ by

$$w_\lambda := (\langle e_\lambda, (\cdot) e_\lambda \rangle \otimes \mathrm{id}) \mathcal{T}^\lambda\,, \tag{4.3}$$

or, less formally, $w_\lambda := \langle e_\lambda, \mathcal{T}^\lambda e_\lambda \rangle$.

Proposition 4.1. *The vacuum functional, when viewed as an element of* $\mathcal{A}_q(K) = \mathcal{U}_h(\mathbf{k})^*$, *is equal to* w_λ, *i.e.,*

$$\langle x \rangle = \langle x, w_\lambda \rangle, \quad \text{for all } x \in \mathcal{U}_h(\mathbf{k}). \tag{4.4}$$

It holds true that

$$S\, w_\lambda = w_\lambda{}^*, \quad \varepsilon(w_\lambda) = 1\,, \tag{4.5}$$

and, concerning the dependence on λ, *we have*

$$w_{\lambda_1+\lambda_2} = w_{\lambda_1} w_{\lambda_2} = w_{\lambda_2} w_{\lambda_1}\,. \tag{4.6}$$

Consequently, it is sufficient to determine w_λ *only for the fundamental weights.*

Furthermore, the restriction to K_{0q},

$$\chi := p_0(w_\lambda) \in \mathcal{A}_q(K_0), \tag{4.7}$$

is a unitary character, i.e.,

$$\Delta\chi = \chi \otimes \chi\,, \tag{4.8}$$

$$S\chi = \chi^* = \chi^{-1}\,, \tag{4.9}$$

with $\chi(x) \equiv \langle x, \chi \rangle$.

Now we can define the quantum coherent state Γ in a way quite parallel to the classical case,[8] namely

$$\Gamma := (\mathcal{T}^\lambda)^{-1}(e_\lambda \otimes 1) \in \mathcal{H}_\lambda \otimes \mathcal{A}_q(K)\,. \tag{4.10}$$

As an element from $\mathcal{H}_\lambda \otimes \mathcal{A}_q(K)$, Γ should be considered as a quantum vector-valued function. Extend the Hermitian product $\langle\cdot,\cdot\rangle$ from $\mathcal{H}_\lambda$ to $\mathcal{H}_\lambda \otimes \mathcal{A}_q(K)$ according to

$$\langle v_1 \otimes c_1, v_2 \otimes c_2 \rangle := \langle v_1, v_2 \rangle c_1^* c_2 \in \mathcal{A}_q(K). \tag{4.11}$$

Using the coherent state one can relate to each vector $u \in \mathcal{H}_\lambda$ a quantum function ψ_u living on the big cell as follows

$$\psi_u := w_\lambda{}^{-1}\langle \Gamma, u \otimes 1 \rangle = w_\lambda{}^{-1}(\langle e_\lambda, (\cdot)u \rangle \otimes \mathrm{id})\mathcal{T}^\lambda. \tag{4.12}$$

Clearly $\psi_{e_\lambda} = 1$.

Proposition 4.2. *The mapping* $u \mapsto \psi_u$ *is injective. Since it holds true that*

$$(p_0 \otimes \mathrm{id})\,\Delta\langle \Gamma, u \rangle = \chi \otimes \langle \Gamma, u \rangle\,, \tag{4.13}$$

$$(p_0 \otimes \mathrm{id})\,\Delta w_\lambda = \chi \otimes w_\lambda\,. \tag{4.14}$$

ψ_u *is* K_0*-invariant, for every* $u \in \mathcal{H}_\lambda$. *Moreover,* ψ_u *is antiholomorphic, i.e.,* $\psi_u \in \mathcal{C}^*$.

Denote by $\mathcal{M}_\lambda$ the image of $\mathcal{H}_\lambda$ in $\mathcal{C}^*$. Thus $\mathcal{M}_\lambda$ can be turned into a $\mathcal{U}_h(\mathbf{k})$ module with the unit as a cyclic vector and, at the same time, the lowest weight vector. We keep the same symbol τ_λ for this representation. It remains to find an explicit prescription for τ_λ. To this end, we recall that $\mathcal{A}_q(K)$ is a left $\mathcal{U}_h(\mathbf{k})$ module with respect to the action $\mathcal{U}_h(\mathbf{k}) \otimes \mathcal{A}_q(K) \ni (Y, f) \mapsto \xi_Y \cdot f \in \mathcal{A}_q(K)$,

$$\xi_Y \cdot f := (\mathrm{id} \otimes \langle Y, \cdot \rangle)\,\Delta f\,. \tag{4.15}$$

The algebra $\mathcal{A}_q(K_0 \backslash K)$ becomes a submodule.

Proposition 4.3. *The representation* τ_λ *of* $\mathcal{U}_h(\mathbf{k})$ *acts in* $\mathcal{M}_\lambda$ *according to the rule*

$$\tau_\lambda(Y)\psi = w_\lambda{}^{-1}\,\xi_Y \cdot (w_\lambda \psi)\,, \quad \forall Y \in \mathcal{U}_h(\mathbf{k}),\ \forall \psi \in \mathcal{M}_\lambda\,. \tag{4.16}$$

References

1. Y. Soibelman, Orbit method for the algebras of functions on quantum groups and coherent states, *Duke. Math. J.* 70A: 151 (1993)

2. B. Jurčo, P. Šťovíček, Quantum dressing orbits on compact groups, *Commun. Math. Phys.*, 152: 97 (1993)

3. V. Lakshmibai, N. Reshetikhin, Quantum deformations of flag and Schubert schemes, *C. R. Acad. Sci. Paris*, 313, Série I: 121 (1991)

4. H. Awata, M. Nuomi, S. Odake, Heisenberg realization for $U_q(sl_n)$ on the flag manifold, preprint YITP/K-1016 (1993)

5. Y. Soibelman, On quantum flag manifolds, RIMS-780 (1991)

6. E. Taft, J. Towber, Quantum deformation of flag schemes and Grassmann schemes I. A q-deformation of the shape-algebra for $GL(n)$, *J. Algebra*, 142: No. 1 (1991)

7. P. Šťovíček, Quantum Grassmann manifolds, *Commun. Math. Phys.*, 158: 135 (1993)

8. A. M. Perelomov, "Generalized Coherent States and their Applications", Springer, Berlin, Heidelberg, New York (1986)

9. B. Jurčo, On coherent states for the simplest quantum groups, *Lett. Math. Phys.*, 21: 51 (1991)

10. B. Jurčo, P. Šťovíček, Coherent states for quantum compact groups, preprint CERN-TH.7201/94 (1994)

11. V. G. Drinfeld, Quantum groups, *in*: "Proc. ICM Berkeley 1986", AMS (1987)

12. M. Jimbo, Quantum R-matrix for the generalized Toda system, *Commun. Math. Phys.*, 102: 537 (1986)

13. N. Yu. Reshetikhin, L. A. Takhtajan, L. D. Faddeev, Quantization of Lie groups and Lie algebras, *Algebra i analiz*, 1: 178 (1989) (in Russian)

14. M. Rosso, Finite dimensional representations of the quantum analog of the enveloping algebra of a complex simple Lie algebra, *Comun. Math. Phys.*, 117: 581 (1988)

15. G. Lusztig, Quantum deformations of certain simple modules over enveloping algebras, *Adv. Math.*, 70: 237 (1988)

ON THE DEFORMATION OF COMMUTATION RELATIONS *

Władysław Marcinek

Institute of Theoretical Physics, University of Wrocław
Poland

Abstract

Deformed commutation relations and corresponding consistency conditions with braid relations are studied in terms of the so-called Wick algebras. We discuss the construction of such algebras and give some examples.

1. INTRODUCTION

Recently the q-deformed commutation relations for creation and annihilation operators (CAO) have been studied from different points of views by several authors, see for example Refs. 1-6. The commutation relations for Hecke braiding have been studied by Kempf.[7] The deformed commutation relations have been also studied by Bożejko and Speicher.[8,9] It is interesting from the algebraic point of view that all deformations of the commutation relations for CAO can be described in terms of the so-called Wick algebras.[10] Note that Wick algebras are some special examples of R-Weyl algebras[11] (see also Refs. 12,13). In a Wick algebra, there are no relations between creation (or annihilation) operators themselves, but such relations are possible in certain cases.[10] Obviously all such relations should be consistent. Hence we need some additional assumption.[10] In this paper we are going to continue the study of deformed commutation relations and the corresponding consistency conditions in terms of Wick algebras. Our study of deformed commutation relations is based on two operators R and R'. These operators are not arbitrary, they must satisfy some consistency conditions like Wess-Zumino conditions for differential calculus on a quantum plane.[14] Our study is a continuation of previous papers.[15-18]

2. WICK ALGEBRAS

We are going here to develop the concept of Wick algebras for our study of deformed commutation relations. Note that our notion of such algebras is based on the paper of Jorgensen, Schmith, and Werner.[10]

DEFINITION A. *A Wick algebra is an algebra $\mathcal{W}(A)$ generated by two sets of generators x_i and x_i^*, for $i \in I$ i.e. $\mathcal{W}(A) = \mathbf{C} < x_i^*, x_i, i \in I >$ such that*
(i) $*$ is an involution

$$x_i^{**} = x_i, \ (x_i x_j)^* = x_j^* x_i^*,$$

*This work is partially supported by KBN, Grant No 2P 302 087 06

(ii) we have the following relation

$$x_i^* x_j = \mathbf{1}\delta_{ij} + A_{ik}(x_j)x_k^*, \tag{2.1}$$

where A_{ik} are algebra-valued coefficients

$$A_{il}A_{lk} = A_{ik}, \quad A_{ik} = \overline{A}_{ki}. \tag{2.2}$$

Let us consider some examples:

Example 2.1. If $A_{ik}(x_j) = \pm x_j\delta_{ik}$, then we obtain the usual CCR or CAR algebras for bosons or fermions.

Example 2.2. In the case $A_{ik}(x_j) = qx_j\delta_{ik}$, we obtain the well-known q-relations.

Example 2.3. If we have the relation

$$A_{ik}(x_j)x_k^* = \tilde{R}_{ij}^{kl} x_l x_k^*, \tag{2.3}$$

where $\tilde{R} = (\tilde{R})_{ij}^{kl}$ is some linear operator, then we obtain the Wick algebras introduced by Jorgensen, Schmith, and Werner.[10]

Example 2.4. We can assume in addition that the operator $\tilde{R}$ in the above example is Yang-Baxter.

Let us denote by E the linear span of generators $x_i, i \in I$, i.e. $E = lin\{x_i | i \in I\}$. Obviously E is a vector space over the field of complex numbers $\mathbf{C}$. We denote by E^* the linear span of $x_I^*, i \in I$. Assume that we have a pairing $< .,. >: E^* \otimes E \longrightarrow \mathbf{C}$ given by

$$< u^*, v > := \Sigma_{i \in I}\overline{u}_i v_i, \tag{2.4}$$

where $u^* = \Sigma_{i\in I}\overline{u}_i x_i^*$ and $v = \Sigma_{i \in I} v_i x_i$.

For the Wick algebra $\mathcal{W}(A)$ we have a linear mapping (a Wick ordering) w : $\mathcal{W}(A) \longrightarrow TE \otimes TE^*$ defined inductively by the formula

$$w(s \otimes x_i^* \otimes x_j \otimes t) = \delta_{ij} s \otimes t + s \otimes A_{ik}(x_j) \otimes x_k^* \otimes t, \tag{2.5}$$

for $s \in \hat{E}^{\otimes k}$, $u^* \in E^*$, $v \in E$ and $t \in \hat{E}^{\otimes l}$, where $\hat{E}$ means E or E^*. We can use the mapping w to order every monomial in $\mathcal{W}(A)$ in such a way that all elements of E should be moved to the left of all elements of E^*. Hence the Wick algebra $\mathcal{W}(A)$ can be understood as a tensor product $TE \otimes TE^*$ equipped with a multiplication as tensor product modulo the relation

$$x_i^* \otimes x_j = \mathbf{1}\delta_{ij} + A_{ik}(x_j) \otimes x_k^*. \tag{2.6}$$

As a special example of Wick ordering, we can consider two linear mappings $\Psi_{E^*, E^{\otimes k}}$: $E^* \otimes E^{\otimes k} \longrightarrow E^{\otimes k} \otimes E^*$ defined by relation

$$\Psi_{E^*, E^{\otimes k}} := \Pi_{i=1}^k \tilde{R}^{(i)}, \tag{2.7}$$

where $\tilde{R}^{(i)} := id_E \otimes \ldots \otimes \tilde{R} \otimes \ldots id_E$, ($\tilde{R}$ in the i-th position), and $\Psi_{(E^*)^{\otimes k}, E^{\otimes l}} : (E^*)^{\otimes k} \otimes E^{\otimes l} \longrightarrow E^{\otimes l} \otimes (E^*)^{\otimes k}$ defined by

$$\Psi_{(E^*)^{\otimes k}, E^{\otimes l}} := \tilde{R}^{(l)} \circ \ldots \circ \tilde{R}^{(1)} \circ \ldots \circ \tilde{R}^{(k+l-1)} \circ \ldots \circ \tilde{R}^{(k-1)} \circ \tilde{R}^{(k+l)} \circ \ldots \circ \tilde{R}^{(k)}, \tag{2.8}$$

where

$$\tilde{R}(x_i^* \otimes x_j) = A_{ik}(x_j) \otimes x_k^*. \tag{2.9}$$

Lemma A. *The Wick algebra $\mathcal{W}(A)$ can be obtained as a quotient $\mathcal{W}(A) \equiv \mathcal{W}(\tilde{R}) := T(E \oplus E^*)/I_{\tilde{R}}$, where $I_{\tilde{R}}$ is the ideal in $T(E \oplus E^*)$ generated by the following relations*

$$id_{E^* \otimes E} = < .,. > \mathbf{1} + \tilde{R}. \tag{2.10}$$

Let us define an evaluation mapping $ev_k : E^* \otimes TE \longrightarrow TE$ by the formulas

$$ev_1 \equiv < .,. >, \; ev_k := ev_1^{(1)} + ev_{k-1}^{(2)} \circ \tilde{R}^{(1)}. \tag{2.11}$$

It is easy to see that

$$id_{E^* \otimes E^{\otimes k}} = ev_k + \Psi_{E^*, E^{\otimes k}}. \tag{2.12}$$

3. BRAID RELATIONS

Let $R : E \otimes E \longrightarrow E \otimes E$ be a linear operator. If we have

$$<,>^{(2)} \circ \tilde{R}^{(1)} = <,>^{(1)} \circ R^{(2)}, \tag{3.1}$$

then we say that $\tilde{R}$ is partially adjoint to R with respect to $<,>$. Let R be an arbitrary Yang-Baxter operator

$$R^{(1)}R^{(2)}R^{(1)} = R^{(2)}R^{(1)}R^{(2)}, \tag{3.2}$$

and let $\tilde{R}$ be the partially adjoint to R. The Wick algebra defined by the formula (2.10) of Lemma A is said to be a Wick algebra with braid relations and is denoted by $\mathcal{W}(R)$.

Let $R' : E \otimes E \longrightarrow E \otimes E$ be a linear operator.

Definition B. *The quotient space*

$$\mathcal{W}(R, R') := T(E \oplus E^*)/I_{R,R'}, \tag{3.3}$$

where $I_{R,R'}$ is the ideal in $T(E \oplus E^*)$ generated by the relations

$$\begin{gathered} id_{E^* \otimes E} = < .,. > \mathbf{1} + \tilde{R}, \\ id_{E \otimes E} - R' = 0, \;\; id_{E^* \otimes E^*} - (R')^T = 0, \end{gathered} \tag{3.4}$$

and equipped with the consistency conditions

$$\begin{aligned} R^{(1)}R^{(2)}R^{(1)} &= R^{(2)}R^{(1)}R^{(2)}, \\ R'^{(1)}R^{(2)}R^{(1)} &= R^{(2)}R^{(1)}R'^{(2)}, \\ R^{(1)}R^{(2)}R'^{(1)} &= R'^{(2)}R^{(1)}R^{(2)}, \\ (id_{E \otimes E} + R)(id_{E \otimes E} &- R') = 0, \end{aligned} \tag{3.5}$$

is an algebra, called a Wick algebra with additional relations.

Note that the properties of the operator $P_2' := id + R$ are essential for the Wick algebra $\mathcal{W}(\tilde{R})$.[9,10] If the kernel of P_2' is trivial, then we have relations only between elements of the Wick algebra and their conjugates. There are no relations between elements of the same kind. Such additional relations are possible if and only if the kernel of P_2' is nontrivial. Observe that, if the operator R has an eigenvalue $\lambda = -1$, then P_2' has a nontrivial kernel and vice versa.

An algebra $\mathcal{A}$ defined as the quotient $\mathcal{A} \equiv \mathcal{A}(E) := TE/I_{R'}$, where the ideal $I_{R'}$ is given by the relation

$$I_{R'} \equiv gen\{id_{E \otimes E} - R'\} \tag{3.6}$$

is said to be an R'-symmetric algebra over E. This means that $\mathcal{A}$ is an algebra generated by $\{x_i | i \in I\}$ and the relation

$$m(x_i \otimes x_j) = m \circ R'(x_i \otimes x_j). \tag{3.7}$$

On the other hand, the quotient $\mathcal{B} := TE^*/I^*$, where the ideal I is given by

$$I^* \equiv gen\{id_{E^* \otimes E^*} - R'^T\}$$

is an R'^T-symmetric algebra. The algebra $\mathcal{A}$ is a quadratic algebra over E, and the algebra $\mathcal{B}$ can be understood as the dual algebra to $\mathcal{A}$ in a certain sense.[24] The algebra $\mathcal{A}$ is graded:

$$\mathcal{A} = \bigoplus_{n \in N} \mathcal{A}^n, \; \mathcal{A}^0 \equiv \mathbf{C}, \; \mathcal{A}^1 \equiv E. \tag{3.8}$$

Let us define some linear mappings on generators by the following formulas:

$$\begin{array}{ll} \Psi_{E,E}(x_i \otimes x_j) = R(x_i \otimes x_j), & \Psi_{E^*,E}(x_i^* \otimes x_j) = \tilde{R}(x_i^* \otimes x_j), \\ \Psi_{E,E^*}(x_i \otimes x_j^*) = T(x_i \otimes x_j^*), & \Psi_{E^*,E^*}(x_i^* \otimes x_j^*) = R^T(x_i^* \otimes x_j^*), \end{array} \tag{3.9}$$

where the operator T is defined by the relation

$$<,>^{(1)} \circ R^{(2)} \circ T^{(1)} = <,>^{(2)}. \tag{3.10}$$

These mappings may be uniquely extended to arbitrary elements of the algebras $\mathcal{A}$ and $\mathcal{B}$. We have, for example,

$$\Psi_{E,\mathcal{A}^2}(x_i \otimes x_j x_k) = m^{(1)} R^{(2)} R^{(1)}(x_i \otimes x_j \otimes x_k). \tag{3.11}$$

It follows from the R'-commutativity that

$$\Psi_{E,\mathcal{A}^2}(x_i \otimes m(x_j \otimes x_k)) = m^{(1)} R'^{(1)} R^{(2)} R^{(1)}(x_i \otimes x_j \otimes x_k). \tag{3.12}$$

On the other hand, we obtain

$$\begin{aligned} &\Psi_{E,\mathcal{A}^2}(x_i \otimes m(x_j \otimes x_k)) \\ &\quad = \Psi_{E,\mathcal{A}^2}(x_i \otimes m \circ R'(x_j \otimes x_k)) = m^{(1)} R^{(2)} R^{(1)} R'^{(2)}(x_i \otimes x_j \otimes x_k). \end{aligned} \tag{3.13}$$

It follows immediately from the second condition (3.5) that the mapping $\Psi_{E,\mathcal{A}^2}$ is well defined. We denote by $P : TE \longrightarrow \mathcal{A}$ and $P^* : TE^* \longrightarrow \mathcal{A}^*$ the quotient mappings. We have

$$\begin{aligned} P_n &\equiv id + \Psi^{(1)}_{E,E} + \Psi^{(1)}_{E,E} \circ \Psi^{(2)}_{E,E} + \ldots, \\ P^*_n &\equiv id + \Psi^{(1)}_{E^*,E^*} + \Psi^{(1)}_{E^*,E^*} \circ \Psi^{(2)}_{E^*,E^*} + \ldots, \end{aligned} \tag{3.14}$$

for homogeneous components of P and P^* respectively.

Lemma B. *The evaluation mapping ev is braid commutative, i.e. we have*

$$ev_{k+l} = ev_{k+l} \circ \Psi^{(2)}_{E^{\otimes k}, \otimes l} \tag{3.15}$$

on $E^* \otimes E^{\otimes k} \otimes E^{\otimes l}$.

Proof: We prove the statement for $k = l = 1$ only. We have

$$\begin{aligned} ev_2 - ev_2 \circ R' &= (ev_1^{(1)} + ev_1^{(2)} \circ \tilde{R}^{(1)})(id - R'^{(2)}) \\ &= ev^{(1)}((id + R)(id - R'))^{(2)} = 0. \end{aligned}$$

It follows from the braid commutativity of the evaluation mapping ev that it can be projected on the algebra $\mathcal{A}$. Hence the projected evaluation is a mapping $\alpha : E^* \otimes \mathcal{A} \longrightarrow \mathcal{A}$ given by the relation

$$\alpha_k(id_{E^*} \otimes P_k) = P_{k-1} \circ ev_k. \tag{3.16}$$

4. REPRESENTATIONS OF WICK ALGEBRAS

We denote by E the space of one-particle states. First we define a representation $a : \mathcal{W}(A) \otimes TE \longrightarrow TE$ by the formulas

$$a_{x_i} s := x_i \otimes s, \quad a_{x_j^*} s := ev_l(x_j^* \otimes s), \tag{4.1}$$

for generators $x_i \in E$, $x_j^* \in E^*$ and for the state $s \in E^{\otimes l}$. We use the following notation:

$$a_{x_i} \equiv a_i^+, \; a_{x_j} \equiv a_j. \tag{4.2}$$

The vacuum state is defined as usual. We define a Ψ-bracket $[.,.]_\Psi$ by the formula

$$[a_i, a_j^+]_\Psi := [.,.]_\Psi(a_i \otimes a_j^+) := c \circ (id - \Psi)(a_i \otimes a_j^+), \tag{4.3}$$

where c is the composition map, and

$$\Psi(a_i \otimes a_j^+) := a_{\tilde{R}(x_i^* \otimes x_j)}. \tag{4.4}$$

Next we use the notation

$$[a_i, a_j^+]_\Psi s := ev \circ [.,.]_\Psi^{(1)}(a_i \otimes a_j^+ \otimes s) \tag{4.5}$$

for $s \in E^{\otimes l}$.

Theorem A. *We have the following commutation relation*

$$[a_i, a_j^+]_\Psi = \delta_{ij}. \tag{4.6}$$

Proof: We calculate the relation (4.6) as follows

$$\begin{aligned} [a_i, a_j^+]_\Psi f &= ev \circ [.,.]_\Psi(a_i \otimes a_j^+ \otimes f) \\ &= ev \circ c^{(1)} \circ (id - \Psi^{(1)})(a_i \otimes a_j^+ \otimes f) \\ &= ev(c \circ (a_i \otimes a_j^+ - \Psi(a_i \otimes a_j^+)) \otimes f) \\ &= ev_{l+1}(x_i^* \otimes (x_j \otimes f)) - ev_l^{(2)} \circ \tilde{R}^{(1)}(x_i^* \otimes x_j \otimes f) \\ &= ev_1^{(1)}(x_i^* \otimes x_j \otimes f) = \delta_{ij} f, \end{aligned} \tag{4.7}$$

where the relation (2.11) has been used.

Example 4.1. The algebra generated by two generators x^* and x and the relation

$$x^* x = 1 + q x x^*, \tag{4.8}$$

is denoted by $\mathcal{W}_q(1)$, here $-1 < q < 1$ is a parameter of deformation. If we introduce the following state vector $|n >= ([n]!)^{-1/2} x^n$, then we obtain

$$|n >= \frac{(a_x)^n}{([n]!)^{1/2}} |o >, \quad a_{x^*}|n >= ([n]!)^{1/2} |n-1 >, \tag{4.9}$$

where

$$a_{x^*}|0 >= 0, \quad [n] = \frac{1-q^n}{1-q}, \text{and} \quad [n]! = [1][2]\ldots[n].$$

Here $R(x \otimes x) = qx \otimes x$ and the kernel of $P_2' = id + R$ is trivial. Hence we have no additional relations here.[20]

Now we give the representation of the Wick algebra $\mathcal{W}(R, R')$ by the formulas (4.1). Obviously the commutation relations in $\mathcal{W}(R, R')$ should be consistent with the additional relations corresponding to the operator R'. It follows from the definition of our representation and from the proof of Lemma B that

$$\begin{aligned} a_i(id_{E\otimes E} - R')x_k \otimes x_l &= ev_2(x_i^* \otimes (id - R')(x_k \otimes x_l)) \\ &= (ev_1^{(1)} + ev_1^{(2)} \circ \tilde{R}^{(1)})(id - R')(x_i^* \otimes x_k \otimes x_l) \;=\; 0. \end{aligned}$$

We also have the relation

$$a_i(id - \Psi)(a_j^+ \otimes a_k^+) = 0. \tag{4.10}$$

We see that our relations are consistent. Now our representation can be projected on the quotient algebra $\mathcal{A} \equiv TE/I_{R'}$. In this way we obtain

$$a_i^+ f \equiv m(x_i \otimes f), \quad a_i f \equiv \alpha(x_i^* \otimes f) \tag{4.11}$$

for every $f \in \mathcal{A}$.

The Ψ-bracket is given here by the relation

$$[\hat{a}_i, \hat{a}_j]_\Psi := [.,.]_\Psi(\hat{a}_i \otimes \hat{a}_j) := c \circ (id - \Psi)(\hat{a}_i, \otimes \hat{a}_j), \tag{4.12}$$

where $\hat{a}_i$ stands for a_i^* or a_i. In order to obtain the Ψ-bracket for creation and annihilation operators explicitly, we need the formulas

$$\Psi(a_i^+ \otimes a_j^+) := a_{R(x_i \otimes x_j)} \quad, \Psi(a_i \otimes a_j^+) := a_{\tilde{R}(x_i^* \otimes x_j)} \quad, \Psi(a_i \otimes a_j) := a_{R^T(x_i^* \otimes x_j^*)}. \tag{4.13}$$

Observe that in our notation we have, for example, for $f \in E^{\otimes l}$:

$$\begin{aligned} (a_i \circ a_j^+)f &\equiv \alpha_{l+1}(x_i^* \otimes (x_j \otimes f)), \\ (a_i^+ \circ a_j^+)f &\equiv m(x^i \otimes x^j \otimes f), \\ (a_i \circ a_j)f &\equiv \alpha_{l+1}^{(1)} \circ \alpha_l^{(2)}(x_i \otimes x_j \otimes f), \end{aligned} \tag{4.14}$$

Theorem B. *We have on the algebra the following commutation relations for the representation of the Wick algebra* $\mathcal{W}(R, R')$

$$[a_i, a_j^+]_\Psi = \delta_{ij}, \quad [a_i^+, a_j^+]_\Psi = 0, \quad [a_i, a_j]_\Psi = 0. \tag{4.15}$$

Proof: The proof of the first relation (4.15) is analogous to the proof of (4.6). We calculate the second relation (4.15):

$$[a_i^+, a_j^+]_\Psi = [.,.]_\Psi(a_i^+ \otimes a_j^+) = a_{(id_{E\otimes E} - B)(x_i \otimes x_j)} = 0. \tag{4.16}$$

The proof of the third commutation relation (4.15) is obvious.

Example 4.2. The algebra $\mathcal{W}_q(2)$ is generated by four generators x_i^*, x_i, $(i = 1, 2)$, and relations

$$x_i^* x_i = 1 + q x_i x_i^*, \tag{4.17}$$

where $-1 \leq q \leq +1$. We have here $R(x_i \otimes x_j) = -q x_j \otimes x_i$ for $i \neq j$ and

$$\ker P_2' = \ker(id + R) = \begin{cases} \Lambda^{\otimes 2}E & \text{for} \quad q = -1, \\ \emptyset & \text{for} \quad -1 < q < 1, \\ S^{\otimes 2}E & \text{for} \quad q = +1 \end{cases} \tag{4.18}$$

where $\Lambda^{\otimes 2}E$ or $S^{\otimes 2}E$ denote the space of antisymmetric or symmetric tensors over E, respectively. Next we have

$$R'(x_i \otimes x_j) = \begin{cases} -x_j \otimes x_i & \text{for} \quad q = -1, \\ x_j \otimes x_i & \text{for} \quad q = +1 \end{cases} \tag{4.19}$$

Hence, for $q = \pm 1$, we obtain the following additional relations

$$x_1^* x_2^* \pm x_2^* x_1^* = 0, \quad x_1 x_2 \pm x_2 x_1 = 0. \tag{4.20}$$

There are no additional relations for $-1 < q < +1$.

References

1. W. Pusz, *Rep. Math. Phys.* 27:394 (1989)
2. W. Pusz and S.L. Woronowicz, *Rep. Math. Phys* 27:231, (1989)
3. D.B. Fairlie and C.K. Zachos, *Phys. Lett. B* 256:43 (1991)
4. S.P. Vokos, *J. Math. Phys.* 32:2979 (1991)
5. M. Arik, The q–difference operator, the 'quantum hyperplane, Hilbert spaces of analytic funtions and q–oscillators, Istanbul Technical University preprint (1991)
6. S. Majid, Free braided differential calculus, braided binomial theorem and the braided exponential map, preprint DAMTP/93-3 (1993)
7. A. Kempf, *Lett. Math. Phys.* 26:11 (1992)
8. M. Bożejko, R. Speicher, Interpolation between bosonic and fermionic relations given by generalized Brownian motions, preprint FSB 132-691, Heidelberg (1992)
9. M. Bożejko, R. Speicher, Completely positive maps on Coxeter groups, deformed commutation relations, and operator spaces (in preparation)
10. P.E.T. Jørgensen, L.M. Schmith, and R.F. Werner, Positive representation of general commutation relations allowing Wick ordering, preprint (1993)
11. J. C. Baez, R-commutative geometry and quantization of Poisson algebras, Wellesley College preprint (July 1991)
12. A. Borowiec, V. K. Kharchenko, Z. Oziewicz, Calculi on Clifford-Weyl and exterior algebras for Hecke braiding, Conferencia dictata por Zbigniew Oziewicz at Centro de Investigation en Matematicas, Guanajuato, Mexico, Seminario de Geometria, jueves 22 de abril de 1993 and at the Conference on Differential Geometric Methods, Ixtapa, Mexico, September 1993
13. E. E. Demidov, On some aspects of the theory of quantum groups, *Uspekhi Mat. Nauk* 48(6):39 (1993) (Russian)
14. J. Wess and B. Zumino, Covariant differential calculus on the quantum hyperplane, CERN preprint 5697/90 (1990)
15. W. Marcinek, On braid statistics and noncommutative calculus, *Rep. Math. Phys.* 33:117 (1993)

16. W. Marcinek, On unital braidings and quantization, U. Wrocław preprint ITP-847 (1993)

17. W. Marcinek, Noncommutative geometry corresponding to arbitrary braidings, *J. Math. Phys.* 35:2633 (1994)

18. W. Marcinek and R. Ralowski, Particle operators from braided geometry, Proceedings of the XXX Winter School of Theoretical Physics, Karpacz, Poland, February 15-26 (1994)

19. C. Daskaloyannis, *J. Phys. A* 24:L789 (1991)

20. D. I. Fivel, *Phys. Rev. Lett.* 65:3361 (1990)

21. J. C. Baez, *Lett. Math. Phys.* 23:1333 (1991)

22. V.V. Lyubashenko, Vectorsymmetries, *Seminar on Supermanifolds* No 19:1 (1987

23. D. Gurevich, A. Radul and V. Rubstov, Noncommutative differential geometry and Yang-Baxter equation, I.H.E.S./M/91/88 (1988)

24. Y. I. Manin, Quantum groups and noncommutative geometry, University of Montréal preprint (1989)

25. W. Marcinek, Algebras based on Yang-Baxter operators, *Rep. Math. Phys.* 32:1 (1993)

26. W. Marcinek, On colour quantization: relations and parastatistics, *Mod. Phys. Lett.* (1995) (to appear).

THE q-DEFORMED QUANTUM MECHANICS IN THE COHERENT STATES MAP APPROACH

V. Maximov[1] and A. Odzijewicz[2]

[1] Institute of Mathematics, Tver State University
Geljiabova 33, Tver, Russia
[2] Warsaw University Division
Lipowa 41, 15-424 Bialystok, Poland

Abstract

We study q-quantum mechanics in one degree of freedom. Among other things, we discuss the holomorphic representation of the q-deformed Heisenberg-Weyl algebra and its realization by covariant Berezin symbols.

1. COHERENT STATES MAP AND q-CANONICAL COMMUTATION RELATIONS

Let $\mathbf{D}_q$ be the open disc in $\mathbb{C}$ of radius $(1-q)^{-1/2}$ centered at zero, where $-1 < q < 1$, and let $\mathcal{M}$ be a complex separable Hilbert space with orthonormal basis $\{e_n\}, n = 0, 1, 2, \ldots$. We define the analytic map (the coherent states map),[1] $K_q : \mathbf{D}_q \to \mathcal{M}$ by

$$K_q(z) = \sum_{n=0}^{\infty} \left[\frac{(1-q)^n}{(1-q)\cdots(1-q^n)}\right]^{\frac{1}{2}} z^n e_n \, , \tag{1.1}$$

where for $n = 0$ one assumes that the coefficient in front of e_0 is equal to one. Since

$$\langle K_q(z)|K_q(z)\rangle = \sum_{n=0}^{\infty} \frac{(1-q)^n}{(1-q)\cdots(1-q^n)} (\bar{z}z)^n =: \exp_q(\bar{z}z) < \infty \, , \tag{1.2}$$

for $\bar{z}z < (1-q)^{-1}$, the definition is correct. The function $\exp_q$ is called[6] the q-deformation of the exponential function $\exp = \exp_1$. So for $q = 1$ one obtains the standard coherent states map (the Bargmann-Fock coherent states map) and for $q = 0$ one gets the geometric sequence, $\exp_0 \bar{z}z = (1-\bar{z}z)^{-1}$.

Having introduced the coherent states map, one can define an annihilation operator by

$$AK_q(z) := zK_q(z), \qquad \forall z \in \mathbf{D}_q. \tag{1.3}$$

This is a bounded operator, with norm $||A|| = (1-q)^{-1/2}$ for $0 \leq q < 1$ and $||A|| = 1$ for $-1 < q < 0$. The operators so defined satisfy q-Heisenberg commutation relations

$$[A, A^\dagger]_q := AA^\dagger - qA^\dagger A = 1 \tag{1.4}$$

which were studied in Refs. 3-5 and 9-12, for instance.

Quantization, Coherent States, and Complex Structures
Edited by J.-P. Antoine *et al.*, Plenum Press, New York, 1995

Next we define the operators

$$Q = Q^{\dagger} := AA^{\dagger} - A^{\dagger}A, \tag{1.5}$$

$$K := (1-q)A^{\dagger}, \tag{1.6}$$

$$J := K(1 + qQ + q^2Q^2 + \cdots). \tag{1.7}$$

Since $||Q|| = 1$ and $|q| < 1$, the above sequence is summable. Therefore, the operators Q, K, and J belong to the C^*-algebra generated by the q-creation $A^{\dagger}$ and the q-annihilation A operators, i.e. the q-Heisenberg-Weyl algebra or Wick algebra W_q.[5]

One has the relations

$$AJ = 1, \tag{1.8}$$

$$[A, Q]_q = 0, \tag{1.9}$$

$$[Q, A^{\dagger}]_q = 0, \tag{1.10}$$

$$[Q, J]_q = 0. \tag{1.11}$$

The relation (1.8) shows that J is a right inverse to the annihilation operator. This is the reason why we shall call it below a q-integration operator.

2. HOLOMORPHIC REPRESENTATION

In this section we represent the Wick algebra by operators acting in the space $L^2\mathcal{O}(\mathbf{D}_q, d\mu_q)$ of holomorphic functions on the disc $\mathbf{D}_q$, which are square integrable with respect to the q-Gauss measure. This allows us to interprete $A, A^{\dagger}, Q$, and J as q-differentiation, multiplication by z, dilation and q-integration operators, respectively. At the end we will show that $\langle K_q(z)|K_q(v)\rangle = \exp_q(\bar{z}v)$ is a reproducing kernel function for the Hilbert space $L^2\mathcal{O}(\mathbf{D}_q, d\mu_q)$. We shall start from the embedding

$$I : \mathcal{M} \hookrightarrow \mathcal{O}(\mathbf{D}_q),$$

which is defined by

$$I(v)(z) := \langle v|K_q(z)\rangle, \tag{2.1}$$

where $v \in \mathcal{M}$. Since coherent states $K_q(z)$, $z \in \mathbf{D}_q$, form a linearly dense subset in $\mathcal{M}$, the map is an antilinear monomorphism of complex vector spaces. Therefore $I(\mathcal{M})$ inherits the Hilbert space structure with the scalar product given by $\langle I(v)|I(v)\rangle := \langle v|v\rangle$. After a simple calculation, one finds:

$$(I \circ A \circ I^{-1}\varphi)(z) = (\partial_q\varphi)(z), \tag{2.2}$$

$$(I \circ A^{\dagger} \circ I^{-1}\varphi)(z) = z\varphi(z), \tag{2.3}$$

$$(I \circ Q \circ I^{-1}\varphi)(z) = \varphi(qz), \tag{2.4}$$

$$(I \circ J \circ I^{-1}\varphi)(z) = (\textstyle\int_q\varphi)(z), \tag{2.5}$$

where

$$(\partial_q\varphi)(z) := \frac{\varphi(z) - \varphi(qz)}{(1-q)z} \tag{2.6}$$

is the q-derivative and

$$(\textstyle\int_q\varphi)(z) := (1-q)z(\varphi(z) + q\varphi(qz) + \cdots + q^n\varphi(q^nz) + \cdots) \tag{2.7}$$

is the q-integral (see Ref. 6 for a detailed discussion).

Now, let us define the scalar product of holomorphic functions $\varphi, \psi \in \mathcal{O}(\mathbf{D}_q)$ by the q-integration over the disc $\mathbf{D}_q$

$$\langle\psi|\varphi\rangle_q := \int_{\mathbf{D}_q} \overline{\psi(z)}\varphi(z)(\exp_q(qz\bar{z}))^{-1}d\mu_q(z,\bar{z}) \tag{2.8}$$

where

$$d\mu_q(z,\bar{z}) = \frac{1}{2\pi}\sum_{n=0}^{\infty} q^n\delta(x - \frac{q^n}{1-q})dx d\varphi. \tag{2.9}$$

One has

$$\langle I(e_n)|\, I(e_n)\rangle_q = \delta_{nm}$$

and thus the Hilbert space isomorphism

$$I : \mathcal{M} \stackrel{\sim}{\rightarrow} L^2\mathcal{O}(\mathbf{D}_q, d\mu_q),$$

and

$$\varphi(v) = \int_{\mathbf{D}_q} \varphi(z)\exp_q(\bar{z}v)[\exp_q(qz\bar{z})]^{-1}d\mu_q(\bar{z},z). \tag{2.10}$$

Thus one obtains in addition

$$(f(\partial_q)\varphi(v) = \int_{\mathbf{D}_q} \varphi(z)f(\bar{z})\exp_q(\bar{z}v)[\exp_q(qz\bar{z})]^{-1}d\mu_q(z,z), \tag{2.11}$$

where $f \in \mathcal{O}(\mathbf{D}_q)$ is such that the integral of the right hand side of (2.11) exists. If $X \in W_q$ and $\|X\| \leq (1-q)^{-1/2}$, then from (2.10) one gets an integral formula for the operator $\varphi(X)$:

$$\varphi(X) = \int_{\mathbf{D}_q} \varphi(z)\exp_q(\bar{z}X)\cdot[\exp_q(qz\bar{z})]^{-1}d\mu_q(\bar{z}z). \tag{2.12}$$

If $q \to 1$, (2.10) becomes the reproducing property for the Bargmann-Fock representation. If $q = 0$, (2.10) becomes the Cauchy formula and $L^2\mathcal{O}(\mathbf{D}_0, d\mu_0)$ is the Hardy-Lebesgue space.

3. q-DEFORMATION OF THE WEYL-HEISENBERG GROUP AND q-EVOLUTION

We have the following statements. The second one gives the q-analogue of the Weyl-Heisenberg group.

Proposition 1.
Let $f(z) = \sum_{n=0}^{\infty} a_n z^n$ be an analytic function on $\mathbf{D}_q$, and let X and Y be q-commuting elements of the Wick algebra, i.e. $[X,Y]_q = XY - qYX = 0$, and such that $\exp_q(X+Y), \exp_q(X), \exp_q(Y), f(X+Y)$ and $f(Y) \in W_q$. Then

$$f(X+Y) = [\exp_q(Y\partial_q)f](X) \tag{3.1}$$

and

$$\exp_q(X+Y) = \exp_q(X)\exp_q(Y) \tag{3.2}$$

Proposition 2.
The q-deformed Weyl-Heisenberg relation is given by

$$\exp_q(sA)\exp_q(tA^\dagger) = \exp_q(\bar{t}A^\dagger)\exp_q(s\bar{t}Q)\exp_q(sA) \tag{3.3}$$

for $s, t \in \mathbf{D}_q$ *and* $-1 < q < 1$.

Let us consider now the q-evolution in time of the system described by the selfadjoint Hamiltonian $H^\dagger = H \in W_q$. We will assume also that the time is bounded:

$$-\frac{1}{(1-q)||H||} < t < \frac{1}{(1-q)||H||}.$$

Therefore, the state

$$\psi(t) = \exp_q(itH)\psi_0, \tag{3.4}$$

for $\psi_0 \in \mathcal{M}$, is well defined and satisfies the q-analogue of the Schrödinger equation:

$$-i\partial_q\psi(t) = H\psi(t). \tag{3.5}$$

Let $B \in W_q$ and let the operator $B(t)$ be defined by the condition

$$\langle\psi(t)|B\psi(t)\rangle = \langle\psi_0|B(t)\psi_0\rangle, \quad \psi_0 \in \mathcal{M}. \tag{3.6}$$

Then one has

$$B(t) = \exp_q(-itH)B\exp_q(itH)$$

and $B(t)$ satisfies the q-analogue of the Heisenberg equation: :

$$i\partial_q B(t) = [H, B(t)]_q - (1-q)(B(t) - it\mathbf{1})H. \tag{3.7}$$

Since $[H, H]_q \neq 0$, the q-evolution operators $\exp_q(itH)$ do not satisfy the one-parameter group property, which holds only when $q \to 1$.

4. THE STAR PRODUCT OF BEREZIN SYMBOLS. CONCLUDING REMARKS

The mean value of the observable $B \in W_q$ in the coherent state $K_q(z)$, where $z \in \mathbf{D}_q$, defines the function

$$\langle B\rangle(z,\bar{z}) := \frac{\langle K_q(z)|BK_q(z)\rangle}{\langle K_q(z)|K_q(z)\rangle}, \tag{4.1}$$

which is called the covariant Berezin symbol of the operator B.[13,15] Since $K_q : \mathbf{D}_q \to \mathcal{M}$ is holomorphic, the operator B is determined by its covariant symbol. The product of two operators $B, C \in W_q$ defines the product of their symbols:

$$\langle B\rangle *_q \langle C\rangle(z,\bar{z}) = \frac{\langle K_q(z)|BCK_q(z)\rangle}{\langle K_q(z)|K_q(z)\rangle}, \tag{4.2}$$

and thus the Wick algebra W_q may be completely described by the $*_q$-product algebra of covariant Berezin symbols.

Let f, g be real analytic functions on the open domain $\Omega \supset \overline{\mathbf{D}_q}$. Since $||A|| = ||A^\dagger|| = (1-q)^{-1/2}$, or $||A|| = ||A^+|| = 1$, the operators

$$f(A^\dagger, A) = \sum_{\bar{n},m=0}^{\infty} f_{\bar{n}m} A^{\dagger^n} A^m, \tag{4.3}$$

$$g(A^\dagger, A) = \sum_{\bar{k},l=0}^{\infty} g_{\bar{k}l} A^{\dagger^k} A^l, \tag{4.4}$$

given by Wick ordered series, belong to W_q. Their covariant Berezin symbols are given by $\langle f(A^\dagger, A)\rangle = f$ and by $\langle g(A^\dagger, A)\rangle = g$, respectively. In this case, the product formula takes the form

$$(f *_q g)(\bar{z}, z) = \frac{1}{\exp_q(z\bar{z})} f(\bar{z}, \bar{\partial}_q)[g(\bar{z}, z)\exp_q(z\bar{z})]. \tag{4.5}$$

We could extend our considerations to the case $|q| < 1, q \in \mathbb{C}$, or to the case $q^k = 1, k \in \mathbb{N}$. But for complex q, the scalar product $\langle K_q(z)|K_q(z)\rangle$ of the coherent states is not expressed by the q-exponential function. As a consequence, the model loses its elegance and its physical interpretation is unclear. However, the case $q^k = 1$ is of physical interest. Namely, if $q \to 1$, $k = 1$, the q-quantum mechanics corresponds to standard Heisenberg-Schrödinger boson quantum mechanics and all relations and formulas considered in the previous sections correspond to their quantum mechanical counterparts.

For $q \to -1$, $k = 2$, the matrix representatives of A and $A^\dagger$ preserve their meaning and give

$$\begin{aligned} Ae_{2n} &= 0, \quad Ae_{2n+1} = e_{2n}, \\ A^\dagger e_{2n} &= e_{2n+1}, \quad A^\dagger e_{2n+1} = 0, \end{aligned} \tag{4.6}$$

Thus one obtains fermionic anti-commutation relations

$$AA^\dagger + A^\dagger A = 1 \quad A^2 = A^{\dagger^2} = 0. \tag{4.7}$$

Since $A^2 = 0$, we can preserve the notion of the coherent states map $K_{-1} : \mathbf{D}_{-1} \to \mathcal{M}$ in this case if and only if $z^2 = 0$. That is, one has to assume that z is a Grassmann variable. Then

$$K_{-1} = e_0 + ze_1 \tag{4.8}$$

and

$$\langle K_{-1}(z)|K_{-1}(v)\rangle = \exp_{-1}(\bar{z}v) = 1 + \bar{z}v, \tag{4.9}$$

for $z^2 = v^2 = \bar{z}^2 = \bar{v}^2 = 0$. The representation in the Hilbert space $\mathcal{M}$ will be irreducible and $K_{-1}(\mathbf{D}_{-1})$ will be linearly dense in $\mathcal{M}$ iff $\dim_{\mathbb{C}}\mathcal{M} = 2$. So the holomorphic representation makes sense, too. Hence, one can interpret the annihilation operator A and the creation operator $A^\dagger$ as the ∂_{-1}-derivative and multiplication by z respectively.

The Heisenberg-Weyl relation (3.3), the q-Schrödinger equation and q-Heisenberg equation are valid for $q = -1$ also. So we conclude that the model of q-quantum mechanics presented in this letter gives a one-parameter deformation $-1 \le q \le 1$ from boson quantum mechanics to fermion quantum mechanics, which are singular cases of the model. Also the case $q^k = 1$, $k > 2$, can be described in the way proposed above, provided one assumes that $z^k = 0$,.

As a result, one is led to q-quantum mechanics of systems which satisfy parastatistics different from bosonic and fermionic ones.

If one takes $K_{q,h} : \mathbf{D}_{q,h} \to \mathcal{M}$, where

$$\mathbf{D}_{q,h} = \left\{ z \in \mathbb{C} : |z| < \sqrt{\frac{h}{1-q}} \right\} \quad \text{and} \quad K_{q,h}(z) = K_q\left(\frac{z}{\sqrt{h}}\right)$$

for $h > 0$, the annihilation and creation operators change by a factor $\sqrt{h}$ and the deformed Heisenberg commutation relation assumes the form

$$AA^\dagger - qA^\dagger A = h.$$

As a consequence, all the formulas in the paper will be deformed by h according to a dimensional argument. Among others, the deformation of the $*$-product formula (4.5) will be given by

$$(f \star_{(q,h)} g)(\bar{z}, z) = \frac{1}{\exp_q(\frac{\bar{z}z}{h})} f(\bar{z}, h\bar{\partial}_q)\big[g(\bar{z}, z)\exp_q(\frac{z\bar{z}}{h})\big]. \tag{4.10}$$

Thus, one has

$$f \star_{(q,h)} g \underset{q,h\to 0}{\longrightarrow} fg \tag{4.11}$$

and

$$\frac{1}{h}[f \star_{(g,h)} g - g \star_{(g,h)} f] \underset{q,h\to 0}{\longrightarrow} \frac{\partial f}{\partial z}\frac{\partial g}{\partial \bar{z}} - \frac{\partial g}{\partial z}\frac{\partial f}{\partial \bar{z}}. \tag{4.12}$$

From this we see that the model discussed here is a (q,h)-deformation of Hamiltonian mechanics of one degree of freedom. In order to have $\mathbf{D}_{q,h} \underset{q,h\to 0}{\longrightarrow} \mathbb{C}$, where $\mathbb{C}$ is interpreted as classical phase space, one assumes that $h, q \to 0$ in such way that $\sqrt{\frac{h}{1-q}} \to \infty$.

The generalization to the n-dimensional case is possible and we plan to study it in further publications.

References

1. A. Odzijewicz, Coherent states and geometric quantization, *Commun. Math. Phys.* 150:385-413, (1992)

2. A. Odzijewicz, On reproducing kernels and quantization of states, *Commun. Math. Phys.* 114:577-597 (1988)

3. P.E.T. Jørgensen, R.F. Werner, Coherent states of the q-canonical commutation relations, preprint Osnabrück (1993)

4. P.E.T. Jørgensen, R.F. Schmit, R.F. Werner, q-canonical commutation relations and stability of the Cuntz algebra, *Pacific.J.Math.* (to appear)

5. P.E.T. Jørgensen, R.F. Schmit, R.F. Werner, q-relations and stability of C^*-isomorphism classes, *in*: "Algebraic Methods in Operator Theory", Birkhäuser Verlag, Basel (1993)

6. A.U. Klimyk, N.J. Vilenkin, "Representation of Lie Groups and Special Functions", Kluwer, Amsterdam

7. M. Arik, D.D. Coon, Hilbert spaces of analytic functions and generalized coherent states, *J. Math. Phys.* 17:524-527 (1976)

8. P.E.T. Jørgensen, R.F. Schmit, R.F. Werner, Positive representation of general commutation relations allowing Wick ordering (preprint)

9. W. Pusz, Twisted canonical anticommutation relations, *Rep.Math.Phys.* 27:349-360 (1989)

10. W. Pusz, S.L. Woronowicz, Twisted second quantization, *Rep. Math. Phys.* 27:231-257 (1989)

11. S.L. Woronowicz, Quantum $E(2)$ group and its Pontryagin dual, *Lett. Math.Phys.* 23:251-263 (1991)

12. S.L. Woronowicz, Operator equalities related to the quantum $E(2)$ group, *Commun. Math. Phys.* 114:417-428 (1992)

13. F.A. Berezin, General concept of quantization, *Commun. Math. Phys.* 40:153-174 (1975)

14. F.A. Berezin, Models of Gross-Neveu type are quantization of a classical mechanics with nonlinear phase space, *Commun. Math. Phys.* 63:131-153 (1978)

15. V. Maximov, *Dokl.Akad.Nauk. USSR* 4 (1986)

16. V. Maximov, Algebraic representations of the fundamental operations in analysis (to appear).

QUANTIZATION BY QUADRATIC POLYNOMIALS IN CREATION AND ANNIHILATION OPERATORS

Wojtek Słowikowski

Institute of Mathematics, Aarhus University
Ny Munkegade, 8000 Aarhus C
Denmark

Let us fix a number $q = 1$ or -1. To a given Hilbert space $\mathcal{H}, <,>$, we attach an algebra $\Gamma_0\mathcal{H}$ generated by $\mathcal{H}$ and a unity $\emptyset$ called the *vacuum*. We call $\Gamma_0\mathcal{H}$ a *Fock algebra* if the scalar product from $\mathcal{H}$ is extended over $\Gamma_0\mathcal{H}$ in such a way that $<\emptyset,\emptyset>=1$ and, for every $x \in \mathcal{H}$, the operator $a^+(x)$ of multiplication by x admits the adjoint $a(x)$, defined on the whole $\Gamma_0\mathcal{H}$ and annihilating the vacuum, i.e. $< xf, g >=< f, a(x)g >$ for all $f, g \in \Gamma_0\mathcal{H}$ and $a(x)\emptyset = 0$. We assume that $a(x)$ fulfills the q-Leibnitz rule, i.e. $[a(x), a^+(y)]_q =< x, y > I$, where $[A, B]_q = AB - qBA$.

In the case $q = 1$, the algebra $\Gamma_0\mathcal{H}$ is commutative and is called a Bose algebra, whereas the case $q = -1$ makes the generators from $\mathcal{H}$ anticommute and $\Gamma_0\mathcal{H}$ is then called a Fermi algebra.

We denote by $\Gamma\mathcal{H}$ the completion of $\Gamma_0\mathcal{H}, <,>$ and we write $\mathcal{H}^n$ for the closure of the linear span of $\{x_1 \cdots x_n : x_1, ..., x_n \in \mathcal{H}\}$.

Take an orthonormal basis $\{e_n\}$ in $\mathcal{H}$. To each operator $A \in \mathcal{B}(\mathcal{H})$, we assign an operator

$$d\Gamma A = \Sigma_{j=1}^{\infty} a^+(Ae_n)a(e_n),$$

which is the unique extension of A to a derivation in $\Gamma_0\mathcal{H}$, and hence does not depend on the choice of $\{e_n\}$. For $A, B \in \mathcal{B}(\mathcal{H})$, we have

$$[d\Gamma A, d\Gamma B] = d\Gamma[A, B],$$

i.e. the transformation $A \to d\Gamma A$ is a Lie algebra homomorphism.

Denote by $\mathcal{L}^q_{hs}(\mathcal{H})$ the space of all Hilbert-Schmidt conjugate linear operators $L : \mathcal{H} \to \mathcal{H}$ such that $L' = qL$, where L' denotes the real adjoint to L.

We define the quadratic polynomial

$$h_L = \Sigma_{j=1}^{\infty} e_n(Le_n) \in \mathcal{H}^2$$

and observe that, for $K, L \in \mathcal{L}^q_{hs}(\mathcal{H})$,

$$< h_L, h_K >= 2q \text{ tr } KL.$$

Then, to each $L \in \mathcal{L}^q_{hs}(\mathcal{H})$, we assign the operator

$$a^+(h_L) : \Gamma_0\mathcal{H} \to \Gamma\mathcal{H}$$

of multiplication by h_L. The adjoint $a(h_L)$ of $a^+(h_L)$ is well defined on $\Gamma_0\mathcal{H}$.

Quantization, Coherent States, and Complex Structures
Edited by J.-P. Antoine *et al.*, Plenum Press, New York, 1995

Fix $\alpha = 1, -1$ and let ι be a conjugate linear mapping such that $\iota^2 = \alpha I$. Let further P be an orthogonal projection in $\mathcal{H}$. For $X \in \mathcal{B}(\mathcal{H})$ let

$$\begin{aligned} P\{X\} &= \iota P X^*(I-P) + q\alpha(I-P)XP\iota \\ P\langle X\rangle &= (I-P)X(I-P) - \alpha\iota P X^* P\iota \in \mathcal{B}(\mathcal{H}). \end{aligned}$$

To $X \in \mathcal{B}(\mathcal{H})$ with $P\{X\} \in \mathcal{L}^q_{hs}(\mathcal{H})$, we assign the transformation

$$: d\Gamma_P X := \frac{1}{2}(a^+(h_{P\{X\}}) - a(h_{P\{qX*\}}))) + d\Gamma P\langle X\rangle$$

of $\Gamma_0\mathcal{H}$ into $\mathcal{H}$.

Theorem 1. *For $X, Y \in \mathcal{B}(\mathcal{H})$ such that $P\{X\}, P\{Y\} \in \mathcal{L}^q_{hs}(\mathcal{H})$, we have*

$$[: d\Gamma_P X :, : d\Gamma_P Y :] =: d\Gamma_P[X,Y] : +\frac{1}{2} q \operatorname{tr} P\langle[[X,P],[P,Y]]\rangle I.$$

Hence $: d\Gamma_P \cdot :$ yields the central extension and for $q = -1$ we get the Schwinger term $\operatorname{tr} P\langle[[X,P],[P,Y]]\rangle$.

We shall apply the theorem to produce simultaneously Kac's wedge bosonization constructed in Chapter 14 of Ref. 1 and Wakimoto's construction from Ref. 2.

Let us fix an orthonormal basis $\{e_n : n \in \mathbf{Z}\}$ in $\mathcal{H}$ and let P be the orthogonal projection on the closure of the span of $\{e_n : n \leq 0\}$. Furthermore, let T_n, $n \in \mathbf{Z}$, be the discrete unitary group of translations $T_n e_k = e_{k+n}$, $k, n \in \mathbf{Z}$, *i.e.*

$$T_n x = \Sigma_{k\in\mathbf{Z}} < e_k, x > e_{k+n}.$$

Corollary 2. *The operators $: d_P T_n :$ considered on $\Gamma_0\mathcal{H}$ obey the CCR, i.e.*

$$[: d\Gamma_P T_n :, : d\Gamma_P T_m :] = q\delta_{n,-m} n,\ n > 0,\ m < 0.$$

References

1. V. Kac, "Infinite Dimensional Lie Algebras", Cambridge University Press (1990)

2. M. Wakimoto, Fock representations of the affine Lie algebra A, *Commun.Math.Phys.* 104:605-609 (1986).

ON DIRAC TYPE BRACKETS

Yurii M. Vorobjev[1*] and Ruben Flores Espinoza[2*]

[1] Department of Applied Mathematics
Moscow Institute of Electronics & Mathematics
B.Vuzovsky per., 3/12, Moscow 109028, Russia
[2] Department of Mathematics, University of Sonora
Rosales y Transversal, 83000 Hermosillo, Sonora, Mexico

Abstract

We investigate the class of Poisson structures with a transversally maximal Lie algebra of infinitesimal automorphisms. We describe such Poisson structures in terms of singular 2-forms and in terms of some universal de Rham cohomology classes of symplectic leaves.

1. INTRODUCTION

We consider a class of *regular, degenerate Poisson* structures with *transversally* maximal Lie algebra of infinitesimal automorphisms (or Poisson vector fields). This class naturally arises in the deformation and cohomology theory of Poisson brackets[1-8] and includes, for example, the *Dirac bracket* and some of its generalization.[7,9]

Let M be a symplectic manifold and $A^1, \ldots, A^r \in C^\infty(M)$ be a set of independent functions such that the matrix $\Delta = ((\Delta^{ij})) \equiv ((\{A^i, A^j\}))$ of pairwise Poisson brackets on M is nondegenerate everywhere. Then the standard Dirac bracket on M is given by the formula

$$\{f, g\}_{\mathrm{DIR}} = \{f, g\} + \sum_{1 \leq i,j \leq r} \Delta_{ij}\{A^i, f\}\{A^j, g\}, \tag{1.1}$$

where $\Delta_{is}\Delta^{sj} = \delta_i^j$. It is clear that the functions A^j are the *Casimir* functions relative to (1.1) and the corresponding symplectic leaves Ω coincide with the level sets of these functions. Consider the set of independent vector fields on M

$$z_i = \sum_{1 \leq j \leq r} \Delta_{ij}\, \mathrm{ad}(A^j), \qquad i = 1, \ldots, r, \tag{1.2}$$

where $\mathrm{ad}(A^j)$ is the *Hamiltonian field* of A^j with respect to the original Poisson structure on M. We claim[4,7,8] that $z_1, \ldots, z_r$ are infinitesimal automorphisms of (1.1), *transverse* to symplectic leaves Ω at each point.

*Research partially supported by Conacyt grant 489100-5-4106 E and Russian Foundation for Fundamental Research

The direct generalization of this situation is the following. For a given regular Poisson manifold $\mathcal{N}$, there exists a subbundle $\mathcal{H} \subset T\mathcal{N}$ *complementary* to the *symplectic foliation* of $\mathcal{N}$ and possessing a local basis of sections (in some neighborhood of every point in $\mathcal{N}$), which are infinitesimal automorphisms. According to Ref. 5, we will call such a Poisson structure *transversally constant.*

We are interested in criteria for brackets in the interior terms of a given Poisson structure to be transversally constant. To do this, we use the approach suggested in Refs. 3,4 for computing the Poisson cohomologies. In particular, we are also interested in the existence of a closed 2-form, compatible in a natural sense with a given Poisson structure. Such a problem arises, for instance, in geometric quantization under the construction of prequantization on Poisson manifolds[10,12,13] and under Hamiltonian regularizations of singular Lagrangian systems.[11]

2. TRANSVERSALLY CONSTANT POISSON STRUCTURES

Let $\mathcal{N}$ be a *regular Poisson manifold* with Poisson bracket

$$\{f,g\} = \langle df, \Psi\, dg\rangle, \qquad f,g \in C^\infty(\mathcal{N}),$$

where the bivector field $\Psi : T^*\mathcal{N} \to T\mathcal{N}$ satisfies the Jacobi identity and

$$\dim(\mathrm{Ker}\,\Psi(\xi)) = r \quad \forall \xi \in \mathcal{N}.$$

Then the integrable plane distribution $\Pi \subset T\mathcal{N}$ with fibers

$$\Pi(\xi) \overset{\mathrm{def}}{=} \mathrm{Ker}\,\Psi(\xi)^\perp, \qquad \xi \in C^\infty(\mathcal{N}),$$

generates the *characteristic foliation* of $\mathcal{N}$ by *symplectic leaves* Ω. Denote by $\Gamma(\Pi)$ the space of sections in the subbundle Π. Thus, $\Gamma(\Pi)$ is the Lie subalgebra of the Lie algebra $\mathcal{X}(\mathcal{N})$ of all vector fields on $\mathcal{N}$.

We say that a 2-form $\mu : T\mathcal{N} \to T^*\mathcal{N}$ on $\mathcal{N}$ is *compatible* with the Poisson structure on $\mathcal{N}$, if

$$\Psi \circ \mu \circ \Psi = \Psi. \tag{2.1}$$

In other words, the restriction of μ to each symplectic leaf Ω coincides with the *symplectic structure* (the *Kirillov form*). A given compatible 2-form μ induces the subbundle $\mathcal{H}_\mu \subset T\mathcal{N}$ with fibers

$$\mathcal{H}_\mu(\xi) = \{u \in T_\xi\mathcal{N} : i_u\mu \text{ annihilates } \Pi(\xi)\},$$

which are transversal to $\Pi(\xi)$ at each point. It is clear that $\mathrm{Ker}_\xi\,\mu \subseteq \mathcal{H}_\mu(\xi)$ and $\dim \mathcal{H}_\mu(\xi) = r$, i.e., $\mathcal{H}_\mu$ is the complementary subbundle to Π in $T\mathcal{N}$.

Conversely, for any subbundle $\mathcal{H}$ complementary to Π, we can choose a compatible 2-form μ such that $\mathrm{Ker}_\xi\,\mu = \mathcal{H}(\xi)$. So, there always exists a compatible 2-form on $\mathcal{H}$, since there always exists a complementary subbundle $\mathcal{H}$.

If, in addition to (2.1), a 2-form μ satisfies the condition

$$\Psi \circ (L_X\mu) \circ \Psi = 0 \tag{2.2}$$

for each (local) section X in $\mathcal{H}_\mu$, then we say that μ is *transversally compatible.*

Note that, if a compatible 2-form μ is *closed* on $\mathcal{N}$, then condition (2.2) holds automatically. For example, in the case of Dirac bracket (1.1), we can choose for μ the *symplectic structure on M*. Sufficient conditions of the existence of a closed, compatible 2-form on $\mathcal{N}$ were obtained in Ref. 10, in the case when a symplectic foliation of $\mathcal{N}$ is a fiber bundle. The problem of defining Poisson brackets on presymplectic manifolds was investigated in Ref. 11. Another interesting problem of the same type concerning the construction of new Poisson structures on a given Poisson manifold by using 2-forms was considered in Ref. 7.

Our goal is to describe criteria for the existence of a transversally compatible 2-form on $\mathcal{N}$ in natural terms of a given Poisson structure. We start with an algebraic criterion.

Recall that a vector field $z \in \mathcal{X}(\mathcal{N})$ is called an *infinitesimal automorphism* of the Poisson structure on $\mathcal{N}$ if its (local) flow preserves Ψ,

$$\exp_*(tz)\Psi(\xi) = \Psi(\exp(tz)\xi) \quad \forall \xi \in \mathcal{N}.$$

Denote by $\mathcal{P}$ the Lie algebra of all infinitesimal automorphisms on $\mathcal{N}$. Single out the Lie subalgebra $\mathcal{P}_0$ of all infinitesimal automorphisms, *tangent* to Π at each point. Then $\mathcal{P}_0$ includes at least the *Hamiltonian fields* $\mathrm{ad}(f) = -\Psi\, df$, $f \in C^\infty(\mathcal{N})$, and

$$[\mathcal{P}, \mathcal{P}_0] \subset \mathcal{P}_0,$$

i.e., $\mathcal{P}_0$ is an *ideal* in $\mathcal{P}$. So, we have the *quotient* Lie algebra $\mathfrak{g}(\mathcal{N}) = \mathcal{P}/\mathcal{P}_0$.

On the other hand, consider the Lie algebra $V(\Pi)$ of all vector fields $v \in \mathcal{X}(\mathcal{N})$ whose (local) flows preserve the characteristic foliation Π,

$$\exp_*(tv)\Pi(\xi) = \Pi(\exp(tv)\xi) \quad \forall \xi \in \mathcal{N}.$$

It is clear that $\Gamma(\Pi) \subset V(\Pi)$ and $\Gamma(\Pi)$ is an ideal in $V(\Pi)$. Denote by $\mathfrak{n}(\Pi) = V(\Pi)/\Gamma(\Pi)$ the corresponding quotient Lie algebra. Then, the inclusions $\mathcal{P}_0 \subset \Gamma(\Pi)$, $\mathcal{P} \subset V(\Pi)$ induce the natural homomorphism $h\colon \mathfrak{g}(\mathcal{N}) \to \mathfrak{n}(\Pi)$ that makes the following diagram commutative

$$\begin{array}{ccccccccc} 0 & \longrightarrow & \mathcal{P}_0 & \xrightarrow{\subset} & \mathcal{P} & \longrightarrow & \mathfrak{g}(\mathcal{N}) & \longrightarrow & 0 \\ & & \cap\downarrow & & \cap\downarrow & & \downarrow h & & \\ 0 & \longrightarrow & \Gamma(\Pi) & \xrightarrow{\subset} & V(\Pi) & \longrightarrow & \mathfrak{n}(\Pi) & \longrightarrow & 0. \end{array} \tag{2.3}$$

Proposition 2.1. *If there is a transversally compatible 2-form μ on $\mathcal{N}$, then the homomorphism h from diagram* (2.3) *is an isomorphism,*

$$\mathfrak{g}(\mathcal{N}) \approx \mathfrak{n}(\Pi). \tag{2.4}$$

Remark 2.1. The existence of a transversally compatible 2-form implies that the Lie algebra $\mathfrak{g}(\mathcal{N})$ is *possibly maximal.* In particular, if the symplectic foliation of $\mathcal{N}$ is a fibration with base $\mathcal{B}$, then $\mathfrak{g}(\mathcal{N}) \approx \mathcal{X}(\mathcal{B})$. Note that condition (2.4) does not hold, for example, for linear degenerate Poisson structures of constant rank corresponding to *compact* Lie algebras. In this case we always have $\mathfrak{g}(\mathcal{N}) = \{0\}$.[3,4,7]

Remark 2.2. From conditions (2.1), (2.2) it follows that the subbundle $\mathcal{H}_\mu$ transversal to Π, possesses the following property. If a local section X of $\mathcal{H}_\mu$ belongs to $V(\Pi|U)$ then this implies that $X \in \mathcal{P}(\mathcal{N}|U)$. This is just the definition of *transversally constant* Poisson structure introduced in Ref. 5. On the contrary, if $\mathcal{N}$ is transversally constant, then there exists a transversally compatible 2-form μ.

Now, following Refs. 3,4, we formulate a cohomology criterion.

Recall that the Schouten bracket[1] between the bivector field Ψ and a vector field $v \in \mathcal{X}(\mathcal{N})$ is the bivector field $[\![\Psi, v]\!] : T^*\mathcal{N} \to T\mathcal{N}$ given by

$$[\![\Psi, v]\!](df, dg) = \{f, L_v(g)\} - \{g, L_v(f)\} - L_v(\{f, g\}).$$

It is easy to see that, if $v \in V(\Pi)$ then the bivector field $[\![\Psi, v]\!]$ annihilates $\operatorname{Ker}\Psi(\xi) \subset T^*_\xi\mathcal{N}$ $\forall\xi \in \mathcal{N}$. Using this we can define the 2-form $\alpha^v : \Pi \to \Pi^*$ for any $v \in V(\Pi)$ by

$$\alpha^v(\Psi\, df, \Psi\, dg) = [\![\Psi, v]\!](df, dg) \quad \forall f, g \in C^\infty(\mathcal{N}).$$

Note that the restriction α^v_Ω of α^v to each symplectic leaf Ω is the *closed* 2-form and if $v \in \Gamma(\Pi)$, then α^v_Ω is *exact*. So, for each symplectic leaf Ω, we have a linear mapping from the Lie algebra $\mathfrak{n}(\Pi)$ to the de Rham cohomology space $\mathsf{H}^2(\Omega; \mathbb{R})$. The image under this mapping is denoted by

$$\mathfrak{V}^2(\Omega) = \{[\alpha^v_\Omega] : v \quad \text{runs over} \quad \mathfrak{n}(\Pi)\}.$$

Proposition 2.2. *If there is a transversally compatible 2-form μ, then*

$$\mathfrak{V}^2(\Omega) = \{0\} \tag{2.5}$$

for each symplectic leaf Ω. In the case when the symplectic foliation of $\mathcal{N}$ is a fibration, the conditions (2.4) *and* (2.5) *are equivalent.*

So, conditions (2.4), (2.5) give us obstructions to the existence of a transversally compatible 2-form on $\mathcal{N}$. Under additional assumptions we have the following criterion.

Theorem 2.1. *Suppose that condition* (2.5) *holds and the foliation of $\mathcal{N}$ by symplectic leaves Ω is a fibration*

$$\pi : \mathcal{N} \to \mathcal{B} \tag{2.6}$$

with parallelizable base $\mathcal{B}$. Then there exists a set of independent infinitesimal automorphisms $z_1, \ldots, z_r$ on $\mathcal{N}$, transversal to Π at each point,

$$z_1, \ldots, z_r \in \mathcal{P}, \tag{2.7}$$

$$T_\xi\mathcal{N} = \operatorname{Span}\{z_1(\xi), \ldots, z_r(\xi)\} \oplus \Pi(\xi). \tag{2.8}$$

Corollary 2.1. *The subbundle $\mathcal{H} \subset T\mathcal{N}$ with fibers $\mathcal{H}(\xi) = \operatorname{Span}\{z_1(\xi), \ldots, z_r(\xi)\}$ generates the following transversally compatible 2-form on $\mathcal{N}$:*

$$\mu_\sigma = \mu_0 + \pi^*\sigma. \tag{2.9}$$

Here the 2-form μ_0 is fixed by (2.1) *and by the condition:*

$$\operatorname{Ker}_\xi \mu_0 = \mathcal{H}(\xi), \tag{2.10}$$

and σ is any 2-form on the base $\mathcal{B}$.

3. CLOSED EXTENSIONS

We will suppose below that the hypotheses of Theorem 2.1 are satisfied. We are interested in the existence of a 2-form σ on $\mathcal{B}$ such that a 2-form μ_σ of (2.9) is closed on $\mathcal{N}$.

In addition we suppose that the symplectic leaves Ω are simply connected and the base $\mathcal{B}$ of the symplectic fibration (2.6) is *integrably parallelizable*. The last assumption means that we can fix a basis of closed 1-forms $\eta^1, \ldots, \eta^r$ on $\mathcal{B}$. Then we can choose vector fields $z_1, \ldots, z_r$ satisfying (2.7), (2.8) and the following

$$\mathbf{i}_{z_i} \pi^*(\eta^j) = \delta_i^j, \qquad i, j = 1, \ldots, r. \tag{3.1}$$

Lemma 3.1. *The pairwise commutator of vector fields* $z_1, \ldots, z_r$ *satisfying* (2.7), (2.8), (3.1) *has the form*

$$[z_i, z_j] = \Psi\, dh_{ij}, \qquad i, j = 1, \ldots, r, \tag{3.2}$$

where $h_{ij} \in C^\infty(\mathcal{N})$ *are some smooth functions.*

It is clear that the functions h_{ij} are determined from (3.2) up to the addition of any Casimir functions $\pi^* f_{ij}$, where $f_{ij} \in C^\infty(\mathcal{B})$. Let us fix h_{ij} and define

$$c_{ijm} = \sum_{(i,j,m)} L_{z_i}(h_{jm}),$$

where the sum is cyclic. By (3.2) it follows that the set of functions c_{ijm} is completely antisymmetric and c_{ijm} is constant along symplectic leaves Ω. Using this, we define the 3-form on the base $\mathcal{B}$:

$$\mathbf{c} = \sum_{i<j<m} c_{ijm} \eta^i \wedge \eta^j \wedge \eta^m. \tag{3.3}$$

Lemma 3.2. *The 3-form* $\mathbf{c}$ *is closed on* $\mathcal{B}$.

So under the above assumptions we obtain

Theorem 3.1. *Let* $\mathbf{c}$ *be an exact form,*

$$\mathbf{c} = d\mathbf{b} \tag{3.4}$$

with some primitive 2-form $\mathbf{b}$ *on* $\mathcal{B}$. *Then the 2-form* μ_σ *defined by* (2.9) *is closed on* $\mathcal{N}$, *where* μ_0 *is associated with vector fields* $z_1, \ldots, z_r$ *satisfying* (2.7), (2.8), (3.1) *and the 2-form* σ *is given by*

$$\sigma = \sum_{i<j} h_{ij} \pi^*(\eta^i) \wedge \pi^*(\eta^j) - \pi^* \mathbf{b}.$$

Corollary 3.1. *Any closed compatible 2-form on* $\mathcal{N}$ *has the representation*

$$\mu^{\mathbf{a},\beta} = \mu_\sigma + \pi^* \mathbf{a} + d\beta, \tag{3.5}$$

where $\mathbf{a}$ *is a closed 2-form on the base* $\mathcal{B}$, *and* β *is a 1-form on* $\mathcal{N}$ *which belongs to the kernel of the Poisson structure* Ψ,

$$\Psi \beta = 0. \tag{3.6}$$

Remark 3.1. In the case when the symplectic foliation of $\mathcal{N}$ is a fibre bundle, $\pi : \mathcal{N} \to \mathcal{B}$, the results of Ref. 10 give us the following conditions for the existence of a closed compatible 2-form μ: (a) a structure group preserves the symplectic structure on Ω; (b) the base $\mathcal{B}$ and the fibers Ω are simply connected; (c) there exists some integer k $(0 \leq 2k \leq \dim \Omega)$ such that $\mathsf{H}^{2k}(\Omega; \mathbb{R}) = 0$.

Remark 3.2. In the general case, the existence of a closed compatible 2-form on $\mathcal{N}$ is equivalent to the triviality of some class $\mathcal{C}$ in a *relative de Rham cohomology* $\mathsf{H}^3(\mathcal{N}; \Pi)$ with respect to the symplectic foliation Π. If we take any compatible 2-form μ on $\mathcal{N}$, then $d\mu$ vanishes on symplectic leaves Ω and defines the cohomology class $[d\mu] \equiv \mathcal{C}$ which does not depend on the choice of μ. So, condition (3.4) implies that $\mathcal{C} = 0$.

The next natural question is the existence of a closed compatible 2-form $\mu^{\mathbf{a},\alpha}$ on $\mathcal{N}$ of *constant rank*. Let us try to find forms $\mathbf{a}$, α from (3.5), (3.6) providing this property.

In addition to the assumptions of Theorem 3.1 suppose that there is a fibration $\tilde{\pi} : \mathcal{N} \to \tilde{\mathcal{B}}$ with base $\tilde{\mathcal{B}}$ and fibers $\tilde{\Omega}$ which is *transversal* to the symplectic fibration (2.6),

$$\dim \tilde{\Omega} = \operatorname{codim} \Omega, \qquad T_\xi \tilde{\Omega} \cap T_\xi \Omega = \{0\} \quad \forall \xi \in \mathcal{N}.$$

In this situation the main obstruction to the existence of the desired 2-form $\mu^{\mathbf{a},\alpha}$ is the cohomology classes $[\mu_\sigma|_{\tilde{\Omega}}] \in \mathsf{H}^2(\tilde{\Omega}; \mathbb{R})$, where $\mu_\sigma|_{\tilde{\Omega}}$ is the restriction of the closed 2-form μ_σ to a fiber $\tilde{\Omega}$.

Proposition 3.1. *Suppose that there exists a closed 2-form* $\mathbf{a}$ *on* $\mathcal{B}$ *such that the de Rham cohomology class induced by the form* $\mu_\sigma + \pi^*\mathbf{a}$ *on each fiber* $\tilde{\Omega}$ *is trivial,*

$$(\mu_\sigma + \pi^*\mathbf{a})\Big|_{\tilde{\Omega}} \quad \textit{is exact.} \tag{3.7}$$

Then there is a 1-form β *on* $\mathcal{N}$ *satisfying* (3.6) *such that the 2-form* $\mu^{\mathbf{a},\beta}$ *of* (3.5) *possesses the property*

$$\operatorname{Ker}_\xi \mu^{\mathbf{a},\alpha} = T_\xi \tilde{\Omega} \quad \forall \xi \in \mathcal{N}.$$

In particular, if

$$\mathsf{H}^2(\mathcal{B}; \mathbb{R}) = 0, \tag{3.8}$$

then conditions (3.7) *holds automatically and we can choose* $\mathbf{a} \equiv 0$.

In the case where $\mathcal{N}$ has a transversally constant Poisson structure and a symplectic fibration (2.6) with simply connected leaves, by Proposition 3.1, we can obtain the following version of the *splitting theorem*[2] in some neighborhood of every symplectic leaf Ω. There are a tubular neighborhood Σ of Ω in $\mathcal{N}$ and a Poisson isomorphism from Σ to the product $\Omega \times U$, where $U \subset \mathbb{R}^r$ is an open set with zero Poisson structure.

It is of interest to note that two Poisson structures with the same symplectic foliation may have opposite properties in the sense of existence of a closed compatible 2-form. An example of such kind concerning a "deformed" Poisson structure on $so^*(3)$ was considered in Ref. 11. We illustrate this "effect" in the following

Example 3.1. Consider the coalgebra $e^*(3) = \mathbb{R}^3_y \times \mathbb{R}^3_z$, where (y, z) are coordinates associated with the decomposition of the Lie algebra $e(3)$ into the semidirect sum $e(3) = so(3) + \mathbb{R}^3$. The linear Poisson structure on $e^*(3)$ is given by the bivector field

$$\Psi = \frac{1}{2}\Big(y \times \frac{\partial}{\partial y}\Big) \wedge \frac{\partial}{\partial y} + \frac{1}{2}\Big(z \times \frac{\partial}{\partial z}\Big) \wedge \frac{\partial}{\partial y} + \frac{1}{2}\Big(z \times \frac{\partial}{\partial y}\Big) \wedge \frac{\partial}{\partial z}.$$

Consider the 6-dimensional Poisson submanifold

$$\mathcal{N} = \{(y,z) \in e^*(3) : y \cdot z \neq 0\}$$

with the characteristic symplectic fibration

$$T^*\mathbb{S}^2 \to \mathcal{N} \to (\mathbb{R} \setminus \{0\}) \times (0, \infty).$$

Then the 2-tensor field Ψ gives the regular Poisson structure on $\mathcal{N}$ which has no transversally compatible 2-form μ, because the necessary condition (2.5) does not hold. But if we consider the "deformed" Poisson structure

$$\widetilde{\Psi} = \frac{|z|}{y \cdot z}\Psi$$

with the same symplectic fibration, then all hypotheses of Theorem 3.1 are satisfied and, hence, there exists closed compatible 2-form relative to $\widetilde{\Psi}$.

Consider the above results from the viewpoint of geometric quantization. Following Refs. 12,13, we can say that there is a *prequantization* (in the sense of Kostant–Souriau) of the Poisson manifold $(\mathcal{N}, \Psi)$ if there exists a complex line bundle over $\mathcal{N}$ with Hermitian connection ∇ and there exists a vector field $v \in \mathcal{P}$ such that

$$\Psi \circ (\alpha^v + \mathit{curvature}\ \nabla) \circ \Psi = -2\pi i \Psi,$$

where the 2-form α^v is defined above. Suppose that the Poisson structure Ψ satisfies the assumptions of Theorem 3.1 and condition (3.8) holds. Fix the form μ_σ of Theorem 3.1. Then we obtain that there exists a prequantization of $(\mathcal{N}, \Psi)$ iff the cohomology class $[\mu_\sigma]$ is integral, $[\mu_\sigma] \in \mathrm{H}^2(\mathcal{N}; \mathbb{Z})$.

Acknowledgements. The authors wish to thank Prof. M. V. Karasev and Prof. G. Marmo for useful discussions.

References

1. A. Lichnerowicz, *J.Diff. Geom.* 12:253-300 (1977)
2. A. Weinstein, *J.Diff. Geom.* 18:3523-3557 (1983)
3. M. V. Karasev and Yu. M. Vorobjev, *Funct. Anal.* 22:1-9 (1988)
4. M. V. Karasev and Yu. M. Vorobjev, *Springer Lecture Notes in Math.* 1453:271-289 (1990)
5. I. Vaisman, *Ann. Inst. Fourier Grenoble* 40:951-963 (1990)
6. P. Xu, *Ann. Inst. Fourier Grenoble* 42:967-988 (1992)
7. M. V. Karasev and V. P. Maslov, "Poisson Brackets. Geometry and Quantization", AMS, Providence, RI (1993)
8. Yu. M. Vorobjev and R. Flores Espinoza, preprint Math. 94-1, Univ. Sonora, Hermosillo, Sonora (1994)
9. J. Sniatycki, *Ann. Inst. H. Poincaré A* 20:365-372 (1974)
10. M. J. Gotay, R. Lashof, J. Sniatycki and A. Weinstein, *Comment. Math. Helvetici* 58:617-621 (1983)
11. B. A. Dubrovin, M. Giordano, G. Marmo and A. Simoni, *Intern. J. Modern Phys.* 8:3747-3771 (1993)
12. J. Huebschmann, *J. für reine angew. Math.* 408:57-113 (1990)
13. I. Vaisman, *J. Math. Phys.* 32: 3339-3345 (1991).

QUANTUM TRIGONOMETRY AND PHASE-SPACE PROPENSITY

Krzysztof Wódkiewicz[1,2] and Berthold-Georg Englert[3,4]

[1] Instytut Fizyki Teoretycznej, Uniwersytet Warszawski
Hoża 69, Warszawa 00-681, Poland
[2] Center for Advanced Studies University of New Mexico
Albuquerque, New Mexico 87131, USA
[3] Max-Planck-Institut für Quantenoptik
D-85748 Garching, Germany
[4] Sektion Physik, Universität München,
Am Coulombwall 1, D-85748 Garching, Germany

Abstract

Quantum trigonometry, corresponding to the operational detection of the quantum phase of an optical field, is derived from the phase-space propensity.

1. INTRODUCTION

The description of quantum phase fluctuations of optical fields is a nontrivial problem because of the well known difficulties associated with the definition of a Hermitian phase operator.[1,2] In the published literature one can find several competing descriptions of quantum phase fluctuations based on different assumptions and mathematical constructions of the phase operator. Only in the classical limit, i.e., when the electromagnetic fields are strong (have a large number of photons), a semiclassical or a statistical description of phase fluctuations is possible. In order to avoid the difficulty with the definition of the hermitian phase operator Susskind and Glogower[3] introduced two operators S and C corresponding to the classical quantities $\sin\varphi$ and $\cos\varphi$:

$$C = \frac{1}{2}\left(\frac{1}{\sqrt{n+1}}b + b^\dagger\frac{1}{\sqrt{n+1}}\right), \tag{1.1}$$

$$S = \frac{1}{2i}\left(\frac{1}{\sqrt{n+1}}b - b^\dagger\frac{1}{\sqrt{n+1}}\right), \tag{1.2}$$

where b and $b^\dagger$ are the annihilation and creation operators of the single mode and $n = b^\dagger b$ is the photon number operator. However these operators do not commute and the trigonometric unity rule does not hold, it is violated for the vacuum state in the form $(S^2 + C^2)|0\rangle = \frac{1}{2}|0\rangle$. Many new phase operators have been suggested as summarized in the excellent review by Carruthers and Nieto.[4]

It is the purpose of this paper to investigate the quantum trigonometry of a single mode of a harmonic oscillator, using the concept of the phase-space propensity and the associated operational positive operator valued measure (POVM). Our approach

Quantization, Coherent States, and Complex Structures
Edited by J.-P. Antoine *et al.*, Plenum Press, New York, 1995

is motivated by an operational approach to phase fluctuations presented in the recent measurements of the phase properties of optical fields by Noh, Fougères, and Mandel (NFM).[5,6]

2. QUANTUM PROPENSITY

A non-operator approach to phase fluctuations is based on the principle of quantum mechanical propensity.[7] The quantum mechanical propensity follows from the assumption that in realistic measurements one compares, for example, a state described by a density matrix ρ with some well-known reference (filter) state, denoted here by $\rho_{\mathcal{F}}$, which in this paper we select to be the vacuum state of a single mode harmonic oscillator $\rho_{\mathcal{F}} = |0\rangle\langle 0|$. In order to compare physically states in realistic laboratory arrangement we have to "displace" the reference filtering device towards the measured field (for example we have to perform a homodyne detection of a an optical field). The displacement can be given as the Glauber displacement operator $D(\beta) = \exp(\beta a^\dagger - \beta^* a)$. Under such a "motion", the reference state is displaced in the coherent phase-space β by the following unitary transformation:

$$\rho_{\mathcal{F}}(\beta) = D(\beta)\rho_{\mathcal{F}} D^\dagger(\beta), \tag{2.1}$$

where the complex parameter $\beta = re^{i\varphi}$ characterizes the phase-space in which a displacement of the reference state by a radius r and a rotation by an angle φ is taking place. We note that, for the selected filtering state, $\rho_{\mathcal{F}}(\beta) = |\beta\rangle\langle\beta|$ is just the projection operator into a coherent state $|\beta\rangle$.

We define the quantum mechanical propensity as the probability density that the state ρ differs from the displaced reference state $|\beta\rangle$ by the angle φ. This propensity is represented by the following expression:

$$\Pr(\varphi) = \mathrm{Tr}\{\rho\mathcal{F}(\varphi)\}, \tag{2.2}$$

where

$$\mathcal{F}(\varphi) = \frac{1}{\pi}\int_0^\infty dr\ r\ |re^{i\varphi}\rangle\langle re^{i\varphi}|. \tag{2.3}$$

Such an operational approach to phase fluctuations has been exemplified by NMF in the recent measurements of the phase properties of optical fields. Here the filter $\mathcal{F}$ accounts for the beam splitters, mirrors, and photon counters used in the homodyne detection. As an illustration of this propensity, we now turn to the operational phase difference of two monochromatic electromagnetic waves determined by measuring its sine and cosine simultaneously in an appropriately designed interferometer. Such a device has been used in the recent NFM experiments[5,6] for a measurement of the quantum phase properties of a low-intensity laser, relative to a high intensity classical field (local oscillator). The experimental data are summarized in the so-called "phase distribution", which is nothing but the propensity density $\Pr(\varphi)$ for the classical phase variable φ that NFM associate operationally with the phase properties of the probe field.

By construction, this propensity (2.2) is periodic, $\Pr(\varphi) = \Pr(\varphi + 2\pi)$, and we normalize it such a way that

$$\int_{(2\pi)} \mathrm{d}\varphi \Pr(\varphi) = 1 \tag{2.4}$$

holds, where the integration covers any φ interval of length 2π.

From this relation we conclude that the propensity density $\Pr(\varphi)$ is given by

$$\Pr(\varphi) = \frac{1}{2\pi}\int_0^\infty \mathrm{dI}\,\langle\beta|\rho|\beta\rangle\,, \tag{2.5}$$

where ρ is the density operator of the photon state of the probe field and $|\beta\rangle$ is a normalized eigenstate of b. Here, $\beta = \sqrt{I}\exp(i\varphi)$ relates the eigenvalue β to the phase variable φ and the intensity I. In the jargon of quantum optics, $\Pr(\varphi)$ is the radially integrated Q function of ρ, and $|\beta\rangle$ is a coherent state or Glauber state.[8] Usually the Q-representation emerges in quantum optics applications in the context of various quasi-distribution functions associated with different orderings of the boson operators b and $b^\dagger$. Here the Q-representation acquires an additional interpretation of a phase-space propensity if and only if the vacuum state is used as a reference state.[9]

From this expression, one can calculate various statistical moments of φ using standard formulas from probability theory. Note that these statistical averages represent quantum fluctuations of the state ρ probed by a device described by a state $|\beta\rangle$. Because of this property, statistical averages of φ should always be interpreted as relative fluctuations with respect to the given reference state. This construction and this interpretation is consistent with the view that, in realistic measurements, we compare the properties of the detected state with a given reference "meter". The classical average of a periodic function $g(\varphi) = g(\varphi + 2\pi)$ is then given by

$$\overline{g(\varphi)} = \int_{(2\pi)} \mathrm{d}\varphi\, \mathrm{g}(\varphi)\Pr(\varphi)\,. \tag{2.6}$$

This number equals the quantum expectation value

$$\overline{g(\varphi)} = \langle G_{\mathcal{F}}\rangle, \tag{2.7}$$

of the corresponding operational operator $G_{\mathcal{F}}(b^\dagger, b)$, which is a function of $b^\dagger$ and b, the creation and annihilation operators for photons in the probe field.[10]

3. OPERATIONAL POVM

The positive-operator-valued-measure or POVM provides a natural description of the quantum world in terms of measurable or observable quantities.[11] We shall show that the operational quantum propensity for the phase measurements determines in a unique way an operational POVM.[10]

The propensity (2.2) determines classical averages of sine and cosine functions as exemplified by

$$\begin{aligned}\overline{\cos\varphi^n} &= \langle C_{\mathcal{F}}^{(n)}\rangle,\\ \overline{\sin\varphi^n} &= \langle S_{\mathcal{F}}^{(n)}\rangle.\end{aligned} \tag{3.1}$$

In view of the linear relations (2.7) and (3.1), these requirements specify a unique set of cosine and sine operators $C_{\mathcal{F}}^{(n)}$ and $S_{\mathcal{F}}^{(n)}$,

$$C_{\mathcal{F}}^{(n)} = \int \mathrm{d}\varphi\, \cos^{\mathrm{n}}(\varphi)\mathcal{F}(\varphi), \tag{3.2}$$

$$S_{\mathcal{F}}^{(n)} = \int \mathrm{d}\varphi\, \sin^{\mathrm{n}}(\varphi)\mathcal{F}(\varphi), \tag{3.3}$$

for the given filter $\mathcal{F}$. The comparison with the standard spectral decomposition of hermitian operators shows that the spectral measure is effectively replaced by the propensity $d\varphi \Pr(\varphi)$, which refers to the filter $\mathcal{F}$ of the measuring device. In one way of looking at quantum measurements,[11] the emphasis is put on the operational POVM. The quantity $d\phi \mathcal{F}(\phi)$ is such a POVM. Here the filter $\mathcal{F}$ accounts for the NFM apparatus and the relations identify the quantum trigonometry of the cosine and sine operational operators.

As a rule, the algebraic properties of the $C_{\mathcal{F}}^{(n)}$ and $S_{\mathcal{F}}^{(n)}$ operators are quite different from those of the powers of $C_{\mathcal{F}}^{(1)}$ and $S_{\mathcal{F}}^{(1)}$. In particular, a factorization is typically impossible, so that, for instance, $C_{\mathcal{F}}^{(2)}$ does not equal $(C_{\mathcal{F}}^{(1)})^2$.

4. QUANTUM TRIGONOMETRY

The relations (2.7) and (3.1) allow to map the classical trigonometry onto the corresponding quantum trigonometry associated with the NFM experiment. This can be accomplished with the help of operational operators corresponding to the Fourier basis. These operators are called *phasors*. The phasor basis $E_{\mathcal{F}}^{(n)}$ is thus identified by the defining property

$$\overline{\exp(in\varphi)} = \left\langle E_{\mathcal{F}}^{(n)} \right\rangle \tag{4.1}$$

for $n = 0, \pm 1, \pm 2, \ldots$. The reality of the propensity density $\Pr(\varphi)$ implies that $E_{\mathcal{F}}^{(-n)}$ is the adjoint of $E_{\mathcal{F}}^{(n)}$, and $E_{\mathcal{F}}^{(0)} = 1$ is an immediate consequence of the normalization (2.4).

The elements of the phasor basis are the cornerstone of quantum trigonometry, because all trigonometric functions are just weighted sums of these fundamental operators. As an example, we have for the cosine (sine) and the cosine^2 (sine^2) functions the following operational definitions:

$$\begin{aligned} C_{\mathcal{F}}^{(1)} &= \frac{1}{2}(E_{\mathcal{F}}^{(1)} + E_{\mathcal{F}}^{(-1)}), \\ C_{\mathcal{F}}^{(2)} &= \frac{1}{4}(E_{\mathcal{F}}^{(2)} + 2E_{\mathcal{F}}^{(0)} + E_{\mathcal{F}}^{(-2)}). \end{aligned} \tag{4.2}$$

In fact, using relation (4.1) one can infer the entire quantum trigonometry from the operational phasors.

From the operational propensity (2.2) or from the POVM (2.7) analysis, one can derive the phasors (4.1) in normally ordered form, compactly presented as [10,12]

$$E_{\mathcal{F}}^{(n)} = \frac{(n/2)!}{n!} : M(n/2, n+1, -b^\dagger b) : b^n \tag{4.3}$$

for $n = 0, 1, 2, \ldots$, where M denotes the confluent hypergeometric function, and the pair of colons indicates normal ordering of the operators $b^\dagger$ and b, that is: all $b^\dagger$s to the left of all bs. A particularly nice form of the basic phasors is [12]

$$E_{\mathcal{F}}^{(n)} = \frac{(b^\dagger b + n/2)!}{(b^\dagger b + n)!} b^n \quad \text{for} \quad \text{n} = 0, 1, 2, \ldots ; \tag{4.4}$$

it is perhaps best suited for the construction of the quantum trigonometry associated with a classical observable $g(\varphi)$. For instance, the operators (4.2) appear as

$$\begin{aligned} C_{\mathcal{F}}^{(1)} &= \frac{1}{2}\left(b^\dagger \frac{(b^\dagger b + 1/2)!}{(b^\dagger b + 1)!} + \frac{(b^\dagger b + 1/2)!}{(b^\dagger b + 1)!} b\right), \\ C_{\mathcal{F}}^{(2)} &= \frac{1}{2} + \frac{1}{4}\left(b^{\dagger 2} \frac{1}{b^\dagger b + 2} + \frac{1}{b^\dagger b + 2} b^2\right). \end{aligned} \tag{4.5}$$

Similar expressions are obtained for the sine operator:

$$
\begin{aligned}
S_{\mathcal{F}}^{(1)} &= \frac{1}{2i}\Big(b^{\dagger}\frac{(b^{\dagger}b+1/2)!}{(b^{\dagger}b+1)!} - \frac{(b^{\dagger}b+1/2)!}{(b^{\dagger}b+1)!}b\Big), \\
S_{\mathcal{F}}^{(2)} &= \frac{1}{2} - \frac{1}{4}\Big(b^{\dagger 2}\frac{1}{b^{\dagger}b+2} + \frac{1}{b^{\dagger}b+2}b^{2}\Big). \qquad (4.6)
\end{aligned}
$$

Note that, due to the operational character of these cosine operators, they differ considerably from the Susskind-Glogower operators (1.1), which are intrinsic in character. The operational cosine and sine functions satisfy the trigonometric unity rule.

$$C_{\mathcal{F}}^{(2)} + S_{\mathcal{F}}^{(2)} = 1. \qquad (4.7)$$

5. GLAUBER'S PHASE-SPACE QUANTUM TRIGONOMETRY

The normally ordered phasors given by (4.3) are very useful for the calculations of the phase-space quantum trigonometry if Glauber's diagonal P-representation of the density operator is used:

$$\rho = \int d^2\beta P(\beta)|\beta\rangle\langle\beta|. \qquad (5.1)$$

In this formula, the diagonal distribution function $P(r,\phi) = P(\beta)$ is the quantum mechanical quasi-distribution corresponding to the state ρ. In this case (3.1) takes the following form:

$$\langle C_{\mathcal{F}}^{(n)}\rangle = \int d^2\beta\, P(\beta)\, \mathcal{C}^{(n)}(\beta), \qquad (5.2)$$

$$\langle S_{\mathcal{F}}^{(n)}\rangle = \int d^2\beta\, P(\beta)\, \mathcal{S}^{(n)}(\beta), \qquad (5.3)$$

where $\mathcal{C}^{(n)}$ and $\mathcal{S}^{(n)}$ are Glauber's phase-space integral representations of $C_{\mathcal{F}}^{n}$ and $S_{\mathcal{F}}^{n}$, respectively. We thus get a phase-space representation of any arbitrary function of the sine and cosine operators in terms of c-number functions.[13]

The sine and cosine functions are given in this case by the following formula:

$$\mathcal{C}^{(1)}(\beta) = \mathcal{A}\cos\varphi, \qquad (5.4)$$

$$\mathcal{S}^{(1)}(\beta) = \mathcal{A}\sin\varphi, \qquad (5.5)$$

where $|\beta|e^{i\varphi}$ denotes the eigenvalue of b. The field amplitude $\mathcal{A}$ depends on the mean number of photons of the coherent state $|\beta|^2 =< n >$, and is given by the following formula:

$$\mathcal{A}(< n >) = \sqrt{\frac{< n >}{\pi}} \int_{-\infty}^{\infty} d\eta\, \frac{\exp(-\frac{<n>\eta^2}{1+\eta^2})}{(1+\eta^2)^2}. \qquad (5.6)$$

For $< n >= 0$, the sine and cosine amplitude is equal to zero as it has to be expected for a vacuum state. For large values of $< n >$, the amplitude approaches the well-known classical behavior for a strong signal, $\mathcal{A}(\infty) = 1$.

The second powers of the sine and cosine functions are given in by the following formulas:

$$\mathcal{C}^{(2)}(\beta) = \cos^2\varphi + \frac{1-2\cos^2\varphi}{2|\beta|^2}(1-e^{-|\beta|^2}), \qquad (5.7)$$

$$\mathcal{S}^{(2)}(\beta) = \sin^2\varphi - \frac{1-2\cos^2\varphi}{2|\beta|^2}(1-e^{-|\beta|^2}). \qquad (5.8)$$

The phase-space cosine and sine functions satisfy the trigonometric unity rule.

$$\mathcal{C}^{(2)}(\beta) + \mathcal{S}^{(2)}(\beta) = 1. \tag{5.9}$$

These trigonometric functions exhibit the classical behavior $\mathcal{C}^{(2)} = \cos^2\varphi$ and $\mathcal{S}^{(2)} = \sin^2\varphi$ for strong signals, whereas $\mathcal{C}^{(2)}$ and $\mathcal{S}^{(2)}$ tend to $\frac{1}{2}$ for the vacuum.

References

1. P. A. M. Dirac, *Proc. Roy. Soc. London A* 114:243 (1927)
2. F. London, *Z. Phys.* 37:915 (1926); *ibid.* 40:193 (1927)
3. L. Susskind, J. Glogower, *Physics* 1:49 (1964)
4. P. Carruthers and M. N. Nieto, *Rev. Mod. Phys.* 40:411 (1968)
5. J. W. Noh, A. Fougères, and L. Mandel, *Phys. Rev. A* 45:424 (1992)
6. J. W. Noh, A. Fougères, and L. Mandel, *Phys. Rev. A* 46:2840 (1992)
7. K. Wódkiewicz, *Phys. Rev. Lett.* **52**, 1064 (1984); *Phys. Lett. A* 115:304 (1986)
8. See, for example, D. F. Walls and G. J. Milburn, "Quantum Optics", Springer-Verlag, Berlin (1994)
9. D. Burak and K. Wódkiewicz, *Phys. Rev. A* 46:2744 (1992)
10. B.-G. Englert, K. Wódkiewicz, to be published
11. See, for example, P. Busch, P. J. Lahti, and P. Mittelstaedt, " The Quantum Theory of Measurement ", Springer-Verlag, Berlin (1991)
12. B.-G. Englert, K. Wódkiewicz, and P. Riegler, to be published
13. P. Riegler and K. Wódkiewicz, *Phys. Rev. A* 49:1387 (1994).

NONCOMMUTATIVE SPACE-TIME IMPLIED BY SPIN

Stanisław Zakrzewski

Department of Mathematical Methods in Physics
Warsaw University, Hoża 74
00-682 Warsaw, Poland

Abstract

This is a short report on our recent study[1] of extended phase spaces for spinning particles. We emphasize the intriguing noncommutativity of space-time, arising by passing from canonical to 'covariant' position variables in our model.

1. INTRODUCTION

Souriau[2] has defined (relativistic) *elementary systems* as mechanical systems whose space of 'motions' ('l'espace des mouvements') is a symplectic manifold with a transitive symplectic action of the Poincaré group (Wigner's philosophy for classical particles). By the momentum mapping theory, transitive actions correspond to coadjoint orbits of the Poincaré group.

We propose[1] a similar algorithm to construct *extended phase spaces* in which motions are solutions of equations of motion, not only abstract points of coadjoint orbits. Like the 'space of motions', an extended phase space is defined as a symplectic 'transitive' space, the transitivity this time being understood with respect to the pair *group + space-time* rather than to the group alone (we 'represent' not only the infinitesimal generators of the group but also functions on space-time, to deal directly with space-time localization).

The simplest example of an extended phase space is provided by the cotangent bundle T^*M to the Minkowski space-time M, together with the action of the Poincaré group G, the (connected) group of affine transformations of M leaving invariant the Lorentz metric g. The action of G on T^*M, implied by the action of G on M, has a canonical momentum mapping $J\colon T^*M \to \mathfrak{g}^*$ ($\mathfrak{g}^*$ is the dual of the Lie algebra $\mathfrak{g}$ of G). Taking inverse images of coadjoint orbits by J leads to a decomposition of T^*M into coisotropic submanifolds (mass shells), whose characteristic foliation determines the (phase) trajectories of the elementary system with a given mass and without spin. T^*M is therefore an extended phase space for massive spinless particles. In order to characterize general extended spaces, we extract three essential properties of the above example (we set $P := T^*M$):

1. P is a Hamiltonian G-space, in other words,

$$\boxed{\text{a complete Poisson map } J\colon P \to \mathfrak{g}^* \text{ is given}} \tag{1.1}$$

2. P is fibered over M (with coisotropic fibers), i.e.

$$\boxed{\text{a complete Poisson map } \pi\colon P \to M \text{ is given}} \tag{1.2}$$

3. the following covariance holds: $X_P(\pi^* f) = \pi^*(X_M f)$, or, equivalently,

$$\boxed{\{J^*X, \pi^* f\} = \pi^*(X_M f)} \tag{1.3}$$

for $X \in \mathfrak{g}$, $f \in C^\infty(M)$. Here X_M (or X_P) denotes the fundamental vector field of the action of G on M (or P), corresponding to $X \in \mathfrak{g}$, and $\pi^* f$ is the pullback of f by π (similarly, J^*X is the pullback of X by J, where X is treated as a linear function on $\mathfrak{g}^*$). Of course for $P = T^*M$, π is the cotangent bundle projection.

We recall that $\mathfrak{g}^*$ is naturally a Poisson manifold. The Poisson structure on M is zero. A Poisson map is said to be *complete*,[3] if it sends (by pullback) functions having complete Hamiltonian vector fields on functions with the same property (such functions are called complete).

Note that the above example has also the following 'transitivity' property:

$$\boxed{X_P,\ \mathbf{H}_{\pi^* f} \text{ (with } X \in \mathfrak{g},\ f \in M) \text{ span } TP}. \tag{1.4}$$

Here $\mathbf{H}_h$ denotes the Hamiltonian vector field of the function h.

2. EXTENDED PHASE SPACES

Here is our basic definition.

Definition 2.1. By an *extended phase space of a relativistic particle* we mean a symplectic manifold P satisfying (1.1), (1.2), (1.3) and (1.4).

A complete Poisson map from a symplectic manifold to a given Poisson manifold N is said to be a *symplectic realization*[4] of N. It is said to be *transitive*, if $\mathbf{H}_{\Psi^* h}$ span TP for $h \in C^\infty(N)$.

In order to find all extended phase spaces, we notice that pairs of symplectic realizations (π, J) satisfying (1.1), (1.2) and (1.3) are in 1-1 correspondence with symplectic realizations of a certain Poisson manifold (a similar fact is known in the theory of crossed products).

Proposition 2.2. *Pairs (π, J) satisfying (1.1), (1.2), (1.3) are in 1-1 correspondence with symplectic realizations Ψ of the semi-direct Poisson product $M \rtimes \mathfrak{g}^*$, given by*

$$\Psi = \pi \times J.$$

(π, J) satisfies (1.4) if and only if Ψ is transitive.

Recall[5] that $M \rtimes \mathfrak{g}^*$ is the cartesian product of M and $\mathfrak{g}^*$ equipped with the *semidirect Poisson structure* defined by

$$\{f_1, f_2\} = 0, \qquad \{X_1, X_2\} = [X_1, X_2], \qquad \{X, f\} = X_M f \tag{2.1}$$

for $f_1, f_2 \in C^\infty(M)$, $X_1, X_2 \in \mathfrak{g}$ (we choose the commutator in $\mathfrak{g}$ based on the right-invariant vector fields in order to keep $X \mapsto X_M$ as homomorphism of Lie algebras).

Since transitive symplectic realizations of a Poisson manifold are just (coverings of) its symplectic leaves (this is a generalization of the familiar fact concerning the moment map of a transitive hamiltonian action), we conclude that extended phase spaces are (coverings of) symplectic leaves in $M \rtimes \mathfrak{g}^*$.

The next proposition describes these leaves completely.

We denote by V the subgroup of translations in the Poincaré group G. This is a normal subgroup and

$$L := G/V$$

is the Lorentz group, acting naturally in V, the tangent space of M. Any choice of $x \in M$ allows to identify L with the stabilizing subgroup G_x of G. We denote by $\mathfrak{l}$ and $\mathfrak{g}_x$ the Lie algebras corresponding to L and G_x.

Proposition 2.3. *The natural map*

$$M \rtimes \mathfrak{g}^* \ni (x, \alpha) \mapsto ((x, p), S) \in T^*M \times \mathfrak{l}^*,$$

where p is the restriction of α to V and S is the restriction of α to $\mathfrak{g}_x \simeq \mathfrak{l}$, is a Poisson isomorphism ($T^*M \times \mathfrak{l}^*$ considered with its direct product Poisson structure).

The proof is easily obtained by calculating the Poisson brackets after the transformation $(x, (p, M)) \mapsto ((x, p), S)$, given in terms of coordinates by

$$S_{jk} = M_{jk} - p_j x_k + p_k x_j,$$

using the Poisson brackets in $M \rtimes \mathfrak{g}^*$:

$$\{M_{jk}, M_{ln}\} = M_{jl}g_{kn} + M_{kn}g_{jl} - M_{jn}g_{kl} - M_{kl}g_{jn}, \qquad \{M_{jk}, p_l\} = -p_j g_{kl} + p_k g_{jl}$$

$$\{M_{jk}, x^l\} = -x_j \delta_k{}^l + x_k \delta_j{}^l, \qquad \{p_j, x^l\} = \delta_j{}^l.$$

(We fix an identification $M \simeq V$, which yields $\mathfrak{g} \simeq V \rtimes \mathfrak{l}$ and treat

$$M_{kl} := e_k \otimes g(e_l) - e_l \otimes g(e_k) \in \mathfrak{l} \subset \operatorname{End} V$$

(e_k is a basis in V) as linear functions on $\mathfrak{g}^* \simeq V^* \times \mathfrak{l}^*$.) Variables x^l, p_j denote position and momentum coordinates.

□

Corollary. *Extended phase spaces are of the form*

$$P = T^*M \times \mathcal{O},$$

where $\mathcal{O}$ is a coadjoint orbit in $\mathfrak{l}^*$. They are in one-to-one correspondence with these coadjoint orbits. The trivial coadjoint orbit yields simply T^*M, the extended phase space of a spinless particle, described in the Introduction.

It is convenient to identify $\mathfrak{l}$ with $\mathfrak{l}^*$ using the Killing form. Our 'spin variable' S will then take values in $\mathfrak{l}$. By definition, $\mathfrak{l} \subset \operatorname{End} V$ is the orthogonal Lie algebra of the Lorentz metric g of signature $(1,3)$ in V. We normalize the Killing form as follows

$$< S, S > := \frac{1}{2} \operatorname{tr} S^2.$$

Knowing that $\mathfrak{l} \simeq sl(2, \mathbb{C})$, one can easily see that the value of the complex Killing form,

$$< S, S >_{\mathbb{C}} = < S, S > -i < *S, S >$$

(the Hodge star $*$ endowes $\mathfrak{l}$ with the complex structure), fully specifies the adjoint orbit (if we consider only nontrivial orbits). Therefore, for any complex number z we have the orbit

$$\mathcal{O}_z := \{S \in \mathfrak{l} \setminus \{0\} :< S, S >_{\mathbf{C}} = -z^2\}$$

(z and $-z$ correspond to the same orbit). Each such orbit is of dimension 4. Corresponding extended phase spaces $P_z := T^*M \times \mathcal{O}_z$ are 12-dimensional.

Recall, that given a 'timelike' vector $u \in V$ such that $g(u,u) = 1$, the spin tensor S is equivalently described by two (spacelike) vectors $\vec{\omega}$, $\vec{v}$, orthogonal to u, where

$$\vec{v} := Su, \qquad \vec{\omega} := (*S)u.$$

We have

$$S = (S - Su\wedge_g u) + Su\wedge_g u = - * (\vec{\omega}\wedge_g u) + \vec{v}\wedge_g u,$$

where $R := S - Su\wedge_g u$ is the 'rotation part' of S and we have denoted by

$$x\wedge_g y := x \otimes g(y) - y \otimes g(x) = (\mathrm{id} \otimes g)(x\wedge y)$$

the element of $\mathfrak{l}$ corresponding to $x \wedge y \in \overset{2}{\bigwedge} V$. Using the positive defined three-dimensional metric in the orthogonal complement of u,

$$\vec{\omega}^2 := -g(\vec{\omega},\vec{\omega}), \qquad \vec{v}^2 = -g(\vec{v},\vec{v}), \qquad \vec{\omega}\cdot\vec{v} := -g(\vec{\omega},\vec{v}),$$

we can write the complex Killing form in terms of $\vec{\omega}$, $\vec{v}$ as follows:

$$< S, S >_{\mathbf{C}} = -(\vec{\omega}^2 - \vec{v}^2) - 2i\vec{\omega}\cdot\vec{v} = -(\vec{\omega} + i\vec{v})^2.$$

The orbit $\mathcal{O}_z$ for $z = a + bi$ is then given by

$$\mathcal{O}_z = \left\{ (\vec{\omega},\vec{v}) : \begin{array}{rcl} \vec{\omega}^2 - \vec{v}^2 & = & a^2 - b^2 \\ \vec{\omega}\cdot\vec{v} & = & ab \end{array} \right\}.$$

3. TRAJECTORIES

The extended phase space $P_z = T^*M \times \mathcal{O}_z$ is related to 'spaces of motions' (coadjoint orbits in $\mathfrak{g}^*$) by symplectic reduction, obtained generically by fixing values of spin and mass (there is no 'multiplicity' in this case: $\dim P_z - 2 \cdot 2 = 8$). In order to deal with the most regular situation, we shall consider only the case of a positive squared mass:

$$m^2 := g(p,p) > 0.$$

We identify the momentum $p \in V^*$ with the corresponding vector $g^{-1}(p) \in V$. The spin s is given by the following expression

$$s^2 = -g((JS)u, (JS)u) = \vec{\omega}^2,$$

where $u := \frac{p}{m}$ is the unit vector in the direction of p (recall that the Pauli-Lubanski vector is defined by $W = (JS)p$).

The characteristic foliation on the constraint submanifold of fixed mass and spin is (generically) 2-dimensional. In order to find the leaves of this foliation, it is sufficient to integrate the hamiltonian vector fields of $\frac{1}{2}m^2$ and $\frac{1}{2}s^2$. Since

$$\{\frac{1}{2}m^2, x\} = p, \qquad \{\frac{1}{2}m^2, p\} = 0, \qquad \{\frac{1}{2}m^2, S\} = 0,$$

the mass constraint generates the usual rectilinear motion with conserved p and S and four-velocity u. In order to calculate the Poisson brackets with $\frac{1}{2}s^2$, note, that

$$s^2 = g(S^2u, u) + < S, S > .$$

A simple calculation yields then

$$\{\frac{1}{2}s^2, p\} = 0, \qquad \{\frac{1}{2}s^2, x\} = \frac{1}{m}(S^2u\wedge_g u)u, \qquad \{\frac{1}{2}s^2, S\} = -S^2u\wedge_g u.$$

The last equality is obtained using the Poisson brackets on $\mathfrak{l}$ transported from $\mathfrak{l}$ by the Killing form:

$$\{S_{jk}, S_{ln}\} = -(S_{jl}g_{kn} + S_{kn}g_{jl} - S_{jn}g_{kl} - S_{kl}g_{jn}) \tag{3.1}$$

(here $S_{jk} := g_{jl}S^l_k$, S^l_k – matrix elements of $S \in \operatorname{End} V$). It follows that p and $\vec{\omega}$ (the rotational part of S) are conserved. Since $S^2u = S\vec{v} = R\vec{v} + \lambda u$, we have $S^2u\wedge_g u = R\vec{v}\wedge_g u$, hence

$$\{\frac{1}{2}s^2, \vec{v}\} = -R\vec{v}.$$

This means that $\vec{v}$ simply rotates around the $\vec{\omega}$ axis. Due to the conservation of $\overline{x} := x + \frac{1}{m}Su$, x moves on a circle around the axis passing through $\overline{x}$ in the direction of $\vec{\omega}$:

$$x = \overline{x} - \frac{1}{m}Su = \overline{x} - \frac{1}{m}\vec{v}.$$

It follows that the characteristics (and also their projections on Minkowski space-time) have the form of a 2-dimensional cylinder. Each of these 'two-dimensional' trajectories corresponds to a point of the coadjoint orbit in $\mathfrak{g}^*$ (with the fixed value of m and s) and should represent the 'history' of the elementary system. In a co-moving frame, an observer should see a circle of the radius r with

$$r^2 = \frac{1}{m^2}\left(\vec{v}^2 - \frac{(\vec{\omega}\cdot\vec{v})^2}{\vec{\omega}^2}\right) = \frac{1}{m^2s^2}\left(s^2(s^2 - a^2 + b^2) - a^2b^2\right) = \frac{1}{m^2s^2}(s^2 - a^2)(s^2 + b^2). \tag{3.2}$$

4. SPIN-INVARIANT POSITION

In order to avoid the unusual 2-dimensional trajectories, it is natural to try to fix the spin first and obtain a 10-dimensional reduced symplectic manifold. The mass shell in this 'reduced extended' phase space has of course 1-dimensional characteristics.

We thus consider the submanifold

$$\mathcal{C}_{z,s} = \{\text{spin} = s\} \subset (P_z)_+ = (T^*M)_+ \times \mathcal{O}_z \subset P_z$$

of a fixed spin. Here $(T^*M)_+ = M \times V^*_+$ is the subset of T^*M corresponding to time-like (positive) momenta.

We recall that the characteristics on $\mathcal{C}_{z,s}$ are 'circles' whose projection on M are circles (in the co-moving frame) with radius r given by (3.2). It follows that the spin function is bounded from below on $(P_z)_+$:

$$s \geq |a|.$$

We have the following two cases.

1. $s > |a|$. In this case $r > 0$, hence $\mathbf{H}_{s^2} \neq 0$ (characteristics really exist), $d(s^2) \neq 0$ and $\dim \mathcal{C}_{z,s} = 11$ ($\mathcal{C}_{z,s}$ is coisotropic). The projections of characteristics on M are 'circles' (not points), therefore the variables x^k **do not pass** to the quotient

$$P_{z,s} := \mathcal{C}_{z,s}/\{\text{circles}\}.$$

Still, the 'renormalized' position $\overline{x} = x - \frac{1}{m}Su$ is of course well defined on $P_{z,s}$. Since we have coisotropic constraints, the Poisson bracket of $\overline{x}^j$ and $\overline{x}^k$ in $P_{z,s}$ is equal to their Poisson bracket in P_z. Using

$$\{x^j, \frac{1}{m}(Su)^k\} = \frac{1}{m^2}(S^{jk} + 2u^j(Su)^k), \qquad \{(Su)^j, (Su)^k\} = -R^{jk},$$

we obtain

$$\{\overline{x}^j, \overline{x}^k\} = \{x^j + \frac{1}{m}(Su)^j, x^k + \frac{1}{m}(Su)^k\} = \frac{1}{m^2}R^{jk}, \qquad (m^2 \equiv p^2). \tag{4.1}$$

2. $s = |a|$. In this case $\vec{\omega}^2 = a^2$, hence $\vec{v}^2 = b^2$. Since $\vec{\omega} \cdot \vec{v} = ab$, we have $\vec{v} \parallel \vec{\omega}$ and it is easy to see that $\dim \mathcal{C}_{z,s} = 10$. It is not difficult to see that for $a = 0$, $\mathcal{C}_{z,s}$ is coisotropic and $P_{z,s} \simeq T^*M$ (the spinless case). We shall calculate the symplectic form in $P_{z,s}$ on vectors tangent to $\mathcal{C}_{z,s}$ in the case $a \neq 0$.

The tangent space to $\mathcal{C}_{z,s} = \{(x,p,S) :< S, S >_{\mathbf{C}} = -z^2,\ aSp = b(*S)p\} =$

$$= \{(x,p,\vec{\omega},\vec{v}) : g(p,\vec{\omega}) = 0 = g(p,\vec{v}), \vec{\omega}^2 = a^2, \vec{v}^2 = b^2,\ a\vec{v} = b\vec{\omega}\}$$

at a point $(x,p,S) = (x,p,\vec{\omega},\vec{v})$ is given by

$$\left\{(\dot{x},\dot{p},\dot{S}) : \dot{S} = \left(\dot{\omega}, -\left(1 + \frac{b^2}{a^2}\right)R\dot{u} + \frac{b}{a}\dot{\omega}\right),\quad \dot{\omega}\cdot\vec{\omega} = 0\right\}$$

(here $\dot{u}$ is the variation of $u = \frac{p}{m}$ defined by the variation $\dot{p}$ of p and R is the part of S corresponding to $\vec{\omega}$). In order to calculate the symplectic form Ω_S at $S \in \mathcal{O}_z$ on a pair $(\dot{S}_1, \dot{S}_2)$,

$$\Omega_S(\dot{S}_1, \dot{S}_2) =< S, [\dot{S}_1, \dot{S}_2] >=< \dot{S}_2, S_1 > \qquad ([S_1, S] = \dot{S}_1), \tag{4.2}$$

we have to solve the equation

$$[S_1, S] = \dot{S}_1 = \left(\dot{\omega}_1, -\left(1 + \frac{b^2}{a^2}\right)R\dot{u}_1 + \frac{b}{a}\dot{\omega}_1\right),\quad \dot{\omega}_1\cdot\vec{\omega} = 0$$

with respect to $S_1 = (\vec{\omega}_1, \vec{v}_1)$. Using some vector analysis we obtain

$$\vec{\omega}_1 = \frac{1}{a^2}\left(\frac{b}{a}R^2\dot{u}_1 - R\dot{\omega}_1\right), \qquad \vec{v}_1 = \frac{1}{a^2}R^2\dot{u}_1.$$

Inserting this in (4.2) gives

$$\Omega_S(\dot{S}_1, \dot{S}_2) = \frac{1}{a^2}\dot{\omega}_2 \cdot R\dot{\omega}_1 - \left(1 + \frac{b^2}{a^2}\right)\dot{u}_2 \cdot R\dot{u}_1.$$

Adding the canonical symplectic form on T^*M,

$$< \dot{p}_1, \dot{x}_2 > - < \dot{p}_2, \dot{x}_1 >,$$

and treating the full expression as a linear function of $(\dot{x}_2, \dot{p}_2, \dot{S}_2)$, we obtain the following formula for the 'lowering indices' by the symplectic form:

$$(\dot{x}_1, \dot{p}_1, \dot{S}_1)^\flat = \left(\dot{p}_1, -\dot{x}_1 + \frac{1}{m}\left(1 + \frac{b^2}{a^2}\right) R\dot{u}_1, \frac{1}{a^2} R\dot{\omega}_1\right). \tag{4.3}$$

It is easy to see that this map is invertible. This means that $\mathcal{C}_{z,s}$ is a symplectic submanifold, i.e. characteristics on $\mathcal{C}_{z,s}$ are points and

$$P_{z,s} := \mathcal{C}_{z,s}/\{\text{characteristics}\} = \mathcal{C}_{z,s}.$$

Using the inverse of (4.3), one is led to the following Poisson brackets of position coordinates:

$$\{x^j, x^k\} = \frac{1}{m^2}\left(1 + \frac{b^2}{a^2}\right) R^{jk}. \tag{4.4}$$

In particular, when $b = 0$, $\{x^j, x^k\} = \frac{1}{m^2} R^{jk} = \frac{1}{m^2} S^{jk}$ (in this case $R = S$). On the other hand, when $a \to 0$, the symplectic submanifold $\mathcal{C}_{z,s}$ tends to the (previously discussed) coisotropic submanifold, which explains the divergence of the Poisson brackets.

Formulas similar to (4.1) can be found in some places in the literature.[6,7] The closest approach consists in enlarging Souriau's evolution space for a spinning particle to include the variable 'canonically conjugate' to spin.[8]

References

1. S. Zakrzewski, Extended phase space for a spinning particle, preprint, Warsaw 199
2. J.-M. Souriau, "Structure des Systèmes Dynamiques", Dunod, Paris (1970)
3. S. Zakrzewski, Quantum and classical pseudogroups, *Commun. Math. Phys.* 134:347-395 (1990)
4. P. Xu, Classical intertwiner space and quantization, *Commun. Math. Phys.* 164:473-488 (1994)
5. A. Weinstein, Poisson geometry of the principal series and nonlinearizable structures, *J. Diff. Geom.* 25:55-73 (1987)
6. A.J. Hanson and T. Regge, The relativistic spherical top, *Ann. Physics (NY)* 87:498-566 (1974)
7. C. Duval and J. Elhadad, Geometric quantization and localization of relativistic spin systems, *Contemp. Math.* 132:317-330 (1992)
8. C. Duval, private communication.

PART IV

MISCELLANEOUS PROBLEMS OF QUANTUM DYNAMICS

SPECTRUM OF THE DIRAC OPERATOR ON THE SU(2) MANIFOLD AS ENERGY SPECTRUM FOR THE POLYANILINE MACROMOLECULE

Hanna Makaruk

Institute of Fundamental Technological Research
Polish Academy of Sciences
Świętokrzyska 21, 00-049 Warsaw
Poland

Abstract

A link is shown between the purely mathematical consideration of spinor structures on a group manifold and the description of a physical system. The $SU(2)$ manifold is taken as an example. The considered model describes physical systems with an order parameter taking values in the $SU(2)$ group, and predicts energy levels for fermions that satisfy the Dirac equation in this kind of system. The assumptions of the theoretical model are shown to be being fulfilled in the case of protonated polyaniline. The theoretical prediction of the polaron energy levels is compared with experimental data, showing a very good agreement.

1. INTRODUCTION

It has been known for a long time that geometric methods are very effective in some areas of physics. The classical example is the geometric description of gravity. Moreover the geometry of multidimensional space-time seems to be the proper framework for the unified theory of fundamental interactions, from simple Kaluza-Klein theory to supergravity and superstrings.[1] Therefore it is not surprising that the geometric methods from the pure field theory penetrate other areas of physics, in particular the theory of condensed matter.[2−5]

This work can serve as one example showing this influence. In the first, purely geometric, part of the work, we consider the spinor structures and spinor bundles (associated by the Dirac representation) on group manifolds. The example for which this construction is realized is the SU(2) manifold with the natural bi-invariant metric. This method gives also the spectrum of the Dirac operator on the considered manifold. The purpose of the next part of the work was to find experimental data corresponding to a physical system which could be described this way. We are looking for an application illustrating this geometrical construction and for checking its validity. The example has been sought in the area of the theory of condensed matter, because in this area the experimental data are usually more complete than in the traditional areas described by field theory, i.e. in cosmology or the theory of fundamental interactions. In fact in the physics of condensed matter the experimental verification of the theoretical predictions is also possible. In addition this area is not as deeply penetrated by the geometrical

Quantization, Coherent States, and Complex Structures
Edited by J.-P. Antoine *et al.*, Plenum Press, New York, 1995

methods as other branches of physics. Then one could expect more experimental data which seek the interpretation of this kind.

The example found is the conductivity mechanism in a synthetic metal — protonated polyaniline. This synthetic metal has the *metallic* type conductivity.[6] It was experimentally proved by electronic paramagnetic resonance (EPR)[7,8] that the charge carriers in polyaniline protonated by the one proton acids are polarons which are *fermions.* It is experimentally proved that polyaniline exhibits liquid-crystalline order of the *nematic* type caused by the local ordering of the very long rectilinear macromolecules of this polymer.[9] The two ends of these molecules are indistinguishable, which provides an *additional symmetry.* As a result the order parameter in polyaniline takes values in the *the SU(2) manifold.*[10]

Some ideas from multidimensional geometry of Kaluza-Klein type are applied here to the description of the conductivity mechanism in polyaniline. The multidimensional space-time used for this description consists of the usual 4-dimensional space-time and of the internal space which is the SU(2) group manifold. Similarly to the Kaluza-Klein description of particles, we examine multidimensional fermions. Their mass, or equivalently energy, in the external space-time occurs as a result of interaction with the internal space and is expressed by eigenvalues of the Dirac operator defined on the internal space. There is a problem with these eigenvalues in the standard Kaluza-Klein theory, because the proper energy scale in this theory should be of the order of the Planck mass and this is far too much for observable particles. On the other hand, in the multidimensional space-time used in the condensed matter theory, we can assume a milder energy scale appropriate for the particular problem under consideration and obtain reasonable energy spectra of the particles.

In this general framework the eigenproblem for the Dirac operator gives the corresponding energy spectra of the polarons. The first step was to build proper spinor structures and spinor bundles over the internal space, introduce the Dirac operator acting on the sections of the latter bundle and finally find the eigenvalues of the Dirac operator on the SU(2) manifold. Next the theoretical predictions obtained in this way are compared with the experimental data for the energy spectra of polarons in polyaniline. In the summary, a short remark is made about the applicability of a similar geometric framework to describe conductivity in other macromolecular synthetic metals.

2. SPINORS AND THE DIRAC OPERATOR SPECTRUM ON THE SU(2) MANIFOLD

The standard definitions of (s)pinor structures and (s)pinor bundles are used.[11] The basis for all these considerations lies in the following exact sequence of groups:

$$1 \rightarrow \mathbb{Z}_2 \rightarrow Pin(g) \xrightarrow{\rho} O(g) \rightarrow 1 \tag{2.1}$$

$$1 \rightarrow \mathbb{Z}_2 \rightarrow Spin(g) \xrightarrow{\rho} SO(g) \rightarrow 1 \tag{2.2}$$

where O(g) is the orthogonal group for the metric g introduced in the (real) vector space V, $SO(g)$ is the special orthogonal group, $Pin(g)$ and $Spin(g)$ are twofold covers thereof, called the pin and spin group respectively.

Consider a manifold M equipped with a metric tensor g. A subgroup $\Omega \subset O(g)$ of the orthogonal group is the structural group of the principal orthonormal frame bundle $\mathcal{F}_g$ over the manifold and, in the case M is orientable and oriented, this subgroup is contained in $SO(g)$. The (s)pinor structures are principal $Spin(g)$ (resp. $Pin(g)$)

bundles corresponding to the twofold covers given above. Therefore, there exists a principal bundle $\mathcal{F}$ with subgroups $\Sigma \subset Pin(g)$ as structural groups ($\Sigma = \rho^{-1}(\Omega)$) and a bundle morphism $\eta : \mathcal{F} \to \mathcal{F}_g$ such that the following diagram is commutative:

$$\begin{array}{ccccccc} & \mathcal{F} \times \Sigma & \longrightarrow & & \mathcal{F} & & \\ \eta \times \rho & \downarrow & & \eta & \downarrow & \begin{array}{c} \searrow^{\pi} \\ \nearrow_{\pi_g} \end{array} & M \\ & \mathcal{F}_g \times \Omega & \longrightarrow & & \mathcal{F}_g & & \end{array} \tag{2.3}$$

where horizontal arrows mean bundle group actions and π and π_g are bundle projections. In the case M is a Lie group manifold G equipped with one of the natural metrics, a bi-invariant or a left-invariant one (the case of a right-invariant metric is equivalent to this last one), it belongs to the class of pseudo-riemannian homogeneous manifolds. For such manifolds, the construction of orthonormal frame bundles and spinor structures could be achieved along the lines of the theorem formulated and proved by Dąbrowski and Trautman.[11,12]

Homogeneous pseudo-riemannian manifolds M are characterized by a group of isometries G acting transitively on M. As a result, M is diffeomorphic to the manifold G/H, where H is the subgroup of G which stabilizes a point of M. The orthonormal frame bundle $\mathcal{F}_g$ is then diffeomorphic to the following bundle:[11]

$$\begin{array}{ccc} G/N & \longleftarrow & H/N \\ \downarrow & & \\ G/H & & \end{array} \tag{2.4}$$

where N is a subgroup of H which acts non-effectively in the tangent space to M (the action on the tangent space is the induced action).

Although orthonormal frame bundles make sense on all (pseudo-)riemannian manifolds, (s)pinor structures do not necessarily exist and if they do, they may not be unique. In general one should examine some Stiefel-Whitney classes for a given manifold M to decide on these questions. However, in the case of homogeneous (pseudo-)riemannian manifolds, Dąbrowski and Trautman invented another way to discuss these problems, which is, moreover, constructive.[11] They proved the following theorem:

The diagram:

$$\begin{array}{ccccccc} & (G/N)_h \times (H/N)_h & \longrightarrow & & (G/N)_h & & \\ \sigma \times \rho & \downarrow & & \sigma & \downarrow & \begin{array}{c} \searrow \\ \nearrow \end{array} & M \\ & G/N \times H/N & \longrightarrow & & G/N & & \end{array} \tag{2.5}$$

gives the pinor structure on M iff $(H/N)_h \simeq \rho^{-1}(H/N)$, where

. *σ is the projection in the bundle associated to the universal covering bundle for G/N by the homomorphism $h : \Pi_1(G/N) \to \mathbb{Z}_2$*

. $(G/N)_h := \sigma^{-1}(G/N)$

. $(H/N)_h := \sigma^{-1}(H/N)$.

I have applied this theorem in the case $M = SU(2)$ and the metric being bi-invariant. The results were published in Ref. 13 and only a short review is presented here.

Orthonormal frame bundles on $SU(2)$ equipped with the left-invariant (a) and bi-invariant (b) metrics are diffeomorphic to:

$$a)\quad \begin{array}{ccc} SU(2) & \longleftarrow & \{e\} \\ \downarrow & & \\ SU(2) & & \end{array} \qquad\qquad b)\quad \begin{array}{ccc} SO(4) & \longleftarrow & SO(3) \\ \downarrow & & \\ SU(2) & & \end{array} \tag{2.6}$$

The spinor structures on $SU(2)$ ($SU(2)$ is orientable) are diffeomorphic to:
a) in the case of the left-invariant metric:

$$\begin{array}{ccc} SU(2) \times \mathbb{Z}_2 & \longleftarrow & \mathbb{Z}_2 \\ \downarrow & & \\ SU(2) & & \end{array} \tag{2.7}$$

b) in the case of the bi-invariant metric:

$$\begin{array}{cccc} Spin(4) = & SU(2) \times SU(2) & \longleftarrow & SU(2) = Spin(3) \\ & \downarrow & & \\ & SU(2) & & \end{array} \tag{2.8}$$

The spinor structure in the bi-invariant case is equivalent to the spinor structure on the 3-dimensional sphere with the metric induced from the 4-dimensional Euclidean space, in which it is embedded as a hypersurface. As a result also spinor bundles, the Dirac operator and the form of its spectrum are identical in these two cases. The spectrum of the Dirac operator for the 3-dimensional sphere is of the form

$$\pm\, m_o(l + \frac{1}{2}), \qquad l \in \mathbb{N} - \{0\}, \tag{2.9}$$

and the same result holds for the $SU(2)$ manifold, where m_o is a constant defining an energy scale in the theory.

Interesting features of the spectrum are: nonexistence of the zero energy level in the spectrum and appearance of negative energies (the minus sign). The physical interpretation of the negative energies in the system is that they are energies of antiparticles. In the case of polyaniline, one could expect polarons of negative and positive sign (charge carriers of the type of "electrons" and "holes") and this agrees with the description of polyaniline.[14,15] The spectrum mathematically takes values from $-\infty$ to ∞, but in real physical systems one could expect only a few levels near zero, for fermions with excitation energy low enough to be bound in the system.

3. PHYSICAL EXAMPLE: ENERGY SPECTRUM OF POLARONS IN POLYANILINE

Looking for a proper application of the mathematical results presented above in the description of condensed matter, one should seek materials which are characterized by an order parameter with values in the $SU(2)$ manifold.[10]

There could be various systems with this property. One of the possibilities is the system consisting of very long rectilinear molecules the ends of which are indistinguishable. Among such substances are some nematic liquid crystals, which are typically built of polymeric molecules. This limits our attention to such materials. We want to use our results concerning the spectrum of the Dirac operator on the $SU(2)$ manifold. Therefore, we concentrate on the description of fermionic excitations in the system, polarons in our case. The considered systems are nematic liquid crystals built of conducting polymers. Our physical example is polyaniline, which is a material of the type described above, for which there exists rather extensive experimental data. The problem of the intra-chain mechanism in polyaniline has been considered by the author.[17]

First of all we should prove there is an order in conducting polyaniline, which is described by the order parameter with values in the $SU(2)$ manifold. Polyaniline is a synthetic metal, quasi one-dimensional conductor. The distinguished direction of conductivity is along the long rectilinear polymer chains. Polyaniline is a nematic liquid crystal as was proved in Ref. 9 and, since the ends of macromolecules (typically built of about $10^4 - 10^5$ monomers) are indistinguishable, the order parameter for such type of materials takes values in the group $SU(2)$. The existence of fermionic polarons is proved by EPR methods.[7,8] In the model we assume that the interaction of the polarons with the polymeric chains is effectively described by the interaction of polarons with the order parameter, which in the case if polyaniline takes values in the SU(2) manifold. These facts and assumptions lead us to the energy spectrum for polarons in polyaniline described by the formula (2.9). As a result, energies of interlevel transitions in polyaniline are expected to be integer multiples of an energy unit. Their ratios are then expressible by ratios of integers. For polyaniline the spectra were measured by many authors.[14,15,16] Spectral lines are 1, 1.5, 2, 3, 4 eV within some experimental limits. The line 1.5 eV was identified by Y.Cao, P.Smith and A.J. Heeger[14] to be connected with hopping of polarons among the polymeric chains. This line should not be described by our model then. The only lines that could be expected to be described by the model are *1, 2, 3, 4* eV. The energies themselves are expressed by integer values. By accident the energy unit coincides with the energy unit of our model if we assume the validity of the latter. After this identification, all spectral lines are in good agreement with the model. All this explains the intramolecular energy spectra of polarons in polyaniline within the geometric model introduced above. It means that the properties of a polymeric chain as a whole and the local order of the chains are more important for the conducting fermions then interaction with local chemical groups in this chain. One can predict that, in another synthetic metal with an order parameter in the same space, the ratios of the fermion energy levels have to be the same.

4. SUMMARY

The example of protonated polyaniline has been considered in detail, to illustrate the idea that a mathematical construction could serve as a model of a physical system. The theory (the Dirac operator spectrum over the group manifold) predicts exact ratios of fermion energy levels in this system. The author works at present on an extension of this research to spinor structures over other group manifolds and spectra of polarons in other synthetic metals, and on finding other physical systems with internal symmetry of this kind and with free fermions.

One final remark is in order. Starting from purely mathematical construction and proceeding to a specific physical system, sometimes one has a chance to describe some crucial properties of the physical system which are not easy to find when one starts from experimental data and looks for a generalization.

5. ACKNOWLEDGEMENTS

I would like to thank the organizers of the Workshop, especially Prof. A. Odzijewicz, for the hospitality and giving me the opportunity to present this lecture and for nice atmosphere during the Workshop. I would also like to thank Profs. J. Śniatycki, W. Słowikowski and I. Kanatchikov for interesting discussions.

References

1. M. Green, J.H. Schwarz and E. Witten, "Superstring Theory", Cambridge U.P., Cambridge (1986)
2. L.D. Landau, V.L. Ginzburg, Towards theory of superconductivity, *ZhETPh*, (in Russian), 20: 1064 (1950)
3. D.J. Amit, "Field Theory, the Renormalization Group and Critical Phenomena", 2nd edition, World Scientific, Singapore (1984)
4. R. Owczarek, Knotted vortices and superfluid phase transition, *Modern Physics Letters*, B7: 1523 (1993)
5. R. Owczarek, On geometric methods in description of quantum fluids, in this volume
6. A. Proń, J. Laska, J.E. Österholm, P. Smith, *Polymer*, 34: 4235 (1993)
7. M. Łapkowski, Effects of the nature of electrolyte on the properties of unpaired spins in polyaniline, *Synthetic Metals*, 35: 183 (1990)
8. M. Łapkowski, E.M. Geniés, Evidence of two kinds of spin in polyaniline from in situ EPR and electrochemistry; Influence of the electrolyte composition, *J. Electroanal. Chem.*, 279: 157 (1990)
9. Y. Cao, P. Smith, Liquid-crystalline solutions of electricity conducting polyaniline, *Polymer*, 34: 3139 (1993),
10. A. Holz, Topological properties of static and dynamic defect configurations in ordered liquids, *Physica*, A182: 240 (1992)
11. L. Dąbrowski, A. Trautman, Spinor structures on spheres and projective spaces, *J. Math. Phys.* 27: 2022 (1986)
12. L. Dąbrowski, A. Trautman, Spinor structures on homogeneous riemannian spaces, SISSA preprint, 23/87/EP, Trieste, (1987)
13. H. Makaruk, Spinor description of charged particles on the SU(2) group, *J. Tech. Phys.*, 35: 75 (1994)
14. Y. Cao, P. Smith, A.J. Heeger, Spectroscopic studies of polyaniline in solution and in spin-cast films, *Synthetic Metals*, 32: 263 (1989)
15. S Stafström, B. Sjögren, J.L. Brédas, Study of the photoinducted absorption spectrum of polyemeraldine, *Synthetic Metals*, 29: E219 (1989)
16. J. Laska, "Synthesis and Properties of Polyaniline Doped by the Diesters of the Phosphoric Acid", Ph.D. thesis (in Polish), Wydawnictwo Politechniki Warszawskiej, Warszawa, (1994)
17. H. Makaruk, Dirac description of energy levels of polarons in polyaniline, *Modern Physics Letters B*, 9: 543 (1995).

ON GEOMETRIC METHODS IN THE DESCRIPTION OF QUANTUM FLUIDS

Robert Owczarek

Institute of Fundamental Technological Research
Polish Academy of Sciences
Świętokrzyska 21, 00-049 Warsaw
Poland

Abstract

Some geometric ideas concerning the description of superfluid helium are presented. Results of application of knot theory to dense systems of quantum vortices are shown. Suggestions concerning applications of Kähler geometry in the description of superfluid helium by means of coadjont orbits of volume preserving diffemorphism groups are made.

This paper is devoted to the presentation of some recent geometric ideas in the description of superfluid helium. It is divided into two parts. In the first part, we show some results obtained by applying knot theory to dense systems of vortices in superfluid helium. In the second part, we present some ideas as to when and how one can introduce Kähler geometry in the description of coadjoint orbits of volume preserving diffeomorphism groups.

1. KNOT THEORY IN THE DESCRIPTION OF SUPERFLUID HELIUM

The model of vortices used previously[1] considers them as solutions of equations for a complex scalar field coupled to a $U(1)$ gauge field. The gauge field is concentrated in a very narrow region of the vortex core. Therefore, the effective description of rarely distributed vortices is in terms of the scalar field alone and the effective description of dense systems of vortices is in terms of the $U(1)$ gauge field $A(\bar{x}, t) = A_\mu dx^\mu$. Before formulating the problem of how to find a proper action for such a system, let me recall some historical background.

In the 1950's Feynman suggested that the superfluid phase transition is caused by proliferating vortex rings. However, many models based on this picture were not successful in obtaining proper characteristics of the phase transition. In contrast, RGT techniques, which do not use such a picture, proved more powerful.[2] Nevertheless, recent experiments performed by the group of Prof. Mc Clintock, showed that very dense systems of vortices are present in superfluid helium close to the critical point.[3] This observation is a strong support for the Feynman picture. On the other hand, we know that in such dense systems of vortices there should occur processes of reconnection, which lead to the appearance of complicated linked and knotted vortex structures in

Quantization, Coherent States, and Complex Structures
Edited by J.-P. Antoine *et al.*, Plenum Press, New York, 1995

the system. We expect, therefore, that the proper action we are looking for should contain information about the topology of the system. Recently, I proposed this action to be

$$S(A) = \frac{1}{4\pi} \int_M A \wedge dA, \tag{1.1}$$

where M is a leaf (space-like or time-like) of a 3+1 foliation of the space-time.[4] I assume M to be a 2+1 D space-time obtained by projection along one of the spatial dimensions. This picture has a lot in common with the one shown in the paper of Rasetti and Regge on canonical approach to quantization of vortices in superfluid helium,[5] which had also important influence on development of the presented approach. Witten's work[6] on a quantum field theoretical representation of knot invariants had also a strong influence on this approach. The quantity $S(A)$, called helicity, is conserved in a classical ideal fluid[7] and it is equal to the Gauss winding number, but in the quantum case it undergoes fluctuations, which are likely to be described by the following Feynman path integral

$$Z = \int DA\, e^{iS(A)}, \tag{1.2}$$

where DA is a measure for the A fields.[4] In the low momentum approximation, one represents this quantity in terms of an auxilliary $(2+1)D$ spinor field coupled to the $U(1)$ gauge field.[4,8,9] To extract the behavior of the excitations of the system connected with fluctuations of topology, one integrates out the gauge field. Standard procedures applied for the field theory obtained this way lead to the thermodynamic partition function, which is the partition function for a large system of Ising models in thermodynamic equilibrium. As a result, heat capacity singularity of superfluid helium is identical to that of the $2\,D$ Ising model[9] and this is the equivalence physicists have been looking for for a long time.[5]

2. KÄHLER GEOMETRY FOR COADJOINT ORBITS OF DIFFEOMORPHISM GROUPS

In the second part of this paper I investigate some ideas concerning open questions in the coadjoint orbit approach to geometric quantization of fluids. There is an interesting approach to the description of fluids in terms of coadjoint orbits of diffeomorphism groups.[10,11] The configuration space of the fluid is identified with $G = S\ \mathit{Diff}\ M$ and the reduced phase space of the fluid is a coadjoint orbit of G, equipped with a canonical KKS symplectic structure. Being symplectic, they are of even dimension and there are some open questions concerning the existence of a complex and a Kähler structures defined on them. Especially interesting is the case when vortices are present in the fluid. A positive answer to the question would be a starting point towards a very nice geometric quantization of the fluid in terms of the Kähler polarization. Such a result was announced by Penna and Spera[12] for the case of structureless vortex filaments in $3\,D$ space. IHowever, it turned out to be false and Goldin, Menikoff and Sharp proved the non-existence of any polarization in this case, pointed out the cases of vortex ribbons and vortex tubes as such when a polarization exists and discussed some other configurations.[13–15] It is desirable to explain whether there exists a Kähler polarization in this case and to find a proper tool for investigating the existence of such polarizations in other cases, e.g. for knotted and linked vortex structures, such as discussed above.

I would like to present some ideas on how such a formalism should be built. I will employ for this aim the theory of flag manifolds, loop groups and certain results from twistor theory.

First, I recall the result that coadjoint orbits of a finite dimensional group are flag manifolds and all flag manifolds can be identified as some coadjoint orbits.[16] Let G be a finite dimensional, connected, compact, simple Lie group. Consider the coadjoint action of G on G'^*, the dual of G', the Lie algebra of G. The stabilizer G_w of a point $w \in G'^*$ is the centralizer $C(T)$ of the torus T generated in G by $\exp(tw^*)$, where w^* is dual to w under the duality defined by the Killing metric. In this way the coadjoint orbit is identified with $G/G_w = G/C(T)$ and thus it is a flag manifold. Flag manifolds for $\dim G < \infty$ carry also complex structures and actually Kähler structures, because the complex structures are compatible with the KKS symplectic structures. The description of such a complex structure which is the most convenient for our purposes is by means of certain "root spaces", described as follows:[17] The Lie algebra G' is splitted as

$$G' = H' \bigoplus M', \tag{2.1}$$

where H' corresponds to G_w and M' is a subspace, which is identified with the tangent space in the point of O_w, which is $e\,G_w$ ($e \in G$ is the unit). The existence of a complex structure is equivalent to the existence of a splitting

$$M'^{\mathbb{C}} = M'_+ \bigoplus M'_- \tag{2.2}$$

where: M'_+ and M'_- are nilpotent algebras, M'_+ is of (1,0) and M'_- of (0,1) type.

For $\dim G < \infty$,

$$M'_+ = \bigoplus_{\alpha\in\Delta+} G'_\alpha \tag{2.3}$$

$$M'_- = \bigoplus_{\alpha\in-\Delta+} G'_\alpha \tag{2.4}$$

where G'_α are certain eigenspaces of the ad_w operator. For $\dim G = \infty$ (like $G = S\ \mathit{Diff}\ M$), this construction does not work, because the spectral problem is not well defined. One should look for a different approach.

In the case $\dim G < \infty$, one could build some additional structures. I hope similar constructions could help us in the infinite-dimensional case. Let us consider the loop group $L\,G = Map(S^1, G)$ consisting of smooth maps from S^1 to G and the based loop group ΩG consisting of those maps from LG that start and end at the unit element $e \in G$. ΩG is then a coset space

$$\Omega G = LG/G. \tag{2.5}$$

ΩG is an infinite dimensional counterpart of flag manifolds. In particular it is a Kähler manifold. Its complex structure is given for an element of the complexified tangent space

$$T_x^{\mathbb{C}}(\Omega G) := T_x(\Omega G) \bigotimes \mathbb{C}, \tag{2.6}$$

i.e. for a vector

$$\xi = \sum_{k\neq 0} \xi_k e^{ik\Theta}, \qquad \xi_k \in (G')^{\mathbb{C}} \tag{2.7}$$

by

$$J_x \xi = -i \sum_{k>0} \xi_k e^{ik\Theta} + i \sum_{k<0} \xi_k e^{ik\Theta}. \tag{2.8}$$

This complex structure is compatible with the canonical symplectic structure given by

$$\omega(\xi, \eta) = \frac{1}{2\pi} \int_0^{2\pi} < \xi(\Theta), \eta'(\Theta) >_{G'} d\Theta, \tag{2.9}$$

where $<,>_{G'}$ is the Killing form for G',

$$\omega(J\xi, J\eta) = \omega(\xi, \eta) \tag{2.10}$$

If G has a trivial center, there exists an isometric immersion of flag manifolds (= coadjoint orbits) in ΩG

$$\Gamma : F \longrightarrow \Omega G, \tag{2.11}$$

given on the trivial coset by

$$\Gamma(eG_w) = \gamma, \qquad \gamma : S^1 \longrightarrow G, \qquad \gamma(e^{it}) = \exp(tw), \tag{2.12}$$

and then smeared over O_w by the group multiplication map. This immersion is geodesic and holomorphic, and ΩG serves as a universal flag manifold.[17] The challenge is to try to repeat this or a similar construction in the case of $G = S\ \mathit{Diff}\ M$ and look for its realizability.

There is one more interesting construction.[17] For $\dim G < \infty$, one can construct for a given flag manifold F a symmetric space $N = G/K$, which serves as a base manifold for the canonical fibration $F \to N$. K is a subgroup of G with Lie algebra

$$G' \cap (\bigoplus_{\alpha \in \Delta} G'_{2\alpha}). \tag{2.13}$$

There is a geodesic immersion $\gamma : N \to G$ such that the diagram

$$\begin{array}{ccc} F & \xrightarrow{\Gamma} & \Omega G \\ \downarrow & & \downarrow \\ N & \xrightarrow{\gamma} & G \end{array} \tag{2.14}$$

is commutative, where vertical arrows denote canonical projections.

It is very likely that for $G = S\ \mathit{Diff}\ M$ it will be much easier to deal with the loop group ΩG than the flag manifold itself, to define the complex structure similar to that given above and then to try to construct an appropriate map Γ. If such a Γ map exists, one should construct such a complex structure for F that Γ be holomorphic. Then one should check its compatibility with the KKS form.

The program sketched above is not easy to realize, but it could be interesting to fulfil it, both from the physical and the mathematical points of view.

3. ACKNOWLEDGEMENTS

I would like to thank the organizers of the Workshop, especially Prof. A. Odzijewicz, for giving me the opportunity to present this lecture and for nice atmosphere during the Workshop, Prof. G.A. Goldin for interesting discussions on problems connected with the coadjoint orbit approach to quantization of superfluid helium, Prof. W. Wojtyński for discussion of problems connected with infinite dimensional Lie groups.

This work was supported by the KBN grant N° 2 P302 125 05.

References

1. R. Owczarek, Topological defects in superfluid helium, *Int. J. Theor. Phys.*, 30:1605 (1991)

2. P.C. Hohenberg, Critical phenomena in ^{4}He, *Physica*, B 109&110:1436, (1982)

3. P.C. Hendry et al., to appear in Nature

4. R. Owczarek, Frames and fermionic excitations of vortices in superfluid helium, *J. Phys., Cond. Matter*, 5:8793 (1993)

5. M. Rasetti, T. Regge, Vortices in He II, current algebras and quantum knots, *Physica*, A80:217 (1975)

6. E. Witten, Quantum field theory and the Jones polynomial, *Commun. Math. Phys.*, 121:351 (1989)

7. Z. Peradzyński, *Int. J. Theor.Phys.*, 29:1277 (1990)

8. R. Owczarek, Knotted vortices and fermionic excitations in bulk superfluid helium, *Mod. Phys. Lett.*, B7:1383 (1993)

9. R. Owczarek, Knotted vortices and superfluid phase transition, *Mod. Phys. Lett.*, B7:1523 (1993)

10. V. Arnold, "Mathematical methods of classical mechanics", Graduate Texts in Math. N° 60, Springer, New York (1978)

11. J. Marsden, A. Weinstein, Coadjoint orbits, vortices, and Clebsch variables for incompressible fluids, *Physica*, D7:305 (1983)

12. V. Penna, M. Spera, A geometrical approach to quantum vortices, *J. Math. Phys.*, 30:2778 (1989)

13. G.A. Goldin, R. Menikoff, D.H. Sharp, Diffeomorphism groups and quantized vortex filaments, *Phys. Rev. Lett.*, 58:2162 (1987)

14. G.A. Goldin, R. Menikoff, D.H. Sharp, Quantum vortex configurations in three dimensions, *Phys. Rev. Lett.*, 67:3499 (1991)

15. G.A. Goldin, The diffeomorphism group approach to nonlinear quantum systems, *Int. J.Mod Phys.*, B6:1905 (1992)

16. D. Freed, Flag manifolds and infinite dimensional Kähler geometry, "Infinite dimensional groups and applications", ed. V. Kac, Berlin, Springer, 83-124 (1985)

17. I. Davydov, A.G. Sergeev, Twistor spaces and harmonic maps, *Uspekhi Matematičeskikh Nauk* (in Russian), 48:3 (1993)

GALACTIC DYNAMICS IN THE SIEGEL HALF-PLANE

G. Rosensteel

Physics Department, Tulane University
New Orleans, LA 70118, USA

Abstract

The dynamics of rotating galaxies is modeled by a Hamiltonian Lax system for which the phase space is a homogeneous G-manifold with the Lie group G equal to either the noncompact real symplectic group $Sp(n, R)$ or a maximal parabolic subgroup $GCM(n)$. The dimensions $n = 1, 2, 3$ correspond respectively to breathing mode oscillations, planar rotations, and three-dimensional collective motion. The homogeneous $GCM(3)$-manifolds correspond to the Riemann ellipsoids. The homogeneous G-manifold $Sp(n, R)/U(n)$, where $U(n)$ is the maximal compact subgroup, is a classical complex domain diffeomorphic to the Siegel upper half-plane S_n. Equilibrium galactic radii are determined for S_1 systems.

1. MANY-BODY COLLECTIVE DYNAMICS

The mathematical description of galactic dynamics is, in principle, straightforward. The phase space for a galaxy of A stars is $\mathbf{R}^{6\mathrm{A}}$, and the Hamiltonian is the sum of the kinetic energy of each star, $T = \sum_\alpha p_\alpha^2/2m_\alpha$, plus the gravitational potential energy, $V = -G\sum_{\alpha<\beta} m_\alpha m_\beta/|\mathbf{x}_\alpha - \mathbf{x}_\beta|$, where the sums are over the particle index $\alpha, \beta = 1, \ldots, A$, m_α
denotes the mass of star α, and $\mathbf{x}_\alpha$, $\mathbf{p}_\alpha$ denote the Cartesian position and momentum vectors for star α. The time evolution of the galaxy $(\mathbf{x}_\alpha(t), \mathbf{p}_\alpha(t)) \in \mathbf{R}^{6\mathrm{A}}$ is computed by integrating the Hamiltonian vectorfield given the initial galactic state $(\mathbf{x}_\alpha(0), \mathbf{p}_\alpha(0))$ in phase space.

There are two practical problems with this analysis because the number of stars is, in a word, astronomical, $A \sim 10^{11} - 10^{14}$. First, the computation of the integral curves is intractable even using the fastest parallel processing supercomputer. Second, the state of the system in $\mathbf{R}^{6\mathrm{A}}$ is not observationally measurable. Astronomical observations provide only partial information about average properties of the stellar distributions in position and momentum space. For example, the projection of the axes lengths of an elliptical galaxy upon the two-dimensional plane perpendicular to the line of sight can be observed optically, and the dispersion in the stellar velocity fields can be determined from the doppler broadening of galactic spectral lines. Thus, our knowledge of real galactic states is very crude, and, even if the initial states could be measured, we cannot calculate their time evolutions.

These technical limitations suggest an alternative analysis that focuses upon the observables which, in fact, are really measurable, and ignores the degrees of freedom

Quantization, Coherent States, and Complex Structures
Edited by J.-P. Antoine *et al.*, Plenum Press, New York, 1995

that are, for all practical purposes, unknowable. The relevant measurable observables are the inertia $Q^{\rm L}$, virial momentum $N^{\rm L}$, and kinetic $T^{\rm L}$ tensors that are defined by the quadratic functions of the Cartesian position and momentum variables summed over all stellar masses:

$$
\begin{aligned}
Q^{\rm L}_{ij} &= \sum m_\alpha x_{\alpha i} x_{\alpha j}, \\
N^{\rm L}_{ij} &= \sum x_{\alpha i} p_{\alpha j}, \\
T^{\rm L}_{ij} &= \sum m_\alpha^{-1} p_{\alpha i} p_{\alpha j},
\end{aligned}
\tag{1.1}
$$

where $i, j = 1,2,3$ index the Cartesian components. These smooth phase space functions are called "collective" observables because of the summations over all particles. Note that the linear collective observables are ignored because these are just the dynamically trivial center of mass and the total momentum variables. The superscript L on each collective quadratic tensor indicates that it is a (laboratory) inertial frame quantity measured with respect to the center of mass; the superscript is omitted for the corresponding body-fixed tensors in the rotating principal axis frame. In the continuum approximation, the sums over particles are replaced by integrals over the density distribution, e.g., $Q^{\rm L}_{ij} = \int \rho(x) x_i x_j d^3x$.

Applying the constellation of ideas associated with co-adjoint orbits, Lax pairs, and homogeneous G-manifolds to the quadratic collective tensors enables the construction of a tractable new model of galactic dynamics. It also elucidates the structure of the well known Riemann ellipsoidal model.[1] The mathematical property that supports this application to galactic dynamics is the following: The collective observables span a finite-dimensional Poisson subalgebra of the Lie algebra of smooth functions on $\mathbf{R}^{6\mathrm{A}}$. This subalgebra is isomorphic to the noncompact real symplectic Lie algebra $sp(3, R)$ for which the Poisson bracket relations are as follows:

$$
\begin{aligned}
\left\{N^{\rm L}_{ij}, N^{\rm L}_{kl}\right\} &= \delta_{il} N^{\rm L}_{kj} - \delta_{jk} N^{\rm L}_{il}, \\
\left\{Q^{\rm L}_{ij}, Q^{\rm L}_{kl}\right\} &= 0, \\
\left\{Q^{\rm L}_{ij}, N^{\rm L}_{kl}\right\} &= \delta_{il} Q^{\rm L}_{jk} + \delta_{jl} Q^{\rm L}_{ik} \\
\left\{T^{\rm L}_{ij}, T^{\rm L}_{kl}\right\} &= 0, \\
\left\{N^{\rm L}_{ij}, T^{\rm L}_{kl}\right\} &= \delta_{il} T^{\rm L}_{jk} + \delta_{ik} T^{\rm L}_{jl} \\
\left\{Q^{\rm L}_{ij}, T^{\rm L}_{kl}\right\} &= \delta_{ik} N^{\rm L}_{jl} + \delta_{il} N^{\rm L}_{jk} + \delta_{jk} N^{\rm L}_{il} + \delta_{jl} N^{\rm L}_{ik}.
\end{aligned}
\tag{1.2}
$$

Several subalgebras of the symplectic algebra are relevant to the physical interpretation and mathematical structure of the galactic symplectic model. The general collective motion $gcm(3)$ algebra is the maximal parabolic subalgebra of $sp(3, R)$ spanned by the inertia tensor $Q^{\rm L}_{ij}$ and the virial momentum tensor $N^{\rm L}_{ij}$. The rotational $rot(3)$ subalgebra of $gcm(3)$ is spanned by the inertia tensor and the antisymmetric part of the virial momentum tensor, viz., the angular momentum $L^{\rm L}_k = \epsilon_{ijk} N^{\rm L}_{ij}$. The Lie algebra $gl(3, R)$ of the general linear group $GL(3, R)$ is generated by the virial momentum tensor, and the Lie algebra $so(3)$ of the rotation group $SO(3)$ is generated by the angular momentum. The maximal compact subalgebra is spanned by the angular momentum and the sum of the inertia and kinetic tensors, $(Q + T)^{\rm L}_{ij}$, and is isomorphic to the Lie algebra $u(3)$ of the unitary group $U(3)$.

The $gcm(3) \supset gl(3, R)$ algebra is the dynamical algebra of the Riemann ellipsoidal model and the $rot(3) \supset so(3)$ algebra corresponds to the Euler rigid body model.

A Riemann ellipsoid is a uniform density fluid with an ellipsoidal boundary whose velocity field is a linear function of the Cartesian position coordinates.[1] Although the exact kinetic tensor T_{ij} is not an element of $gcm(3)$, its linear velocity field value, the collective kinetic tensor, is a function of the algebra generators, $t = {}^tN \cdot Q^{-1} \cdot N$. The Kelvin circulation of a linear velocity field may be expressed in terms of the $gcm(3)$ generators by $C_k = \epsilon_{kij}(Q^{-1/2} \cdot N \cdot Q^{1/2})_{ij}$. In the principal axis frame, the inertia tensor Q is, by definition, diagonal, and its eigenvalues are proportional to the squared axis lengths a_i^2 of the inertia ellipsoid. Hence, in the rotating frame, the circulation simplifies to $C_k = \epsilon_{ijk}(a_j/a_i)N_{ij}$.

To describe planar galaxies, restrict the Cartesian indices in Eq. (1.1) to $i, j = 1, 2$ and consider the symplectic algebra $sp(2, R)$. Two-dimensional Riemann ellipses correspond to the algebra $gcm(2) \subset sp(2, R)$. One-dimensional breathing mode oscillations are described by the observables that are traces of the three-dimensional tensors,

$$\begin{aligned} Q^{\mathrm{L}} &= \sum m_\alpha r_\alpha^2, \\ N^{\mathrm{L}} &= \sum \mathbf{x}_\alpha \cdot \mathbf{p}_\alpha, \\ T^{\mathrm{L}} &= \sum m_\alpha^{-1} p_\alpha^2. \end{aligned} \tag{1.3}$$

These generate the one-dimensional symplectic algebra $sp(1, R) \simeq sl(2, R)$.

In the next section the dynamics of Riemann ellipsoids is presented as a Lax pair.

2. RIEMANN ELLIPSOID LAX DYNAMICS

Time evolution in the classical collective models based upon $rot(3), gcm(3)$, and $sp(3, R)$ is governed by Hamiltonian dynamics of a special type known as a Lax system. Consider first the simple case of $rot(3)$ for which the dynamics corresponds to Euler rigid body rotation. If the inertia ellipsoid is rotating with an angular velocity $\Omega_{ij} = \epsilon_{ijk}\omega_k$ and $L_{ij} = N_{ij} - N_{ji}$ is the angular momentum tensor, then Hamiltonian dynamics is given by

$$\dot{L} = [\Omega, L]. \tag{2.1}$$

In terms of vectors, this equation is the familiar law $\dot{\mathbf{L}} = -\vec{\omega} \times \mathbf{L}$ that determines the precession of the angular momentum vector in the body-fixed frame.

A matrix equation of the form $\dot{X} = [F, X]$ is called a Lax equation and $X - F$ are referred to as a Lax pair.[2] A useful property of any Lax equation is that the trace of any power of X is conserved. Let I_p denote the trace of the pth power of the matrix X.

$$I_p = \frac{1}{p}\mathrm{Tr}(X)^p. \tag{2.2}$$

For any Lax equation, it is evident that every I_p is a constant of the motion,

$$\dot{I}_p = \mathrm{Tr}\left(X^{p-1} \cdot \dot{X}\right) = \mathrm{Tr}\left(X^{p-1} \cdot [F, X]\right) = \mathrm{Tr}\left(X^{p-1}FX - X^pF\right) = 0. \tag{2.3}$$

In the case of the Euler equation, $I_2 = -\mathbf{L} \cdot \mathbf{L}$ is the negative of the squared length of the angular momentum vector. If p is odd, then I_p is zero. If $p > 2$ is even, then I_p is a function of the squared length of the angular momentum vector. Thus, there is only one independent invariant among the Lax invariants.

The coefficients of the characteristic equation for X are combinations of the traces of powers of X itself. Hence, the trace powers determine uniquely the eigenvalues of X, which consequently are invariants themselves. For the Euler equation, a real antisymmetric 3×3 matrix L has three eigenvalues: $0, \pm i\sqrt{\mathbf{L}\cdot\mathbf{L}}$. Thus, there is only one independent eigenvalue, and it is determined by one Lax invariant, viz., I_2.

Suppose that $X(t)$ is a solution to the Lax equation, $\dot{X} = [F, X]$, corresponding to the initial condition $X = X_0$. If $g(t)$ is a smooth curve of invertible matrices satisfying the matrix differential equation $\dot{g} = F \cdot g$ with the initial condition $g = I$, then the solution to the Lax equation is just the isospectral deformation,[3]

$$X(t) = g(t) \cdot X_0 \cdot g(t)^{-1}. \tag{2.4}$$

This is proven using the identity $dg^{-1}/dt = -g^{-1}\dot{g}g^{-1}$. It is now clear at a deeper level why each trace of a power of $X(t)$ is a conserved quantity: The similarity transformation Eq. (2.4) preserves the eigenvalues of the matrix X_0. If Ω is constant, the matrix differential equation $\dot{g} = \Omega \cdot g$ for the Euler equation has the unique solution $g(t) = \exp(\Omega t)$ for the initial condition $g(0) = I$. Thus, $g(t)$ is a curve in the rotation group $SO(3)$, and the isospectral deformation $L(t) = g(t)L_0\, g(t)^{-1}$ describes explicitly the precession of the angular momentum in the body-fixed frame resulting from the rotation $g(t)$ of the intrinsic frame relative to the laboratory frame. Because of the choice of initial conditions for g, L_0 represents the constant angular momentum vector in the inertial laboratory frame.

To present the time evolution for Riemann ellipsoids as a Lax system, suppose the gravitational potential energy in the body-fixed frame $V = V(a_1, a_2, a_3)$ is a smooth function of the axes lengths. Define the Chandrasekhar potential tensor W in the rotating frame to be the diagonal matrix,[4]

$$W_{ij} = -\delta_{ij} a_i \frac{\partial V}{\partial a_i}, \tag{2.5}$$

and, to impose a constraint to constant volume, define the pressure tensor $\Pi = p\,v$ to be the product of the hydrostatic pressure p times the ellipsoid's volume $v = 4\pi a_1 a_2 a_3/3$. Hamiltonian dynamics for Riemann ellipsoids is given as follows[5] :

Theorem. If the inertia ellipsoid is rotating with an angular velocity $\Omega_{ij} = \epsilon_{ijk}\omega_k$, then the Riemann ellipsoid equations of motion are equivalent to the Lax system, $\dot{X} = [F, X]$, where the 6×6 real matrices X and F in the body-fixed frame are given by

$$X = \begin{pmatrix} N & -Q \\ t & -{}^{t}N \end{pmatrix}, \quad F = \begin{pmatrix} \Omega & I \\ (W + \Pi) \cdot Q^{-1} & \Omega \end{pmatrix}. \tag{2.6}$$

Equilibrium Riemann ellipsoids are solutions for which the body-fixed inertia, virial momentum, and angular velocity tensors are constant in time. Hence, an equilibium solution corresponds to $\dot{X} = 0$, or the Lax pair X and F commute, $[F, X] = 0$. The equilibrium solutions have been studied extensively for self-gravitating systems.[1]

The quadratic Lax invariant equals the negative of the squared length of the Kelvin circulation vector, $I_2 = \mathrm{Tr}(N^2 - t \cdot Q) = -C^2$. The higher order Lax invariants are either zero (odd powers) or are functions of the circulation vector's squared length. From the Lax equation for Riemann ellipsoids, one may derive the Lax equation for the precession of the angular momentum, $\dot{L} = [\Omega, L]$.

The phase space for a Riemann ellipsoid obeying the Lax equation is a co-adjoint orbit of the Lie group $GCM(3)$. A faithful matrix representation of $gcm(3)$ is given by the 6×6 real matrices,

$$gcm(3) \simeq \left\{ (\Xi, X) \equiv \begin{pmatrix} X & 0 \\ \Xi & -{}^{t}X \end{pmatrix}, \Xi = {}^{t}\Xi \right\}, \tag{2.7}$$

where $Q^{L}_{ij} \mapsto e_{i+3,j} + e_{j+3,i}$, $N^{L}_{ij} \mapsto e_{ij} - e_{3+j,3+i}$, and e_{km} denotes the 6×6 matrix with 1 at the intersection of row k with column l and zero elsewhere. The Lie group $GCM(3)$ is given by exponentiation,

$$GCM(3) \simeq \left\{ (\Delta, g) \equiv \begin{pmatrix} g & 0 \\ \Delta g & {}^{t}g^{-1} \end{pmatrix}, \Delta = {}^{t}\Delta, g \in Gl(3, R) \right\}, \tag{2.8}$$

and obeys the semidirect product multiplication rule

$$(\Delta_1, g_1)(\Delta_2, g_2) = (\Delta_1 + {}^{t}g_1^{-1}\Delta_2 g_1^{-1}, g_1 g_2). \tag{2.9}$$

Thus, $GCM(3)$ is isomorphic to a semidirect product of the abelian normal subgroup $\mathbf{R}^6$ of 3×3 real symmetric matrices under addition with the general linear group $Gl(3, R)$.

The following theorem has been proven[6] :

Theorem. Each Riemann ellipsoid phase space is diffeomorphic to a coset space of GCM(3). The coset depends upon the value of the circulation C:

$$\mathcal{O}_C = \begin{cases} GCM(3)/SO(2) \cong \mathbf{R}^{12} \times S^2, & C \neq 0, \quad \dim = 14, \\ GCM(3)/SO(3) \cong \mathbf{R}^{12}, & C = 0, \quad \dim = 12. \end{cases} \tag{2.10}$$

The degenerate orbit is diffeomorphic to 12-dimensional Euclidean space. The generic orbits are diffeomorphic to the Cartesian product $\mathbf{R}^{12} \times S^2$ of Euclidean space with the two-dimensional sphere. The topology of the sphere forces the circulation to be quantized to integer multiples of $\hbar$ in a way parallel to the usual angular momentum quantization. Thus, the spectrum of the squared length of the quantum circulation operator is quantized to $C(C+1)\hbar^2$, where C is a nonnegative integer.[7,8]

3. SELF-GRAVITATING SYMPLECTIC LAX SYSTEMS

Dynamics in the galactic symplectic model is given in the rotating frame by the Lax equation, $\dot{X} = [F, X]$, if, in the Lax matrix X of Eq. (2.6), the linear approximation t to the kinetic energy is replaced by its exact expression T. The Lax equations for the Riemann and symplectic models in the inertial laboratory frame are given by replacing each body-fixed quantity in the pair $X - F$ by its corresponding L- superscripted inertial frame value and also by deleting the angular velocity matrix Ω from F. The inertial frame equations are provided by the Lax pair $X^{L} - F^{L}$.

The symplectic conservation laws are provided by the Lax invariants I_p. The quadratic Casmir invariant of the symplectic algebra is the quadratic Lax invariant, $\mathcal{C}^{(2)} = \mathrm{Tr}(N^2 - Q \cdot T)$. Note that for a linear velocity field, the quadratic symplectic invariant simplifies to the negative of the squared length of the Kelvin circulation vector. The odd order invariants vanish. The quartic symplectic Casimir invariant is the quartic Lax invariant,

$$\mathcal{C}^{(4)} = \mathrm{Tr}\left[(NQ - Q^{t}N)(TN - {}^{t}NT)\right] - 1/2\,\mathrm{Tr}\left[(N^2 - QT)^2\right]. \tag{3.1}$$

There is only one more independent Casimir and Lax invariant $\mathcal{C}^{(6)} = I_6$; the higher order invariants are functionally dependent upon the three independent Casimirs $\mathcal{C}^{(p)} = I_p$ for $p = 2, 4, 6$.

The symplectic Lie group $Sp(3, R)$ is defined by the 6×6 matrices that preserve an antisymmetric nondegenerate form J:

$$Sp(3,R) = \{g \in Gl(6,R) \mid {}^t gJg = J\} \text{ where } J \equiv \begin{pmatrix} 0 & I \\ -I & 0 \end{pmatrix}. \tag{3.2}$$

Its Lie algebra may be characterized by the form J also,

$$sp(3,R) \simeq \{X \in gl(6,R) \mid {}^t XJ + JX = 0\}. \tag{3.3}$$

It is easy to verify that the matrices X, F, X^{L}, and F^{L} are each elements of the symplectic Lie algebra. Suppose that X^L is a solution to the inertial frame symplectic Lax equation,

$$\dot{X}^{\mathrm{L}} = \left[F^{\mathrm{L}}, X^{\mathrm{L}}\right], \tag{3.4}$$

corresponding to the initial condition $X^{\mathrm{L}} = X_0^{\mathrm{L}}$. Let $g(t) \in Gl(6,R)$ be a smooth curve satisfying the matrix differential equation $\dot{g} = F^{\mathrm{L}} \cdot g$, with the initial condition $g = I$. Then, the solution to the symplectic dynamical equation with the given initial conditions is just the group transformation, $X^{\mathrm{L}}(t) = g(t)X_0^{\mathrm{L}}g(t)^{-1}$. Since F^{L} is evidently an element of the symplectic Lie algebra,

$$\begin{aligned} 0 &= {}^t g\left({}^t F^{\mathrm{L}} J + JF^{\mathrm{L}}\right)g \\ &= {}^t(F^{\mathrm{L}}g)Jg + {}^t gJF^{\mathrm{L}}g \\ &= {}^t\dot{g}Jg + {}^t gJ\dot{g} \\ &= \frac{d}{dt}({}^t gJg). \end{aligned} \tag{3.5}$$

Since g equals the identity matrix at the initial time, ${}^t gJg = J$ for all times, and the following theorem is proven[9] :

Theorem. The solutions to the symplectic Lax system are given by a symplectic group transformation $g(t) \in Sp(3,R)$ applied to the initial state

$$X^{\mathrm{L}}(t) = g(t) \cdot X_0^{\mathrm{L}} \cdot g(t)^{-1}, \tag{3.6}$$

where X_0^{L} and X^{L} are elements of the symplectic Lie algebra $sp(3,R)$. The group element $g(t)$ is a solution to the matrix differential equation $\dot{g} = F^{\mathrm{L}}g$ with the initial condition $g = I$ if and only if X^{L} is a solution to the Lax equation (3.4) with the initial condition $X^{\mathrm{L}} = X_0^{\mathrm{L}}$.

Consider the co-adjoint orbit of the symplectic group through the point X^{L},

$$\mathcal{O}_{X^{\mathrm{L}}} = \left\{g \cdot X^{\mathrm{L}} \cdot g^{-1} \mid g \in Sp(3,R)\right\}. \tag{3.7}$$

The co-adjoint orbit is regarded as a surface in the Euclidean symplectic dual space, $sp(3,R)^*$. A manifold that intersect each co-adjoint orbit exactly once is called a "transversal". A transversal $\mathcal{T}$ for the symplectic co-adjoint group action is provided by a three-dimensional surface:

$$\mathcal{T} = \left\{\hat{S} = \begin{pmatrix} 0 & -S \\ S & 0 \end{pmatrix} \in sp(3,R)^* \mid S = \mathrm{diag}(s_1, s_2, s_3)\right\}. \tag{3.8}$$

Transversal points correspond to elementary systems for which the virial momentum tensor vanishes, $N^{\mathrm{L}} = 0$, and the inertia and kinetic tensors are equal and diagonal, $Q^{\mathrm{L}} = T^{\mathrm{L}} = S$. Since the inertia and kinetic tensors are positive-definite, the physically relevant transversal consists of only those points for which S is positive-definite, $s_i > 0$.

An orbit of the transversal point $\hat{S} \in \mathcal{T}$ is diffeomorphic to a coset space of the symplectic group modulo the isotropy subgroup. These isotropy subgroups may be proven to be subgroups of the unitary group,

$$U(3) \simeq \left\{ \begin{pmatrix} U & -V \\ V & U \end{pmatrix} \in Sp(3,R) \mid U + iV \in U(3) \right\}, \tag{3.9}$$

and, thereby, the coset spaces are given explicitly as follows[13–15] :

Theorem. The galactic symplectic phase spaces are diffeomorphic to coset spaces of $Sp(3,R)$:

$$\mathcal{O}_S = \begin{cases} Sp(3,R)/[U(1)\times U(1)\times U(1)], & s_i \text{ distinct}, \quad \dim = 18, \\ Sp(3,R)/[U(2)\times U(1)], & s_1 = s_2 \neq s_3, \quad \dim = 16, \\ Sp(3,R)/U(3), & s_1 = s_2 = s_3, \quad \dim = 12. \end{cases} \tag{3.10}$$

The minimal dimension degenerate orbit, $Sp(3,R)/U(3)$, is diffeomorphic to the $GCM(3)$ phase space with vanishing circulation,$GCM(3)/SO(3)$. The diffeomorphism is given by the mapping $gSO(3) \mapsto gU(3)$ for $g \in GCM(3)$. This mapping is well-defined since $GCM(3) \cap U(3) = SO(3)$, and it is surjective since $Sp(3,R) = GCM(3) \cdot U(3)$.

4. SIEGEL HALF-PLANE DYNAMICS

The 12-dimensional degenerate phase space $Sp(3,R)/U(3) \simeq GCM(3)/SO(3)$ of the galactic symplectic model is diffeomorphic to the Siegel upper half-plane of 3×3 symmetric complex matrices whose imaginary part is positive-definite,

$$S_3 = \{ z \in M_3(\mathbf{C}) |\ {}^t z = z, y = \ Im z > 0 \}. \tag{4.1}$$

The symplectic group acts transitively on the Siegel half-plane by linear fractional transformations

$$g \cdot z \equiv (az+b)(cz+d)^{-1}, \tag{4.2}$$

for $z \in S_3$ and

$$g = \begin{pmatrix} a & b \\ c & d \end{pmatrix} \in Sp(3,R). \tag{4.3}$$

Note that the unitary group $U(3)$ is the isotropy subgroup at $z = i$ and, hence, the diffeomorphism of S_3 with the coset space $Sp(3,R)/U(3)$ is given by the mapping $gU(3) \mapsto g \cdot i$. The tensor observables forming the symplectic algebra are given on the Siegel half-plane by

$$\begin{aligned} Q &= s(y + xy^{-1}x) \\ N &= sxy^{-1} \\ T &= sy^{-1}, \end{aligned} \tag{4.4}$$

where $z = x + iy \in S_3$.

For the one-dimensional breathing mode theory of galactic oscillations, the space S_1 is the usual upper half-plane $\{z = x + iy \in \mathbf{C}, y > 0\}$. The symplectic structure is given by the form

$$\omega = -\frac{s}{2} y^{-2} dx \wedge dy, \tag{4.5}$$

and a smooth function $H(x, y)$ on the upper half-plane defines a vector field

$$X_H = (dH)^\sharp = -\frac{2}{s} y^2 \left(\frac{\partial H}{\partial y} \frac{\partial}{\partial x} - \frac{\partial H}{\partial x} \frac{\partial}{\partial y} \right). \tag{4.6}$$

Hence, the Poisson bracket of two smooth functions $f(x, y)$ and $g(x, y)$ is given by

$$\{f, g\} = -\frac{2}{s} y^2 \left(\frac{\partial f}{\partial x} \frac{\partial g}{\partial y} - \frac{\partial g}{\partial x} \frac{\partial f}{\partial y} \right). \tag{4.7}$$

The vector fields corresponding to the symplectic generators are

$$\begin{aligned} X_Q &= 2(x^2 - y^2) \frac{\partial}{\partial x} + 4xy \frac{\partial}{\partial y} \\ X_N &= 2 \left(x \frac{\partial}{\partial x} + y \frac{\partial}{\partial y} \right) \\ X_T &= 2 \frac{\partial}{\partial x}, \end{aligned} \tag{4.8}$$

Consider the harmonic oscillator Hamiltonian that describes galactic oscillations,

$$H(x, y) = \frac{1}{2}(T + Q) = \frac{s}{2} y^{-1} (1 + x^2 + y^2) \tag{4.9}$$

The constant energy curves are the solutions to $E = H(x, y) = sc$, where $c > 1$ is a constant, and are the circles centered on the y-axis at $(0, c)$ with radius $\sqrt{c^2 - 1} \geq 0$,

$$x^2 + (y - c)^2 = c^2 - 1. \tag{4.10}$$

A simple application of the algebraic method is provided by the gcm(1) prediction for equilibria with respect to breathing mode oscillations. Let $H(\mathbf{x}_\alpha, \mathbf{p}_\alpha)$ be the self-gravitating Hamiltonian on $\mathbf{R}^{6A}$. If $p = (\mathbf{x}_\alpha, \mathbf{p}_\alpha)$ is a point in phase space corresponding to a $Sp(1, R)$ transversal point, $\sum \mathbf{x}_\alpha \cdot \mathbf{p}_\alpha = 0$ and $\sum m_\alpha r_\alpha^2 = \sum m_\alpha^{-1} p_\alpha^2 = s > 0$, then the energy of the $GCM(1)$ orbit point $(\Delta, g) \cdot p$ is calculated to be

$$E(\Delta, g) = g^{-1} E(p) + \frac{s}{2}(g^{-2} - g^{-1}) + \frac{s}{2} \Delta^2 g^2. \tag{4.11}$$

Equilibrium when $E(p) < s$ is attained at the minimized energy: $\Delta = 0$ and

$$g = \frac{2}{1 - E(p)/s}. \tag{4.12}$$

The equilibrium monopole moment is, therefore,

$$Q = g^2 s. \tag{4.13}$$

I would like to thank E. Ihrig and T. Dankova for helpful suggestions. This material is based upon work supported by the National Science Foundation under Grant No. PHY-9212231.

References

1. S. Chandrasekhar, "Ellipsoidal Figures of Equilibrium, Yale University Press, New Haven (1969)
2. P. Lax, *Commun. Pure Appl. Math.* 21:467 (1968)
3. Ju. Moser, *Adv. Math.* 16:197 (1975)
4. James Binney and Scott Tremaine, "Galactic Dynamics", Princeton University Press, Princeton (1987)
5. G. Rosensteel, Lax representation of Riemann ellipsoids, *Appl. Math. Lett.* 6:55 (1993)
6. G. Rosensteel, Rapidly rotating nuclei as Riemann ellipsoids, *Ann. Phys. (N.Y.)* 186:230 (1988)
7. G. Rosensteel and E. Ihrig, Geometric quantization of the $CM(3)$ Model, *Ann. Phys. (N.Y.)* 121:113 (1979)
8. G. Rosensteel and E. Ihrig, Geometric quantization of Riemann rotors, *in*: "Geometric Quantization and Coherent States Methods", S. Twareque Ali, I.M. Mladenov, and A. Odzijewicz (eds.), World Scientific, Singapore (1993)
9. G. Rosensteel, Self-gravitating symplectic systems, *Astrophys. J.* 416:291 (1993).

GRADED CONTRACTIONS OF $so(4,2)$

J. Tolar[1] and P. Trávníček[2]

[1] Department of Physics and Doppler Institute
Faculty of Nuclear Sciences and Physical Engineering
Czech Technical University
Břehová 7, 115 19 Prague 1
Czech Republic
[2] Institute of Atmospheric Physics
Czech Academy of Sciences
Boční II čp. 1401, 141 31 Prague 4
Czech Republic

Abstract

All $\mathbb{Z}_2 \times \mathbb{Z}_2$-graded contractions preserving the space isotropy with grading induced by the Π (space inversion) and Θ (time reversal) automorphisms of the Lie algebra $so(4,2)$ are listed. Some properties of these contractions are discussed.

1. INTRODUCTION

$\mathbb{Z}_2 \times \mathbb{Z}_2$-graded contractions of the kinematical groups of space-time were the subject of a previous article.[1] There the new method[2] based on the preservation of a grading through the contraction was studied in the case of the complex Lie algebra B_2 and its real forms. These contractions are physically interesting because the non-compact real forms $o(4,1)$ and $o(3,2)$ correspond to the de Sitter and the anti–de Sitter groups respectively, and among the contracted Lie algebras those of the Poincaré and the Galilei group are present. As a result, the classification of the 10-dimensional kinematical groups emerged under the very natural assumptions of space isotropy and preservation of the $\mathbb{Z}_2 \times \mathbb{Z}_2$-grading induced by the Lie algebra automorphisms of space inversion Π and time reversal Θ (see also Refs. 3,4).

We find it worthwhile and also interesting from the physical point of view to extend the scope of our investigation to the *conformal group of space-time* because of its overall importance, especially in quantum field theories as the symmetry of theories of massless particles.[5] As a matter of fact, already I. E. Segal[6] gave the very first definition of a contraction yielding an approximate (asymptotic) symmetry from the initial exact symmetry on an example of the conformal Lie algebra $so(4,2)$.

In order that the results of this paper could be compared with the preceding ones,[1] we dwell on the assumptions of the space isotropy and of preservation of the $\Pi \times \Theta$–grading.

Quantization, Coherent States, and Complex Structures
Edited by J.-P. Antoine *et al.*, Plenum Press, New York, 1995

2. NOTION OF GRADED CONTRACTIONS

We summarize the basic concepts and also introduce some useful notations.

Definition 1: *Lie algebra graded by an Abelian finite group.*[2] Let G be a finite Abelian group. Let us suppose that the Lie algebra $\mathcal{L}$ is decomposed as a linear space into a direct sum of (grading) subspaces

$$\mathcal{L} = \bigoplus_{i\in G} \mathcal{L}_i, \tag{2.1}$$

such that the commutation relations in $\mathcal{L}$ have the graded structure, i.e. for every choice of elements $x \in \mathcal{L}_i$ and $y \in \mathcal{L}_j$ we have

$$[x, y] = z, \tag{2.2}$$

where z belongs to the grading subspace $\mathcal{L}_{i+j}$, as long as the commutator differs from zero. For simplicity of notation we write

$$0 \neq [\mathcal{L}_i, \mathcal{L}_j] \subseteq \mathcal{L}_{i+j}, \quad i, j \in G. \tag{2.3}$$

Then the Lie algebra $\mathcal{L}$ is called *graded* by the finite Abelian group G.

Definition 2: *The graded contraction of a Lie algebra.*[2] The *G-graded contraction* $\mathcal{L}^\varepsilon$ of $\mathcal{L}$ is defined as a Lie algebra which has the same linear space structure as $\mathcal{L}$, but with modified commutation relations

$$0 \neq [\mathcal{L}_i, \mathcal{L}_j]_\varepsilon \stackrel{def}{=} \varepsilon_{i,j}\, [\mathcal{L}_i, \mathcal{L}_j] \subseteq \varepsilon_{i,j}\, \mathcal{L}_{i+j}, \quad i, j \in G, \tag{2.4}$$

where the contraction parameters $\varepsilon_{i,j}$ (zero or not) are such that the Jacobi identity is never violated. This requirement leads to the so called *first basic set of contraction equations* [2]

$$\varepsilon_{i,j}\varepsilon_{k,i+j} = \varepsilon_{j,k}\varepsilon_{i,j+k}, \quad i, j, k \in G \tag{2.5}$$

which the contraction parameters must satisfy. For contractions of the complex or real Lie algebras, the parameters $\varepsilon_{i,j}$ are complex or real respectively.

We introduce the following notation for the commutator of elements of a Lie algebra:

$$a \quad \overset{b}{\boxed{\;c\;}} \quad \Leftrightarrow \quad [a, b] = c. \tag{2.6}$$

Moreover we will use the following abbreviations :

$$\begin{aligned}
[\mathbf{a}, \mathbf{b}] = \mathbf{c} \quad &\Leftrightarrow \quad [a_i, b_j] = \varepsilon_{ijk}\, c_k, \\
[\mathbf{a}, \mathbf{b}] = d \quad &\Leftrightarrow \quad [a_i, b_j] = \delta_{ij}\, d, \\
[\mathbf{a}, d] = \mathbf{b} \quad &\Leftrightarrow \quad [a_i, d] = b_i, \qquad\qquad i, j, k \in \{1, 2, 3\},
\end{aligned} \tag{2.7}$$

where δ_{ij} is the Kronecker symbol ($\delta_{ij} = 1$ for $i = j$ and $\delta_{ij} = 0$ for $i \neq j$) and ε_{ijk} is totally skew–symmetric Levi-Cività tensor with $\varepsilon_{123} = 1$.

3. THE LIE ALGEBRA $so(4, 2)$

3.1. $\Pi \times \Theta$-Graded Structure

There are two natural commuting involutive automorphisms of the conformal group of the Minkowski space-time: the space inversion Π and the time reversal Θ, which can be represented by the matrices

$$\Pi \stackrel{def}{=} \mathrm{diag}\,(-1, -1, -1, +1, +1, +1), \quad \Theta \stackrel{def}{=} \mathrm{diag}\,(+1, +1, +1, -1, +1, +1), \tag{3.1}$$

acting on the Lie algebra $so(4,2)_{K_\mu}$ by similarity transformations.

Now each of the involutive automorphisms $T \equiv \Pi$ and $T \equiv \Theta$ has the eigenvalues $+1$ and -1,

$$T\,x\,T^{-1} = (-1)^\alpha x, \qquad x \in \mathcal{L},\ \alpha \in \{0,1\}, \tag{3.2}$$

hence automatically induces $\mathbb{Z}_2$ grading of the Lie algebra $so(4,2)_{K_\mu}$. So the two similarity transformations by $T \equiv \Pi$ and $T \equiv \Theta$ induce the following $\mathbb{Z}_2 \times \mathbb{Z}_2$ grading of $so(4,2)_{K_\mu}$:

$$\begin{aligned} so(4,2)_{K_\mu} &= \bigoplus_{\alpha\beta\in\{0,1\}} \mathcal{L}_{\alpha\beta} = \mathcal{L}_{00} \oplus \mathcal{L}_{01} \oplus \mathcal{L}_{10} \oplus \mathcal{L}_{11} = \\ &= \operatorname{span}\{\mathbf{J},\ D\} \oplus \operatorname{span}\{p_4,\ q_4\} \oplus \operatorname{span}\{\mathbf{p},\ \mathbf{q}\} \oplus \operatorname{span}\{\mathbf{M}\}. \end{aligned} \tag{3.3}$$

3.2. $\Pi \times \Theta$-Graded Contractions

Let us now rewrite the commutation relations of $so(4,2)_{K_\mu}$ in the form modified by the real contraction parameters $\varepsilon_{i,j}$, $i,j \in \{a \equiv 00,\ b \equiv 01,\ c \equiv 10,\ d \equiv 11\}$:

	$\mathbf{J}$	D	p_4	q_4	$\mathbf{p}$	$\mathbf{q}$	$\mathbf{M}$
$\mathbf{J}$	$-\varepsilon_{a,a}\,\mathbf{J}$	$+\varepsilon_{a,a}\,0$	$+\varepsilon_{a,b}\,0$	$+\varepsilon_{a,b}\,0$	$-\varepsilon_{a,c}\,\mathbf{p}$	$-\varepsilon_{a,c}\,\mathbf{q}$	$-\varepsilon_{a,d}\,\mathbf{M}$
D		$+\varepsilon_{a,a}\,0$	$-\varepsilon_{a,b}\,q_4$	$-\varepsilon_{a,b}\,p_4$	$-\varepsilon_{a,c}\,\mathbf{q}$	$-\varepsilon_{a,c}\,\mathbf{p}$	$+\varepsilon_{a,d}\,0$
p_4			$+\varepsilon_{b,b}\,0$	$+\varepsilon_{b,b}\,D$	$-\varepsilon_{b,c}\,\mathbf{M}$	$+\varepsilon_{b,c}\,0$	$-\varepsilon_{b,d}\,\mathbf{p}$
q_4				$+\varepsilon_{b,b}\,0$	$+\varepsilon_{b,c}\,0$	$+\varepsilon_{b,c}\,\mathbf{M}$	$-\varepsilon_{b,d}\,\mathbf{q}$
$\mathbf{p}$					$-\varepsilon_{c,c}\,\mathbf{J}$	$-\varepsilon_{c,c}\,D$	$-\varepsilon_{c,d}\,p_4$
$\mathbf{q}$						$+\varepsilon_{c,c}\,\mathbf{J}$	$-\varepsilon_{c,d}\,q_4$
$\mathbf{M}$							$+\varepsilon_{d,d}\,\mathbf{J}$

(3.4)

If all $\varepsilon_{i,j}$, $i,j \in \{a,b,c,d\}$, are equal to 1, we have the commutation relations of $so(4,2)_{K_\mu}$ in the non-contracted form (trivial contraction).

It is convenient to store the information about the zeros of non-contracted commutation relations in a (symmetric) matrix κ.[2] In many cases the grading of a Lie algebra is non-generic, i. e. some commutators $[\mathcal{L}_i, \mathcal{L}_j]$, $i,j \in G$ are identically zero before contraction. Then the corresponding elements of κ are set to zero. This was the case of the $\Pi \times \Theta$-grading of $o(5)$.[1] But now the commutation relations determine a matrix κ *without zeros*

$$\kappa = \begin{pmatrix} 1 & 1 & 1 & 1 \\ & 1 & 1 & 1 \\ & & 1 & 1 \\ & & & 1 \end{pmatrix}. \tag{3.5}$$

In Ref. 2, also the case of the generic $\mathbb{Z}_2 \times \mathbb{Z}_2$-graded structure was considered. It was shown that there are 40 inequivalent solutions of the basic set of contraction equations (2.5)

$$\varepsilon_{i,j}\varepsilon_{k,i+j} = \varepsilon_{j,k}\varepsilon_{i,j+k}, \qquad i,j,k \in \mathbb{Z}_2 \times \mathbb{Z}_2. \tag{3.6}$$

The scaling transformation φ

$$l_i \overset{\varphi}{\longrightarrow} l_i' = a_i l_i, \qquad a_i \in \mathbb{R}, \qquad l_i \in \mathcal{L}_i, \qquad i \in \{a,b,c,d\}, \tag{3.7}$$

of the basis of the Lie algebra changes its commutation relations into

$$[l_i, l_j] = l_{i+j} \overset{\varphi}{\longrightarrow} [l_i', l_j'] = \frac{a_i a_j}{a_{i+j}}\, l_{i+j}', \qquad i,j \in \{a,b,c,d\}. \tag{3.8}$$

In this way the non–zero contraction parameters ε_{ij}, $i,j \in \{a,b,c,d\}$, which differ from zero can be normalized to ± 1.[2]

We will consider two different results of $\Pi \times \Theta$-graded contractions of $so(4,2)_{K_\mu}$ as physically equivalent, if their generators have the same physical meaning even if they differ by a multiplier. In this sense the generators $\mathbf{J}$, H, $\mathbf{P}$ and $\mathbf{K}$ of $o(5)$ in Ref. 1 have the same physical meaning as the generators $\mathbf{J}$, $(p_4 + q_4)$, $(\mathbf{p} + \mathbf{q})$ and $\mathbf{M}$ of $so(4,2)_{K_\mu}$.

In order to preserve the usual transformation properties of the physical generators also in the contracted Lie algebras, we shall study the $\Pi \times \Theta$-graded contractions of $so(4,2)$ under the following *physical assumptions:*

(i) Space isotropy: the rotational invariance of space means that the form of the commutation relations involving $\mathbf{J}$ remains unchanged after contraction.[3]

(ii) Parity and time reversal transformations are retained as involutive automorphisms of the Lie algebras after contraction.

In accordance with the physical assumption (i), we keep the contraction parameters $\varepsilon_{a,a}$, $\varepsilon_{a,c}$ and $\varepsilon_{a,d}$ equal to 1. This assumption restricts the first basic set of contraction equations (2.5) (or (3.6)) to the set of 8 independent equations

$$
\begin{aligned}
\varepsilon_{a,b} &= \varepsilon_{a,b}\varepsilon_{a,b}, & \varepsilon_{b,d} &= \varepsilon_{a,b}\varepsilon_{b,d}, & \varepsilon_{b,b} &= \varepsilon_{b,d}\varepsilon_{b,c},\\
\varepsilon_{b,b} &= \varepsilon_{a,b}\varepsilon_{b,b}, & & & \varepsilon_{c,c} &= \varepsilon_{b,c}\varepsilon_{c,d},\\
\varepsilon_{b,c} &= \varepsilon_{a,b}\varepsilon_{b,c}, & \varepsilon_{c,d} &= \varepsilon_{a,b}\varepsilon_{c,d}, & \varepsilon_{d,d} &= \varepsilon_{c,d}\varepsilon_{b,d}.
\end{aligned}
\tag{3.9}
$$

From these the $\Pi \times \Theta$-graded contractions of $so(4,2)$ were computed:

$$
\varepsilon^D = \begin{pmatrix} 1 & \cdot & 1 & 1 \\ & \cdot & \cdot & \cdot \\ & & \cdot & \cdot \\ & & & \cdot \end{pmatrix}, \tag{3.10}
$$

$$
\varepsilon^1 = \begin{pmatrix} 1 & 1 & 1 & 1 \\ & \hat{\varepsilon}_{bb} & \hat{\varepsilon}_{bc} & \hat{\varepsilon}_{bb}\hat{\varepsilon}_{bc}^{-1} \\ & & \hat{\varepsilon}_{bc}\varepsilon_{cd} & \varepsilon_{cd} \\ & & & \hat{\varepsilon}_{bb}\varepsilon_{cd}\hat{\varepsilon}_{bc}^{-1} \end{pmatrix}, \tag{3.11}
$$

$$
\varepsilon^3 = \begin{pmatrix} 1 & 1 & 1 & 1 \\ & \cdot & \hat{\varepsilon}_{bc} & \cdot \\ & & \hat{\varepsilon}_{bc}\varepsilon_{cd} & \varepsilon_{cd} \\ & & & \cdot \end{pmatrix}, \tag{3.12}
$$

$$
\varepsilon^{2,4} = \begin{pmatrix} 1 & 1 & 1 & 1 \\ & \cdot & \cdot & \varepsilon_{bd} \\ & & \cdot & \varepsilon_{cd} \\ & & & \varepsilon_{cd}\varepsilon_{bd} \end{pmatrix}; \tag{3.13}
$$

the hat over the ε-s means that the corresponding ε_{ij} is never equal to zero ($\hat{\varepsilon}_{ij} \neq 0$).

Because of the first five of the equations (3.9), the contraction parameter $\varepsilon_{a,b}$ may be contracted only discretely to zero. This case gives the *discrete contraction* (3.10). When we keep $\varepsilon_{a,b}$ equal to 1, then the first five equations (3.10) are trivial. The remaining ones give the three types (3.11)-(3.13) of $\Pi \times \Theta$-graded contractions of $so(4,2)_{K_\mu}$.

4. LIST OF $\Pi \times \Theta$–GRADED CONTRACTIONS OF $so(4,2)$

The resolution of (3.9) over $\{-1,0,1\}$ yields 28 different contraction matrices. Under the scaling transformation (3.7) involving a change of the signs of $\varepsilon_{b,c}$, $\varepsilon_{b,d}$ and $\varepsilon_{c,d}$, we find that among 28 solutions, only 15 contraction matrices are independent.

The $\varepsilon^{R1d\alpha}$-contractions, $\alpha \in \{1,2,3\}$, give two pairs of isomorphic Lie algebras.

- $\alpha = 3$: In the case of the contraction ε^{R1d3} we obtain the form of the commutation relations of $so(4,2)$ Lie algebra, which we denote as $so(4,2)_{-K_\mu}$. There is another way, in which it is possible to obtain the form $so(4,2)_{-K_\mu}$ of the commutation relations of $so(4,2)$ Lie algebra, concretely by exchanging the physical role of the generators p_μ and q_μ, $\mu \in \{1,2,3,4\}$:

$$P_\mu = p_\mu + q_\mu \quad \longrightarrow \quad {P'}_\mu = q_\mu + p_\mu \equiv P_\mu, \tag{4.1}$$
$$K_\mu = p_\mu - q_\mu \quad \longrightarrow \quad {K'}_\mu = q_\mu - p_\mu \equiv -K_\mu, \tag{4.2}$$

which physically corresponds only to a change of the sign of the generators of the special conformal transformations. Then we can consider the forms $so(4,2)_{K_\mu}$ and $so(4,2)_{-K_\mu}$ of the commutation relations of $so(4,2)$ Lie algebra as physically equivalent.

- $\alpha = 1$: If the metric in the matrices $(A.5)$ is $g_{ab} = \mathrm{diag}(+1,\ +1,\ +1,\ +1,\ +1,\ -1)$ and we simultaneously keep the notation introduced by $(A.5)$, then we obtain the same commutation relations as under the contraction ε^{R1d1}. This fact implies that we obtain the commutation relations of $so(5,1)$. We denote this form by $so(5,1)_{K_\mu}$.

- $\alpha = 2$: By exchanging the role of p_μ and q_μ, $\mu \in \{1,2,3,4\}$ in $so(5,1)_{K_\mu}$ we obtain the commutation relations in the same form as under the contraction ε^{R1d2} and denote this form by $so(5,1)_{-K_\mu}$. As in the case of the pair of $so(4,2)$, Lie algebras we consider the forms $so(5,1)_{K_\mu}$ and $so(5,1)_{-K_\mu}$ of the commutation relations of $so(5,1)$ as physically equivalent.

The resulting contracted Lie algebras can be divided into 4 families.

- **Family 1:** These $\Pi \times \Theta$–graded contractions of $so(4,2)_{K_\mu}$ consist of the contractions ε^{R1} (trivial contraction), ε^{R1d1}, ε^{R1d2}, and ε^{R1d3}.

$$\varepsilon^{R1} = \begin{pmatrix} 1 & 1 & 1 & 1 \\ & 1 & 1 & 1 \\ & & 1 & 1 \\ & & & 1 \end{pmatrix} \sim so(4,2)_{K_\mu}, \quad \varepsilon^{R1d1} = \begin{pmatrix} 1 & 1 & 1 & 1 \\ & -1 & 1 & -1 \\ & & 1 & 1 \\ & & & -1 \end{pmatrix} \sim so(5,1)_{K_\mu},$$

$$\varepsilon^{R1d2} = \begin{pmatrix} 1 & 1 & 1 & 1 \\ & 1 & -1 & -1 \\ & & -1 & 1 \\ & & & -1 \end{pmatrix} \sim so(5,1)_{-K_\mu}, \quad \varepsilon^{R1d3} = \begin{pmatrix} 1 & 1 & 1 & 1 \\ & -1 & -1 & 1 \\ & & -1 & 1 \\ & & & 1 \end{pmatrix} \sim so(4,2)_{-K_\mu}.$$

- **Family 2** consists of contractions which have one discrete alternative, i.e. of ε^{R2}, ε^{R2d}, ε^{R3}, ε^{R3d}, ε^{A1} and ε^{A1d}. The pairs ε^{R3}, ε^{R3d} and ε^{A1}, ε^{A1d} of the graded contractions give the (physically) same results.

$$\varepsilon^{R2} = \begin{pmatrix} 1 & 1 & 1 & 1 \\ & \cdot & \cdot & 1 \\ & & \cdot & 1 \\ & & & 1 \end{pmatrix} \sim [so(3,1)_{\mathbf{J,M}} \oplus T_{1D}] \ni\!\!\!\!+ [T_{4P_\mu} \oplus T_{4K_\mu}],$$

$$\varepsilon^{R2} = \begin{pmatrix} 1 & 1 & 1 & 1 \\ & \cdot & \cdot & 1 \\ & & \cdot & 1 \\ & & & 1 \end{pmatrix} \sim [so(4)_{\mathbf{J,M}} \oplus T_{1D}] \ni\!\!\!\!+ [T_{4P_\mu} \oplus T_{4K_\mu}],$$

$$\varepsilon^{R3} = \begin{pmatrix} 1 & 1 & 1 & 1 \\ & \cdot & 1 & \cdot \\ & & 1 & 1 \\ & & & \cdot \end{pmatrix} = \varepsilon^{R1d3} \bullet \varepsilon^{R3d} \sim so(4,1)_{\mathbf{J},\mathbf{P},\mathbf{K},D} \mathrel{\ni\!\!\!\!+} [T_{3\mathbf{M}} \oplus T_{1P_4} \oplus T_{1K_4}],$$

$$\varepsilon^{A1} = \begin{pmatrix} 1 & 1 & 1 & 1 \\ & 1 & 1 & 1 \\ & & \cdot & \cdot \\ & & & \cdot \end{pmatrix} = \varepsilon^{R1d3} \bullet \varepsilon^{A1d} \sim [so(3)_{\mathbf{J}} \oplus so(2,1)_{P_4,K_4,D}] \mathrel{\ni\!\!\!\!+} [T_{3\mathbf{M}} \oplus T_{3\mathbf{P}} \oplus T_{3\mathbf{K}}].$$

The operation $\bullet$ is the composition of the matrices defined in Ref. 2 by the multiplication of corresponding elements.

- **Family 3** consists of the contractions ε^{R4}, ε^{A2} and ε^{A3}. Let us denote

$$A = so(3)_{\mathbf{J}} \oplus T_{1D}, \quad B = (T_{1P_4} \oplus T_{1K_4}), \quad C = T_{3\mathbf{M}}, \quad D = (T_{3\mathbf{P}} \oplus T_{3\mathbf{K}}). \tag{4.3}$$

Then we have

$$\varepsilon^{R4} \sim A \mathrel{\ni\!\!\!\!+} [(B \mathrel{+\!\!\!\!\in} C) \oplus D], \ \varepsilon^{A2} \sim A \mathrel{\ni\!\!\!\!+} [(D \mathrel{+\!\!\!\!\in} C) \oplus B], \ \varepsilon^{A3} \sim A \mathrel{\ni\!\!\!\!+} [(C \mathrel{+\!\!\!\!\in} D) \oplus B], \tag{4.4}$$

where

$$\varepsilon^{R4} = \begin{pmatrix} 1 & 1 & 1 & 1 \\ & \cdot & \cdot & \cdot \\ & & \cdot & 1 \\ & & & \cdot \end{pmatrix}, \varepsilon^{A2} = \begin{pmatrix} 1 & 1 & 1 & 1 \\ & \cdot & \cdot & 1 \\ & & \cdot & \cdot \\ & & & \cdot \end{pmatrix}, \varepsilon^{A3} = \begin{pmatrix} 1 & 1 & 1 & 1 \\ & \cdot & 1 & \cdot \\ & & \cdot & \cdot \\ & & & \cdot \end{pmatrix}. \tag{4.5}$$

- **Family 4** consists of the remaining contractions ε^{A4} and ε^{D}

$$\varepsilon^{A4} = \begin{pmatrix} 1 & 1 & 1 & 1 \\ & \cdot & \cdot & \cdot \\ & & \cdot & \cdot \\ & & & \cdot \end{pmatrix} \sim [so(3)_{\mathbf{J}} \oplus T_{1D}] \mathrel{\ni\!\!\!\!+} [T_{3\mathbf{M}} \oplus T_{4P_\mu} \oplus T_{4K_\mu}],$$

$$\varepsilon^{D} \text{ (see (3.10))} \sim \{[so(3)_{\mathbf{J}} \oplus T_{1D}] \mathrel{\ni\!\!\!\!+} [T_{3\mathbf{M}} \oplus T_{3\mathbf{P}} \oplus T_{3\mathbf{K}}]\} \oplus (T_{1P_4} \oplus T_{1K_4}).$$

5. DISCUSSION OF THE RESULTS

Our results can be summarized in a simple graphical form, Fig. 1. For $so(4,2)$ there are 8 continuous contractions described by 8 vertices of the cube, 1 discrete contraction ε^{D}, and 6 additional discrete contractions $\varepsilon^{R1d\alpha}$, $\alpha \in \{1,2,3\}$, ε^{R2d}, ε^{R3d}, ε^{A1d}.

The comparison with the preceding investigation[1] of B_2 and its real forms shows many similarities, but also two important differences:

1. The matrix κ exhibiting zero commutators, given in (3.5), has no zeros here, whereas there was one zero entry ($\kappa_{a,b} = 0$) for B_2. This led to the extra discrete contraction ε^{D}.

2. Even if the remaining 14 ε-matrices turned out to be the same as in Ref. 1, there is still a second important difference. In Ref. 1, among the contracted Lie algebras there were pairs of isomorphic Lie algebras, however with different assignments of physical generators, hence physically inequivalent. Here no such isomorphisms among the contracted Lie algebras are present.

Finally, we should comment on I. E. Segal's treatment of contraction of $so(4,2)$.[6] His example was constructed with the aim of obtaining the 10-dimensional Poincaré Lie algebra as subalgebra of the 15-dimensional contracted Lie algebra. In our treatment only one contraction has this property, namely ε^{R2}. Among the remaining 5 generators, the dilatation generator D forms together with the Poincaré generators the basis of the Lie algebra of the Weyl group — the largest transformation group of the Minkowski space-time which preserves the causal order.[8]

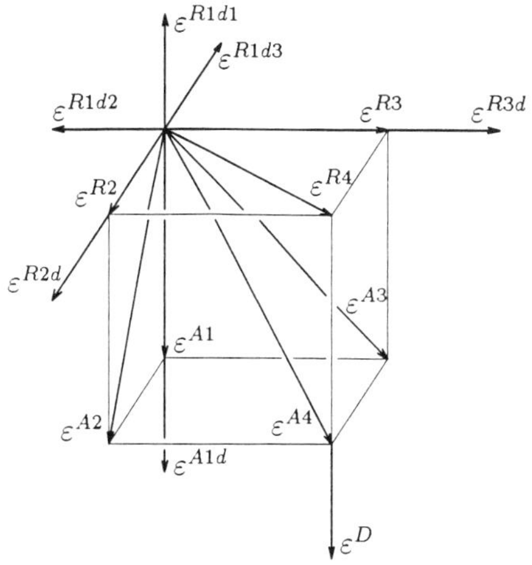

Figure 1: Contraction scheme for all $\Pi \times \Theta$–graded contractions of $so(4,2)$ (except the trivial contraction ε^{R1}).

ACKNOWLEDGEMENTS

J. T. acknowledges the support of the Grant Agency of Czech Republic (contract No. 202/93/1314). P. T. is grateful to his wife Markéta Trávníčková for her great patience and he is still charmed by her beautiful blue eyes.

APPENDIX: PHYSICAL GENERATORS OF $so(4,2)$

The Lie algebra of the conformal group of the Minkowski space–time is formed by the following differential operators

$$
\begin{array}{ll}
\text{Lorentz transformations} & M_{\mu\nu} = i\,(x_\mu \partial_\nu - x_\nu \partial_\mu), \\
\text{translations} & P_\mu = i\,\partial_\mu, \\
\text{special conformal transformations} & K_\mu = i\,(2x_\mu x^\nu \partial_\nu - x^2 \partial_\mu), \\
\text{dilatations} & D = i\,x^\nu \partial_\nu,
\end{array}
\qquad (A.1)
$$

where $\partial_\mu \equiv \dfrac{\partial}{\partial x^\mu}$. They satisfy the famous commutation relations

$$\begin{aligned}
&[M_{\mu\nu}, M_{\sigma\rho}] = i\,(g_{\mu\rho}M_{\nu\sigma} + g_{\nu\sigma}M_{\mu\rho} + g_{\mu\sigma}M_{\rho\nu} + g_{\nu\rho}M_{\sigma\mu}), && [K_\mu, K_\nu] = 0,\\
&[P_\lambda, M_{\mu\nu}] = i\,(g_{\mu\lambda}P_\nu - g_{\nu\lambda}P_\mu), && [D, M_{\mu\nu}] = 0,\\
&[P_\mu, P_\nu] = 0, && [P_\mu, D] = i\,P_\mu,\\
&[K_\lambda, M_{\mu\nu}] = i\,(g_{\mu\lambda}K_\nu - g_{\nu\lambda}K_\mu), && [D, K_\mu] = i\,K_\mu,\\
&[P_\mu, K_\nu] = 2\,i\,(g_{\mu\nu}D - M_{\mu\nu}), && [D, D] = 0,
\end{aligned} \tag{A.2}$$

where $g_{\mu\nu}$ is the metric tensor of the Minkowski space-time. Let us note that the space rotations J_1, J_2, J_3 and the Lorentz boosts M_1, M_2, M_3 can be introduced by

$$J_i = \tfrac{1}{2}\varepsilon_{ijk}M_{jk}, \quad M_i = M_{4i}, \qquad i \in \{1,2,3\}. \tag{A.3}$$

The generators of $so(4,2)$ can be represented by the matrices $\mathbb{L}_{ab}$, $a, b \in \{1, 2, \ldots, 6\}$, with elements defined as

$$(\mathbb{L}_{ab})_i^{\ j} = i(\delta_{ia}\delta^{jb}g_{aa} - \delta_{ib}\delta^{ja}g_{bb}), \tag{A.4}$$

where both δ_{ij} and δ^{ij} are Kronecker symbols ($\delta_{ij} = \delta^{ij} = 1$, $i = j$ and $\delta_{ij} = \delta^{ij} = 0$, $i \neq j$). In (A.4) the upper and the lower indices correspond to the columns and the rows, respectively.

In this article we start from $so(4,2)_{K_\mu}$. The subscript K_μ means that we use the metric $g_{ab} = \mathrm{diag}(+1, +1, +1, -1, +1, -1)$. Moreover, we have multiplied all matrices $\mathbb{L}_{ab}$, $a, b \in \{1, 2, \ldots, 6\}$, by $-i$

$$\begin{aligned}
J_i = -\tfrac{1}{2}\, i\, \varepsilon_{ijk}\mathbb{L}_{jk}\,|_{\hat{g}^+}, \quad M_i = -i\,\mathbb{L}_{4i}\,|_{\hat{g}^+},\\
p_\mu = -i\,\mathbb{L}_{5\mu}\,|_{\hat{g}^+}, \quad q_\mu = -i\,\mathbb{L}_{6\mu}\,|_{\hat{g}^+}, \quad D = -i\,\mathbb{L}_{56}\,|_{\hat{g}^+},
\end{aligned} \tag{A.5}$$

where $i, j, k \in \{1,2,3\}$ and $\mu \in \{1,2,3,4\}$. In the basis (A.5) the commutation relations of $so(4,2)$ have the form (3.4) with all $\varepsilon_{i,j}$, $i, j \in \{a,b,c,d\}$ equal to 1. If we substitute

$$P_\mu = p_\mu + q_\mu, \quad K_\mu = p_\mu - q_\mu, \tag{A.6}$$

$\mu \in \{1,2,3,4\}$, we obtain the commutation relations of $so(4,2)$ in the form (A.2) (up to the multiplicative i on all right hand sides of the commutation relations), i. e. we have

	$\mathbf{J}$	$\mathbf{M}$	$\mathbf{P}$	P_4	D	$\mathbf{K}$	K_4
$\mathbf{J}$	$-\mathbf{J}$	$-\mathbf{M}$	$-\mathbf{P}$	$\cdot$	$\cdot$	$-\mathbf{K}$	$\cdot$
$\mathbf{M}$		$\mathbf{J}$	P_4	$\mathbf{P}$	$\cdot$	K_4	$\mathbf{K}$
$\mathbf{P}$			$\cdot$	$\cdot$	$\mathbf{P}$	$2\,(D - \mathbf{J})$	$2\,\mathbf{M}$
P_4				$\cdot$	P_4	$-2\,\mathbf{M}$	$-2\,D$
D					$\cdot$	$\mathbf{K}$	K_4
$\mathbf{K}$						$\cdot$	$\cdot$
K_4							$\cdot$

Lie algebra of Lorentz group

Lie algebra of Poincaré group

Lie algebra of Weyl group

Lie algebra of conformal group of Minkowski space-time

. (A.7)

References

1. M. de Montigny, J. Patera and J. Tolar, Graded contractions and kinematical groups of space-time, *J. Math. Phys.* 35: 405-425 (1994)

2. M. de Montigny and J. Patera, Discrete and continuous graded contractions of Lie algebras and superalgebras, *J. Phys. A: Math. Gen.* 24:525-547 (1991)

3. H. Bacry and J.-M. Lévy-Leblond, Possible kinematics, *J. Math. Phys.* 9:1605-1614 (1968)

4. H. Bacry and J. Nuyts, Classification of ten-dimensional kinematical groups with space isotropy, *J. Math. Phys.* 27:2455-2457 (1986)

5. I. T. Todorov, Infinite-dimensional Lie algebras in conformal QFT models, in *Lecture Notes in Physics* No. 261, pp. 387–443, Springer-Verlag, Berlin (1986)

6. I. E. Segal, A class of operator algebras which are determined by groups , *Duke Math. J.* 18:221-265 (1951)

7. B. G. Wybourne, "Classical Groups for Physicists", John Wiley, New York (1973)

8. E. C. Zeeman, Causality implies Lorentz group, *J. Math. Phys.* 5:490-493 (1964).

THE BERRY PHASE AND THE GEOMETRY OF COSET SPACES

Ewgenij A. Tolkachev and Arthur A. Tregubovich

B. I. Stepanov Institute of Physics
Byelorussian Academy of Sciences
220072, F. Skaryna avenue 70, Minsk
Republic of Belarus

Abstract

We consider the Berry phase for quantum systems with dynamical symmetry and nondegenerate spectrum. It is shown that the corresponding phase factor is the holonomy group element of the group fibre bundle $G(G/S)$ where G is the dynamical symmetry group and S is its maximal commutative subgroup. The connection between such elements and Cartan structure 1-forms of the bundle is established. The particular case of $SU(3)$ symmetry group is completely investigated.

1. DYNAMICAL SYMMETRY AND GEOMETRIC PHASE

Physical effects connected with the geometric phase that appears in parametric quantum systems with quasi-periodic wave functions are now a subject of intensive investigation.[1] Such an effect can be mathematically expressed by a specific phase factor being acquired by the system's wave function when the corresponding parameters evolve periodically and there are no transitions in the system. This kind of phase factors was discovered by Berry[2] in the case of the evolution of adiabatic parameters and a nondegenerate spectrum. Concretely, let us consider a quantum system with Hamiltonian $H(\vec{R})$, where $\vec{R}$ denotes the set of parameters $\{R_i; i = 1, ..., N\}$ and is formally considered as a vector in $\mathbf{R^n}$. Let now $\vec{R}$ evolve periodically and classically, i.e. slowly enough for the adiabatic theorem[3] to be valid:

$$H(\vec{R})|n(\vec{R}) >= E_n(\vec{R})|n(\vec{R}) > \ . \tag{1.1}$$

The equation (1.1) is quasi-stationary because $\vec{R} = \vec{R}(t)$ is a function of time. Comparing (1.1) with the non-stationary Schrödinger equation

$$i\frac{\partial}{\partial t}|\Psi(t) >= H(\vec{R}(t))|\Psi(t) >, \tag{1.2}$$

one can obtain the solution of the problem for $\vec{R}(0) = \vec{R}(T)$ at some T

$$|\Psi_n(T) >= exp(-i\Phi_n(T) + i\gamma_n(C))\ |n(\vec{R}(0)) >, \tag{1.3}$$

Quantization, Coherent States, and Complex Structures
Edited by J.-P. Antoine *et al.*, Plenum Press, New York, 1995

where

$$\gamma_n(C) = \oint_C \vec{A}_n\,(\vec{R})d\vec{R}; \quad \Phi_n(t) = \int_0^T E_n(\tau)d\tau, \tag{1.4}$$

$$\vec{A}_n\,(\vec{R}) = i < n(\vec{R})|\nabla_{\vec{R}} n(\vec{R}) >, \tag{1.5}$$

where the integration is performed over the closed contour C followed by $\vec{R}$ in the parameter space. As it can be easily seen from (1.3), $\gamma_n(C)$ depends only on the contour C, but not on the parametrisation of C, i.e. not on the concrete details of the parameter evolution along C. Generally speaking, to calculate the phase under consideration in some concrete case, one has to find the eigenvalues and eigenstates of the Hamiltonian $H(\vec{R})$ and then apply formulae (1.4), (1.5). This is possible only in a few simple cases. Nevertheless there exists a class of Hamiltonians for which it is possible to simplify the calculation procedure. These Hamiltonians possess a so-called dynamical symmetry, i.e. they are in some way related to finite Lie groups.

In this paper we consider the case in which the Hamiltonian is an element of some Lie group. It is shown that the Berry phase of such Hamiltonians can be expressed by using the geometric characteristics of the manifold defined by the symmetry group and the given energy level. This expression can be represented as an integral of some combination of Cartan structure 1-forms on the manifold. These 1-forms are written in the local coordinates which are to be connected with the initial set of parameters.

Let the Hamiltonian $H(\vec{R})$ be an element of a Lie algebra **g** corresponding to the Lie group G. At an arbitrary instant t, it can be diagonalized by the unitary transformation $U(t)$

$$H(t) = U(t)\ D(t)\ U^\dagger(t), \tag{1.6}$$

where $D(t) = \mathrm{diag}\big(E_1(t), \ldots, E_n(t)\big)$. It follows from the adiabatic theorem that $U(t)$ plays the role of the evolution operator in this case:

$$|\Psi_n(t) >= U(t)\ |\Psi_n(0) > .$$

Since $\vec{R}(t)$ evolves cyclically, we have at the instant T :

$$H(T) = H(0) = U(T)\ H(0)\ U^\dagger(T),$$

so that

$$\big[U(T), H(0)\big] = 0$$

and

$$U(T) = \exp\big(i\ a_\alpha(C)\ \Lambda_\alpha\big), \tag{1.7}$$

where Λ_α commutes with $H(0)$ and $a_\alpha(C)$ are some path dependent coefficients. The set of generators $\{\Lambda_\alpha\}$ is a subalgebra $\mathbf{s} \subset \mathbf{g}$. Here we restrict ourselves to a nondegenerate spectrum, thus **s** is a maximal commutative subalgebra in **g** and the corresponding group S is a maximal torus in G. The space of states is thus geometrically equivalent to the group fiber bundle $G(G/S, S)$ and the Berry phase is completely defined by the parallel transport in it.

2. GEOMETRIC PHASE FACTORS AND PARALLEL TRANSPORT ON COSET SPACES

Geometrical properties of the parallel transport in such cases are well known. To apply the general theory, one has to find out what is the relation between two connections, on the group G and on the bundle $G(G/S, S)$. For this purpose, let us represent $\mathbf{g}$ as

$$\mathbf{g} = \mathbf{s} \oplus \mathbf{m}. \tag{2.1}$$

It is easy to check that

$$[\mathbf{s}, \mathbf{m}] \subset \mathbf{m}, \tag{2.2}$$

so one can conclude that $(AdS)\mathbf{m} \subset \mathbf{m}$.

This fact allows us to apply the theorem about group fiber bundles[4] which states that, in such a case, a connection 1-form on the fiber bundle is the $\mathbf{s}$-component of the canonical 1-form on G with respect to decomposition (2.1). This connection is invariant under the left action of G. The holonomy group $\Phi(e)$ at the point e (identity in G) consists of the $\mathbf{s}$-components of the commutators $\left[X, Y\right]$ where $X, Y \in \mathbf{m}$. To find the element of $\Phi(e)$ for an arbitrary closed curve, let us consider the parallel transport in $G(G/S, S)$. Let x and ξ be the sets of local coordinates in the base manifold and in the fiber, respectively, and let θ be the canonical left-invariant 1-form on G. This 1-form can be expressed in terms of the 1-form basis $\{\theta^a\}$ of the cotangent space $T^*_e(G\ )$ (a summation over repeated indices is taken):

$$\theta = \theta^a T_a; \quad a = 1, \ldots, N, \tag{2.3}$$

where T_a are the generators of G and

$$\theta^a(T_b) = \delta^a_b.$$

The 1-form θ obeys the Maurer-Cartan equation

$$d\theta = -\frac{1}{2}\ \theta \wedge \theta, \tag{2.4}$$

or, using the basis $\{\theta^a\}$ explicitly,

$$d\theta^a = -\frac{1}{2}\ C^a_{\ bc}\ \theta^b \wedge \theta^c\ , \tag{2.5}$$

where $C^a_{\ bc}$ are the structure constants of G. For any vector $A \in T_{\vec{R}}(G)$ defined by the set of coordinates $\{R^a\}$, the following relation holds:

$$\theta(A) = A,$$

or, in other words,[5]

$$g(\vec{R} + d\vec{R})\, g^{-1}(\vec{R}) = \tilde{\omega}^k(x, dx, \xi, d\xi)\ X^k + \theta^\alpha(x, dx, \xi, d\xi)\ \Lambda_\alpha, \tag{2.6}$$

where

$$X_k \in \mathbf{g} \setminus \mathbf{s},\ \Lambda_\alpha \in \mathbf{s}\ ;\ k = 1, \ldots N_1;\ \alpha = 1, \ldots N_2;$$
$$N_1 = \dim(\mathbf{g} \setminus \mathbf{s}),\ N_2 = \dim\ \mathbf{s},\ N_1 + N_2 = N.$$

We consider horizontal curves, so that $d\xi = 0$, and choose the coordinates in the fiber such that $\xi = 0$ at the initial point. The last equation completely defines the parallel transport in $G(G/S)$. Let now $x(t)$ be a curve in G/S. It is well known that, if $A(t)$ is the tangent vector at $x(t)$, then

$$U\left(g(t)\right) = T\ \exp\left(-i \int_0^t A(\tau) d\ \tau\right), \tag{2.7}$$

where $A(t)$ is chosen to be hermitian and T denotes T-ordering. It follows from (2.6), (2.7) that

$$U\left(g\left(\vec{R}(T)\right)\right) \equiv U(T) = \mathcal{P}\exp\left(-i\oint_C \tilde{\omega}^k\, X_k - i\oint_C \theta^\alpha\, \Lambda_\alpha\right). \tag{2.8}$$

The quasi-periodicity condition gives

$$\oint_C \tilde{\omega}^k = 0,$$

and we obtain finally

$$U(T) = \exp\left(-i\oint_C \theta^\alpha\, \Lambda_\alpha\right), \tag{2.9}$$

where the P-ordering is removed because all Λ_α's commute.

Our purpose is to find the connection between θ^α and the metric properties of the coset manifold G/S. It can be done in the following way. It is well known[4] that parallel transport in an arbitrary affine connection manifold can be described in terms of the Cartan structure 1-forms which obey the equations

$$d\omega^i = \omega^i{}_k \wedge \omega^k + \frac{1}{2}\, T^i{}_{jk}\, \omega^j \wedge \omega^k \tag{2.10}$$

$$d\omega^i{}_j = \omega^i{}_k \wedge \omega^k{}_j + \frac{1}{2}\, R^i{}_{jkl}\, \omega^k \wedge \omega^l, \tag{2.11}$$

where $T^i{}_{jk}$ and $R^i{}_{jkl}$ are the components of the torsion and curvature tensors. To find the correspondence between two sets of 1-forms $\tilde{\omega}^i, \theta^\alpha$ and $\omega^i, \omega^i{}_j$, let us consider the left quasi-regular representation of G which acts in the algebra of differentiable functions on G/S as follows

$$L_g \cdot f(x) = f\left(g^{-1}\cdot x\right);\ f:\ G/S \to \mathbf{R},\ x \in G/S,\ g \in G.$$

The differential of the function f can be written as

$$df(x) = i\left(\tilde{\omega}^k\, X_k + \theta^\alpha\, \Lambda_\alpha\right) f(x), \tag{2.12}$$

where X_k and Λ_α are supposed to be first order differential operators acting in the algebra of differentiable functions. Using the Poincaré identity and collecting the coefficients of X_k and Λ_α, we obtain equations for θ_α and $\tilde{\omega}^k$:

$$d\theta^\alpha = \frac{1}{2}\, C^\alpha{}_{kl}\, \omega^k \wedge \omega^l, \tag{2.13}$$

$$d\tilde{\omega}^i = \frac{1}{2}\, C^i{}_{jk}\, \omega^j \wedge \omega^k + C^i{}_{\alpha l}\, \theta^\alpha \wedge \omega^l, \tag{2.14}$$

where the relations

$$C^\alpha{}_{\beta\gamma} = C^i{}_{\alpha\beta} = C^\alpha{}_{\beta i} = 0$$

are used. The first two types of structure constants vanish because of the commutativity of **s** and the last one does in view of the condition (2.2). Now we introduce the set of 1-forms

$$\omega^i{}_j = C^i{}_{\alpha j}\, \theta_\alpha. \tag{2.15}$$

In terms of ${\omega^i}_j$, the equations (2.13), (2.14) can be rewritten as

$$d\tilde{\omega}^i = {\omega^i}_k \wedge \tilde{\omega}^k + \frac{1}{2}\, {C^i}_{jk}\, \tilde{\omega}^j \wedge \tilde{\omega}^k, \qquad (2.16)$$

$$d{\omega^i}_j = {\omega^i}_k \wedge {\omega^k}_j + \frac{1}{2}\, {R^i}_{jkl}\, \tilde{\omega}^k \wedge \tilde{\omega}^l, \qquad (2.17)$$

where the Jacobi identity has been used in the form

$${C^i}_{k\beta}\, {C^k}_{j\alpha} = {C^i}_{k\alpha}\, {C^k}_{j\beta}.$$

Comparing (2.13), (2.14) with (2.16), (2.17), one can conclude that the forms $\tilde{\omega}^i, {\omega^i}_j$ describe parallel transport in G/S and the torsion and curvature tensors equal

$${T^i}_{jk} = {C^i}_{jk}\ ; \quad {R^i}_{jkl} = {C^i}_{\beta j}\, {C^\beta}_{kl}\ .$$

To obtain the final result, one has to reexpress θ^α in terms of ${\omega^i}_j$. For this purpose we introduce

$$\theta_\alpha = {C_{\alpha i}}^j\, {\omega^i}_j. \qquad (2.18)$$

One can check that

$$\theta_\alpha = g_{\alpha\beta}\, \theta^\beta, \qquad (2.19)$$

where the tensor

$$g_{\alpha\beta} = {C^i}_{\alpha j}\, {C^j}_{\beta i}$$

is introduced. Supposing $g_{\alpha\beta}$ to be nondegenerate, we get

$$\theta^\alpha = g^{\alpha\beta}\, \theta_\beta, \qquad (2.20)$$

where $g^{\alpha\beta}$ is the inverse tensor to $g_{\alpha\beta}$. From (2.18), (2.20), follows the expression for the cyclic evolution operator

$$U(T) = \exp\left(-i\, {A^\alpha_k}^{\,l} \oint_C {\omega^k}_l\, \Lambda_\alpha\right), \qquad (2.21)$$

where

$${A^\alpha_k}^{\,l} = g^{\alpha\beta}\, {C_{\beta k}}^l.$$

The important point is that the tensor $g_{\alpha\beta}$ must be nondegenerate. In the case where G is semisimple, it is easy to see that this happens because $g_{\alpha\beta}$ coinsides with the restriction of the Killing-Cartan tensor on the Cartan subalgebra. The quantities

$$I_\alpha = {A^\alpha_k}^{\,l} \oint_C {\omega^k}_l$$

are nothing but the Anandan angles introduced in Ref. 6.

3. THE BERRY PHASE FOR A THREE-LEVEL SYSTEM

As shown above, the geometric phase factor can be expressed in terms of Cartan 1-forms which characterize the metrical properties of the manifold G/S and structure constants which characterize the algebraic properties of $\mathbf{g}$ modulo $\mathbf{m}$. Nevertheless, if the connection coefficients are not known in explicit form, it is also possible to calculate the Berry phase purely algebraically by calculating the canonical left-invariant form on the group G. In this section we demonstrate this possibility in the simple case of the $SU(3)$ symmetry group.

Let us consider the following Hamiltonian:

$$H(t) = R_a(t)\ \lambda_a, \quad a = 1, \dots, 8, \tag{3.1}$$

where λ_a are the Gell-Mann matrices and R_a are parameters which are supposed to be good functions of time. It is more convenient to rewrite (3.1) in the form

$$H(t) = U(t)\ H_0(t)\ U^\dagger(t), \quad H_0(t) = \mathrm{diag}\left(E_1(t), E_2(t), E_3(t)\right). \tag{3.2}$$

We define the $U(t)$ parametrisation in the following way:

$$U(t) = V_3(t)\ V_2(t)\ V_1(t), \quad V_i(t) = \exp\left(\xi_i(t)\ J_+^{(i)} - \xi_i^*(t)\ J_-^{(i)}\right), \tag{3.3}$$

where $\xi_i(t)$ are functions to be connected with $R_a(t)$ and $J_\pm^{(i)}$ are the Cartan-Weyl basis elements

$$J_\pm^{(1)} = \frac{1}{2}(\lambda_1 \pm i\lambda_2), \ J_\pm^{(2)} = \frac{1}{2}(\lambda_4 \mp i\lambda_5), \ J_{\pm-}^{(3)} = \frac{1}{2}(\lambda_6 \pm i\lambda_7), \tag{3.4}$$

$$H_1 = \lambda_3, \quad H_2 = \frac{1}{2}\left(\lambda_3 + \sqrt{3}\ \lambda_8\right), \quad H_3 = \frac{1}{2}\left(-\lambda_3 + \sqrt{3}\ \lambda_8\right), \tag{3.5}$$

with the following commutation relations

$$\left[J_\pm^i, J_\pm^j\right] = \mp\varepsilon_{ijk}\ J_\mp^k, \quad \left[H_i, J_\pm^j\right] = \pm C_{ij}, \quad \left[J_+^k, J_-^k\right] = H_k, \tag{3.6}$$

with the matrix C

$$C = \begin{pmatrix} 2 & -1 & -1 \\ 1 & -2 & 1 \\ -1 & -1 & 2 \end{pmatrix}$$

and all other commutators equal to zero. From the expansion (3.3) follows the expression for the canonical 1-form

$$U^{-1}\ dU = V_1^{-1}\ dV_1 + V_1^{-1}\left(V_2^{-1}\ dV_2\right)\ V_1 + (V_2\ V_1)^{-1}\ \left(V_3^{-1}\ dV_3\right)\ (V_2\ V_1). \tag{3.7}$$

As each of the matrices V_i parametrizes a copy of the coset space $SU(2)/U(1)$, one can apply the expression for $V^{-1}\ dV$ which was obtained in Refs. 7-9. Namely, if the matrix V has the form shown in (3.3) with parameter ξ, then

$$V^{-1}\ dV = \tilde{\omega}_+(\xi)\ J_+ + \tilde{\omega}_-(\xi)\ J_- + \tilde{\omega}_0(\xi)\ J_3, \tag{3.8}$$

where $J_\pm, J_3$ are convenient symbols for $SU(2)$ generators and the 1-forms $\omega_\pm, \omega_0$, equal

$$\tilde{\omega}_+ = \frac{d\alpha}{1+|\alpha|^2}, \quad \tilde{\omega}_- = -\tilde{\omega}_+^*, \quad \tilde{\omega}_0 = \frac{\alpha^*\ d\alpha - \alpha d\alpha^*}{1+|\alpha|^2}, \quad \alpha = -\tan|\xi|\ e^{-i\phi}. \tag{3.9}$$

To find the 1-form $U^{-1}\ dU$ from that, it is sufficient to use the expression (3.7) and the formula

$$e^A\ B\ e^{-A} = e^{ad\ A}\ B, \quad (ad\ A)\cdot B = [A, B].$$

After simple calculations we obtain

$$U^{-1}\ dU = \sum_{k=1}^{3}(\omega_+^{(k)}\ J_+^{(k)} + \omega_-^{(k)}\ J_-^{(k)} + \omega_0^{(k)}\ H_k), \tag{3.10}$$

where the 1-forms $\omega_{\pm}^{(k)}$ are irrelevant for our purposes and the 1-forms $\omega_0^{(k)}$ equal

$$\begin{aligned} \omega_0^{(1)} &= \tilde{\omega}_0(\xi_1) + \varphi_0^1, \\ \omega_0^{(2)} &= \cos^2|\xi_1|\,\tilde{\omega}_0(\xi_2) + \cos^2|\xi_2|\,\sin^2|\xi_1|\,\tilde{\omega}_0(\xi_3), \\ \omega_0^{(3)} &= \sin^2|\xi_1|\,\tilde{\omega}_0(\xi_2) + \cos^2|\xi_2|\,\cos^2|\xi_1|\,\tilde{\omega}_0(\xi_3), \end{aligned} \tag{3.11}$$

where

$$\varphi_0^1 = f(\xi_1,\xi_2)\,\tilde{\omega}_+(\xi_3) + f^*(\xi_1,\xi_2)\,\tilde{\omega}_-(\xi_3) + g(\xi_1,\xi_2)\,\tilde{\omega}_0(\xi_3), \tag{3.12}$$

$$f(\xi_1,\xi_2) = \frac{\xi_1\xi_2^*}{2|\xi_1||\xi_2|}\sin||\xi_2|\,\sin(2|\xi_1|), \quad g(\xi_1,\xi_2) = -\sin^2|\xi_2|\,\cos 2|\xi_1|.$$

Using the quasi-periodicity condition, it is now possible to write the expression for the cyclic evolution operator $U(T)$

$$U(T) = \exp\left(i\sum_{k=1}^{3}(\oint_C \omega_0^{(k)}\;H_k\right). \tag{3.13}$$

We introduce on each coset space $SU(2)/U(1)$ coordinates

$$\xi_k = -\frac{\theta_k}{2}\,e^{-i\phi},$$

where θ_k, ϕ_k are polar and azimuthal spherical angles. These angles can be easily connected with the matrix elements H_{ij} of the Hamiltonian, which are, of course, linear combinations of R_a :

$$\tan(\theta_1/2) = \frac{\left|M_{12}^{(1)}\right|}{\left|M_{11}^{(1)}\right|}, \quad \tan(\theta_2/2) = \frac{\left|M_{13}^{(1)}\right|}{\left|M_{12}^{(1)}\right|}, \quad \tan^2(\theta_3/2) = \frac{1+\tan^2(\theta_2/2)}{A-\tan^2(\theta_1/2)}, \tag{3.14}$$

$$\phi_1 = \arg M_{12}^{(1)}, \quad \phi_2 = \arg M_{13}^{(1)}, \tag{3.15}$$

$$\sin(\phi_1-\phi_2+\phi_3) = -\frac{Im\; e^{i\phi_2}M_{31}^{(3)}}{M_{33}^{(3)}\,\tan(\theta_1/2)\tan(\theta_3/2)}, \tag{3.16}$$

where

$$A = \left(\left|M_{31}^{(2}\right|^2 + \left|M_{32}^{(2)}\right|\right) / \left(\left|M_{33}^{(2)}\right|^2\right)$$

and $M_{jk}^{(i)}$ is the algebraic supplement corresponding to the element $\Lambda_{jk}^{(i)}$ of the matrix

$$\Lambda_{(i)} = H - E_i \cdot I$$

(I denotes the identity operator).

References

1. A. Shapere and F. Wilczek, "Geometric Phases in Physics", World Scientific, Singapore (1989)
2. M. V. Berry, *Proc. Roy. Soc. A*, 392:45 (1987)
3. A. Messiah, "Quantum Mechanics", vol.2, North-Holland, Amsterdam (1970)
4. S. Kobayashi and K. Nomizu, "Foundations of Differential Geometry", vol. 1 , Interscience, New York (1963)

5. M. K. Volkov and V. N. Pervushin, "Essentially Nonlinear Quantum Theories, Dynamical Symmetries and Meson Physics", Atomizdat, Moscow (1978) (in Russian)

6. J. Anandan, *Phys.Lett. A,* 129:201 (1988)

7. E. A. Tolkachev and A. Ya. Tregubovich, *in:* "Topological Phases in Quantum Theory", p.119; B. M. Markovski and S. I. Vinitsky (eds.), World Scientific, Singapore (1989)

8. S. Chaturvedi, M. Sriram and V. Srinivasan, *J. Phys. A,* 20:L1071 (1987)

9. G. Giavarini and E. Onofri, *J. Math. Phys.,* 30:659 (1989).

PARTICIPANTS

S. Twareque Ali, Montreal
J-P. Antoine, Louvain-la-Neuve
S. Berceanu, Paris
A. Bette, Stockholm
F. Cantrijn, Gent
G. Chadzistakos, Prague
I. Dmitriyeva, Odessa
G. G. Emch, Gainesville
J. Gancarzewicz, Kraków
J-P. Gazeau, Paris
G. A. Goldin, New Brunswick, N.J.
M. J. Gotay, Honolulu
J. Grabowski, Warszawa
C. Gross, Darmstadt
W. Hann, Wrocław
M. Horowski, Białystok
P. Jakóbczak, Kraków
I. Kanatchikov, Aachen
M. V. Karasev, Moscow
M. Kozak, Białystok
M. B. Kozlov, Moscow
M. Krivoruchenko, Moscow
A. K. Kwaśniewski, Białystok
M. De Leon, Madrid
W. Lisiecki, Białystok
H. Makaruk, Warszawa
W. Marcinek, Wrocław
G. Marmo, Napoli
V. Maximov, Moscow
S. Mehdi, Paris
I. M. Mladenov, Sofia
M. Modugno, Firenze
W. Mulak, Wrocław
E. Mundt, Berlin
J. Myszewski, Warszawa
P. Nattermann, Clausthal-Zellerfeldt
K-H. Neeb, Darmstadt
E. Novikova, Moscow
A. Odzijewicz, Białystok
P. Oellers, Mannheim
R. Owczarek, Warszawa
A. M. Perelomov, Moscow
W. Piechocki, Warszawa
E. Prugovečki, Toronto
S. Rauch-Wojciechowski, Linköping
M. A. Robson, Cambridge
G. Rosensteel, New Orleans
A. Schmitt, Berlin
T. Schmitt, Berlin
D. J. Simms, Dublin
W. Słowikowski, Aarhus
J. Sniatycki, Calgary, Alberta
P. Šťovíček, Prague
A. Strasburger, Warszawa
J. Szmigielski, Saskatoon
J. Tafel, Warszawa
J. Tolar, Prague
A. Tregubovich, Minsk
G. M. Tuynman, Lille
O. Viskov, Moscow
Y. M. Vorobjev, Moscow
K. Wódkiewicz, Warszawa
W. Wojtyński, Warszawa
S. Zakrzewski, Warszawa

INDEX